T0237096

Springer Series in Statistics

Springer Series in Statistics (SSS) is a series of monographs of general interest that discuss statistical theory and applications.

The series editors are currently Peter Bühlmann, Peter Diggle, Ursula Gather, and Scott Zeger. Peter Bickel, Ingram Olkin, and Stephen Fienberg were editors of the series for many years.

More information about this series at http://www.springer.com/series/692

Göran Kauermann • Helmut Küchenhoff •
Christian Heumann

Statistical Foundations, Reasoning and Inference

For Science and Data Science

 Springer

Göran Kauermann
Department of Statistics
LMU Munich
Munich, Germany

Helmut Küchenhoff
Department of Statistics
LMU Munich
Munich, Germany

Christian Heumann
Department of Statistics
LMU Munich
Munich, Germany

ISSN 0172-7397 ISSN 2197-568X (electronic)
Springer Series in Statistics
ISBN 978-3-030-69829-4 ISBN 978-3-030-69827-0 (eBook)
https://doi.org/10.1007/978-3-030-69827-0

Mathematics Subject Classification: 62-01, 62-07

This Springer imprint is published by the registered company Springer Nature Switzerland AG.
The registered company address is: Gewerbestrasse 11, 6330 Cham, Switzerland

To Ursula and the boys
G.K.

Preface

The following book was developed for a two-semester course on "*Statistical Reasoning and Inference*" in the master's programme *Data Science* at LMU Munich. The master's programme puts emphasis on the combination of statistics and computer science and its fusion to the new discipline of data science. We consider the material to be generally applicable to similar master's programmes, where the emphasis on statistics is sufficiently strong. The book also provides a general overview of statistical concepts and ideas, and, where necessary in our view, we also go into technical details.

The first chapter provides a general discussion on why statistics and statistical ideas are of essential relevance in the field of data science. This chapter also contrasts the book from alternatives which focus more on statistical and machine learning. The subsequent Chaps 2–5 can be considered as a condensed introduction to statistical estimation theory. We do cover both, frequentist as well as Bayesian approaches. Putting it differently, we introduce likelihood-based models in the same strength as explaining numerical methods such as Monte Carlo Markov Chain in Bayes models. Chapter 6 discusses statistical testing, which is related in the same way to uncertainty quantification using confidence intervals, as well as Bayesian inference and classification. These first six chapters provide the core syllabus and are certainly more focussed on theoretical and conceptual ideas but less on applications. Chapter 7 looks at the wide field of regression models, and while this chapter is more applied, it certainly does not cover the field as it should with respect to a general education in a data science programme. In our opinion, regression is a fundamental concept in statistics and data science which should be treated in a separate lecture/course; hence, the material needs to be included in a separate book, some of which we refer to in our book. However, a statistics book without touching regression is also inappropriate in our view. We therefore have included a chapter on regression, but we kept it short and refer to specialised material in this field.

The remaining chapters of the book provide a potpourri of statistical ideas and concepts which are in our view essential for data scientists and data analysts. Quite central is the question of quantification of uncertainty based on numerical methods. Bootstrapping is one technical solution to this, which is discussed in depth in

Chap. 8. Moreover, model selection is crucial in general and we discuss this point in depth in Chap. 9. In Chap. 10, we go beyond single dimensions and introduce up-to-date methods to model multivariate data. This includes ideas of copulas and extends towards modelling extreme values. A general question in the area of data science is whether the sheer quantity of data allows to accept reduced quality of the data. Looking at different aspects such as missing data and biased data, for instance, we discuss these questions in Chap. 11. Finally, the last chapter puts emphasis on how to draw causal conclusions from available data, where we also propose the use of experimental data.

Some of the exercises in the book make use of data sets, which we provide on the book's webpage: https://github.com/SFRI-SDS-lmu/book_first_edition/.

The writing of the book was an enjoyable experience which showed us how rich the discipline of statistics is. It made clear that statistics plays an important and central role in data science, and building bridges between the "traditional" fields of statistics and computer science is necessary and beneficial. We are grateful to our students, who read the material carefully, found several mistakes and provided very useful suggestions. In particular, through the digital semesters due to the COVID-19 pandemic, they studied the material even more carefully. Our special thanks go to Katharina Hechinger, Jason Jooste as well as Elisa Noltenius for carefully reading, improving and latexing the material. We also thank Christoph Striegel and Giacomo De Nicola for contributing to the exercise sections.

All in all, we enjoyed composing the material, shaping it and putting it into this book format. We hope that our readers will find the book interesting and useful for their work with data.

Munich, Germany   Göran Kauermann
April 2021   Helmut Küchenhoff
                                                                         Christian Heumann

Contents

Chapter 1
Introduction

Even coincidence is not unfathomable;
it has its regularity.

(NOVALIS, 1772–1801)

1.1 General Ideas

We need more well-trained data scientists. The information era and its ever-growing masses of data make such a statement abundantly clear. Inevitably, our first hurdle in discussing the topic is to even define such a fluid and wide-ranging field. However, when pressed for a definition we, hesitantly, might suggest the symbiosis of statistics and informatics as a good working definition. This very symbiosis is an ever-relevant theme within data science, with the two sides often offering differing vocabularies, approaches and perspectives on identical phenomena. More than two decades ago, Cleveland (2001) defined data science as the discipline of retrieving information from data and, on remarkably prescient grounds, called for the expansion of the technical areas of statistics towards the field of informatics. With the current gold-rush surrounding all things "Big Data", it is becoming clear that this collaboration between statistics and informatics is more necessary than ever. In this book, we focus on the statistical aspects of data science and intend to build a statistical foundation for data scientists. We will emphasise statistical thinking and methods as well as the mathematical principles of statistical inference. In this respect, we do not intend to provide a book full of tools to be used in data analysis, be it by data scientists or others. Our aim is to instead present a solid foundation in statistics, which in turn will allow the reader to truly understand and apply statistical thinking and inference to data analysis in general. We are convinced that data scientists most certainly need deep knowledge of statistical learning, as covered, for instance, in Friedman et al. (2009) or more recently in Hastie et al. (2015). However, the material in our book starts earlier, with the general principles and foundational statistics that are necessary to understand and apply common data analytic tools.

A central goal in statistics is the analysis of data. While this sounds like the goal of data science in general, we want to distinguish between two different, and sometimes competing, approaches to data analysis. One is nowadays called machine learning, in the sense that algorithms are applied to a (novel) dataset in order to draw conclusions from it. Stereotypically and quite simplifyingly spoken, these algorithms can be considered as some kind of "black box", where data is inserted on one side and condensed information comes out on the other side. Machine learning is clearly driven by algorithms. The results of such algorithms could be group indicators, if the intention is to cluster observations, or predictions, if the intention is to get from input variables to one or multiple output variables. While machine learning is a sound and powerful approach to distilling information from data, it is not the typical approach that a statistician would pursue. Again, using a stereotypical description, a statistician intends to divide the data into a **systematic component** and a **stochastic component**. The latter, the stochastic component, is the key ingredient in statistical data analysis. It is based on the assumption that, no matter what data we record at whatever resolution, there is always randomness. With randomness, the statistician can bundle all variations that cannot be explained, all things that will behave differently from experiment to experiment, from individual to individual or from measurement to measurement, into a single package. This stochastic component in the data is what statisticians are interested in and what they try to model. This belief in a stochastic component in the data is the key difference between this and the aforementioned machine learning approach.

The reader might rightfully ask why the stochastic part of the data is more relevant or interesting to the statistician than its deterministic counterpart. The answer is that the stochastic part is not itself the end-product of a data analysis, but its vehicle and essential component. In fact, statisticians primarily use stochastic models, not because the stochastic structure is the main goal of the data analysis, but using stochastic models allows the two components in the data to be separated from each other. The systematic component is usually of primary interest for the data scientist, but the stochastic component allows for the separation of the two and simultaneously allows for the quantification of uncertainty.

Let us demonstrate the two approaches with a very simplistic example. Assume an internet retailer wants to analyse data on money spent per sale. An algorithmic approach could be to take the numbers of recent sales and calculate, for instance, the mean expenditure per sale for a given time window. The model-based approach, on the other hand, would start differently and first find a suitable stochastic model for the data, arguing that expenditures vary from sale to sale and are not deterministic. Using a simple model, like the normal distribution model, one might now be interested in the mean (or as we will call it later, the expectation) of the expenditure. This mean value is a parameter of the stochastic model, which can be estimated from the data. That is, we use the data to get information about the (typically) unknown parameters of the stochastic model. A reasonable and plausible estimate for the true mean value is, in turn, the arithmetic mean from the data. In this simple example, the algorithmic and the model-based approaches come to the same output, namely the arithmetic mean. The reasoning as to why this output is useful, however, is completely different between the two approaches. At first sight, this might sound confusing. The reader might even consider this distinction pedantic. We must stress, however, that statisticians approach data in a completely different way from computer scientists. Statisticians also apply algorithms. However, these algorithms have no justification in and of themselves. Instead, the algorithms are only applied in order to discern unknown parameters in stochastic models. Furthermore, the stochastic model enables the statistician to make statements about the reliability of a result or prediction. We must also emphasise right from the beginning that we have no intention of ranking the two approaches or stating that one is better than the other. In fact, if the focus is on fast prediction, e.g. in speech or image processing, statistical approaches currently are far behind those of computational and machine learning. We refer the interested reader to Breiman (2001), who contrasts these two different ways of thinking in more depth. That being said, in this book, we exclusively pursue the statisticians' viewpoint and intend to lay a foundation for statistical reasoning and inference and therein to convey the power of deciphering and modelling randomness.

1.2 Databases, Samples and Biases

A typical workday for a data scientist could be the following: a dataset or a database is made available and needs to be analysed with appropriate tools. In this book, we will follow the statistical approach to such a situation, that is, to first pose two questions before any data analysis: where does the data come from and what is the question/hypothesis? This means that automatic analyses of databases in the style of data mining are not necessarily compatible with statistical thinking. The traditional (and certainly slightly artificial) statistical framework is thereby as follows: either a sample is drawn from a population or measurements are taken in an experiment. Both samples and experiments are designed to answer a specific question, be it an experiment to assess whether a particular new medicine has an effect or a poll to get information about preferences for a range of political parties. The data are collected in such a way that the analysis provides an answer to the question. If, in contrast, data is collected without any specific purpose, for instance, data recorded from internet transactions or from a technical process, it is usually called observational data (see e.g. Rosenbaum 2002). In the case of observational data, one needs to question whether analysis of this data could lead to a systematic error because it was recorded for a different purpose. If there are systematic differences between the parameter of interest and the result of the analysis, statisticians call it "bias". To avoid such bias is a central aim of statistical data analysis. While observational data *does* contain useful information, the analyst needs to be careful not to run into a bias-trap, leading to incorrect interpretations. An often cited example in this direction is the Boston street bump app, see Harford (2014). Drivers in Boston could download an app that used the GPS and accelerometer of their smartphone to record the location of bumps and potholes. The idea was that the city council would no longer need to actively search for potholes, but just repair the ones recorded in the app. It turned out, however, that potholes were primarily detected in wealthy neighbourhoods. The explanation for this bias is quite simple. At the time of data collection, smartphones with GPS were expensive and therefore owned by wealthier drivers, while poorer drivers did not generally have access to high-end smartphones. This easily explains the bias in the detection of potholes: they were more often detected in streets where wealthy people drive. While the bias in this example is obvious, the story demonstrates the problems that can occur with observational data in general. Hence, a warning applies when it comes to the analysis of observational data. Do not run into a bias-trap and do not draw invalid conclusions. We will discuss strategies to avoid this kind of pitfalls later in the book.

Beyond bias, there are further problems that may occur with observational data. Assume, for instance, that one has data on the number of sales of a product and the corresponding price at which the product was sold. If the focus is on estimating how price determines demand, one should bear in mind that price and demand are mutually dependent. Therefore, one should first question how the price is fixed in the data. In the real world, the seller will increase the price if an increased demand occurs and the opposite holds for the customer. Demand decreases as price increases. Hence, price and demand are mutually dependent. In this case, it is difficult to draw a (causal) conclusion from observational data. The problem occurs simply because the data on sales do not result from a controlled experiment. As a consequence, the recorded data on sales do not directly allow us to forecast the demand for a fixed price. In fact, we will present results from econometrics (i.e. the scientific discipline of statistics in economics) that explain the dependence of demand on price and show that it is critical, both for the understanding of observational data and for solutions to this problem of interdependency. The above examples should demonstrate that one needs to carefully question how the data have been collected before using them to draw conclusions.

Nowadays, data science is closely related to the analysis of massive data, which is reflected in the term (or buzzword) "Big Data Analytics". Datasets in the world of "Big Data" are high in volume or velocity or both, meaning that they are massive and, as such, that computation is numerically demanding or even infeasible. Sometimes, a database is just too large to upload to a standard desktop computer and even cluster computing might not be a suitable alternative. In this case, statistical thinking and reasoning might be helpful. As described above, a statistician first asks what questions should be answered with the data, before even touching it. This approach can often help with large datasets in so far that it is often unnecessary to analyse them in their entirety, but just to take a subset of them for the analysis. This may sound like a capitulation to the technical challenges involved, but in fact it is not. Assume that you have access to all individuals in a country, let us take Germany as an example, and that you want to gather information about the political preferences of its citizens. Assume that every single individual who is eligible to vote has a smartphone and that an omniscient organisation has access to all telephone numbers. Now, in order to get information about the 60 million people entitled to vote, we could send out 60 million messages and collect 60 million responses using an app specifically designed for this purpose. This is certainly a "Big Data" challenge, resulting in heavy computation. However, the question remains whether this effort is worth it. A statistician would directly answer with "no". Instead, the statistician would suggest running a poll and getting information, not from all individuals, but just from a subset (usually called a sample). If we took this approach, we would not have the preferences of all individuals and thus would not be certain of the result. In fact, the resulting party preferences obtained from this sample would carry some randomness. The results would vary, dependent on which individuals we asked in the population. However, we can bring statistical methodology to bear to quantify this uncertainty, which leads to confidence intervals, i.e. intervals for the true percentage of voters who support each political party. Such polls are well accepted and in fact, even with a small number of respondents (e.g. 1000), the results are valid within the order of +/- 3%. Hence, instead of approaching all individuals via their smartphones, it is more cost-efficient to simply get the information from a small sample of the population. This idea of sampling is a core instrument in statistics and it can also have a large impact in "Big Data" situations. In fact, it may be more cost- (and computing-) efficient, not to analyse an entire massive database, but just to draw a reasonable sample from it. Statisticians have developed quite efficient sampling strategies for this purpose, some of which we will present in this book.

And finally, whenever we talk about bias and sampling, we must also mention missing data. In real-life databases, we are often confronted with the fact that some data entries are missing. This absence may be due to several factors, some of critical importance. For data analysis, missing data is a common problem. A common strategy is therefore to simply ignore missing data, by omitting either observations or entire variables associated with missing entries. It is clear that this strategy carries the risk of producing a biased analysis. Incorporating missing data in the analysis has its own risk of bias and one needs to think carefully about the reasons for this missingness. Are the numbers just missing randomly? Or is their absence somehow related to their value? For example, if one asks individuals to record their monthly income, those with particularly high or low income might be tempted to not answer. Hence, the absence depends on the (unknown) value of the variable. In this case, and many others, if we simply choose to analyse only the people who have provided the information (so called complete cases), our results can be heavily biased because key groups are simply not present in the data. We will discuss problems with missing data in a later chapter of the book and introduce available statistical methodology that helps us to cope with them.

In his article, "Big data: are we making a big mistake" in the Financial Times (see Harford 2014), Harford summarised: "As for the idea that with enough data, the numbers speak for themselves—that seems hopelessly naive in datasets where spurious patterns vastly outnumber genuine discoveries. The challenge now is to solve new problems and gain new answers without making the same old statistical mistakes on a grander scale than ever". This statement and the above ideas have hopefully made clear that data scientists need statistics. It is essential to be able to think statistically and to know the principles of statistical reasoning and inference. The foundation for this much-needed perspective on data science is the content of this book.

Chapter 2
Background in Probability

We begin our work with a summary of many essential concepts from probability theory. The chapter starts with different viewpoints on the very definition of probability and moves on to a number of foundational concepts in probability theory: expected values and variance, independence, conditional probability, Bayes theorem, random variables and vectors, univariate and multivariate distributions and limit theorems. We think a thorough understanding of these topics is absolutely necessary for the coming chapters and if they seem straightforward then simply skimming through this chapter would be appropriate. However, if some of these topics are not already clear then please keep in mind that this chapter is intended more as a reference than as a thorough explanation of the topics at hand. We refer to Heumann and Schomaker (2016) for a more basic introduction to the field of probability theory and statistics. Let us now begin with the deceptively difficult task of defining the meaning of probability itself.

2.1 Random Variables and Probability Models

2.1.1 Definitions of Probability

The foundation of statistical modelling is stochastic models. Data are assumed to be generated by a random process, which can be described by probabilities. In order to define probabilities, the notion of a random experiment is essential. A random experiment is a process for which the outcome is uncertain. The set of all possible outcomes is called sample space and denoted by Ω. For instance, when throwing a die, $\Omega = \{1, 2, 3, 4, 5, 6\}$ and for the length of a person's life, we have $\Omega = [0, \infty)$. Events can be defined as subsets of the sample space Ω. Examples are $A = \{2, 4, 6\}$ ("even number"), $B = \{6\}$ ("six") or $C = (60, \infty)$ ("lifetime longer than 60"). Mathematically, probabilities are functions, which assign a number $P(A)$ between

G. Kauermann et al., *Statistical Foundations, Reasoning and Inference*,
Springer Series in Statistics, https://doi.org/10.1007/978-3-030-69827-0_2

0 and 1 to events A, where $A \subset \Omega$. If $P(A)$ is small, then the event A occurs seldomly and if $P(B)$ is close to 1, then B occurs rather often. Andrey Kolmogorov (1933) developed a system of axioms for probabilities, which still represent the basis of modern probability theory. We define these axioms as follows:

Definition 2.1 (Mathematical Probability Definition) The **probability** P is a function defined on the collection $\mathcal{A}$ of all subsets of a sampling space Ω, which fulfils the following properties.

(i) $0 \leq P(A) \leq 1$ for all $A \in \mathcal{A}$ $\qquad\qquad\qquad\qquad\qquad\qquad$ (2.1.1)

(ii) $P(\Omega) = 1$ $\qquad\qquad\qquad\qquad\qquad\qquad\qquad\qquad\qquad$ (2.1.2)

(iii) $P\left(\bigcup_{i=1}^{\infty} A_i\right) = \sum_{i=1}^{\infty} P(A_i)$ $\qquad\qquad\qquad\qquad\qquad$ (2.1.3)

$\qquad$ for all mutually disjoint sequences of events, i.e. $A_i \in \mathcal{A}, i = 1, \ldots, \infty$

$\qquad$ with $A_i \cap A_j = \emptyset$ for $i \neq j$.

This definition is purely formal and does not give any actual meaning to the term probability. However, it has proven to be both mathematically sound and practically useful. Despite the clear mathematical foundation, a long philosophical debate continues over the true definition of probability itself. We briefly discuss the two most common definitions, as we find them illustrative of the various strategies of statistical inference described in later chapters. The first interpretation of probabilities is based on infinite replications of the random experiment and was introduced by Richard Von Mises (1928).

Definition 2.2 (Frequentist Probability Definition) The **probability** of an event is defined as the limit of the relative frequency of its occurrence in a series of replications of the respective random experiment, that is

$$P(A) := \lim_{n \to \infty} \frac{n_A(n)}{n} \qquad\qquad\qquad (2.1.4)$$

with n as the number of replications and $n_A(n)$ being the number of occurrences of A in n experiments.

The above definition has some shortcomings including, of course, that it is impossible to actually perform an infinite number of replications. Another important definition of probability was proposed by De Finetti (1974) and is based on the idea that uncertainty can also be seen as a lack of information.

Definition 2.3 (Subjective Probability Definition) The **probability** of an event A is defined as the degree of belief of an individual at a certain time with a certain amount of information. It is quantified by imaginary bets. That is, one's degree of belief in A is P(A), if and only if P(A) units of utility is the price at which one would buy or sell a bet that pays 1 unit of utility if A occurs and 0 units if A does not occur.

Consider the example of a (fair) die roll. Before we throw a die, the number of dots it will show is unknown and once we have thrown the die and it lies on the table, we then know the number of dots showing. However, what happens if we throw the die by putting it in a cup? Shaking the cup and putting it upside down on the table fixes the number of dots showing on the die. Hence, the random experiment is realised. However, before we lift the cup, we still do not know what the die shows. With a closed cup on the table the random experiment is over, but the uncertainty about the number of dots still exists until the cup is lifted. Using De Finetti's approach we can still apply a probability model, i.e. the probability that the die under the cup shows six dots on the top is 1/6. This shows that uncertainty can be well expressed in terms of probability statements. Both definitions are essential in interpreting results using statistical models. An overview on defining probability with a more detailed discussion of the philosophical background can be found in Gillies (2000). No matter how probability is defined, Kolmogorov's axioms will always hold. Therefore, we proceed with some results in probability theory that result from these axioms.

2.1.2 Independence, Conditional Probability, Bayes Theorem

Definition 2.4 (Conditional Probability and Independence) Let A and B be two events. Assuming that $P(A) > 0$, the conditional probability of B given A is

$$P(B|A) = \frac{P(A \cap B)}{P(A)}. \tag{2.1.5}$$

Two events A and B are called independent if and only if

$$P(A \cap B) = P(A) \cdot P(B). \tag{2.1.6}$$

The interpretation is that $P(B|A)$ is the probability of B if A is known to have occurred. In the case of rolling a fair die, let us take $B = \{6\}$ to be the event of rolling a 6, which has probability 1/6. When we know that an even number occurred, i.e. $A = \{2, 4, 6\}$, then the probability of B is 1/3, i.e. $P(B|A) = 1/3$. Note also that independence of A and B implies that $P(B|A) = P(B)$, i.e. knowing that A has occurred does not change the probability of B.

Property 2.1 (Law of Total Probability and Bayes Theorem) Let $\Omega = \bigcup_{i=1}^{\infty} A_i$ where $A_1, A_2, A_3, \ldots$ are countable partitions of Ω (i.e. $A_i \cap A_j = \emptyset$ for $i \neq j$.). Then, for each event B it holds that

$$P(B) = \sum_{i=1}^{\infty} P(B|A_i)P(A_i) \tag{2.1.7}$$

$$P(A_j|B) = \frac{P(B|A_j)P(A_j)}{\sum_{i=1}^{\infty} P(B|A_i)P(A_i)}. \tag{2.1.8}$$

Both equations can be directly deduced from Kolmogorov's axioms and Definition (2.1.5). The first equation describes the law of total probability and simply states that the probability of an event can be calculated as the sum of the probabilities of that event conditional on each event of a set of events, as long as that set encompasses all possible events of that partition. The second equation is also known as Bayes Theorem and is attributed to the English reverend and philosopher Thomas Bayes (1702–1761). The main idea of this theorem is to compute conditional probabilities "the other way round". The conditional probabilities $P(A_j|B)$ are calculated from conditional probabilities $P(B|A_j)$ and the (prior) probabilities $P(A_j)$. If the sets A_j characterise some unobservable events or some theories concerning the event B, then Bayes theorem gives us an understanding of what we learn about A_j when we observe B. This is the basis of what is called Bayesian inference, which is discussed later in Chap. 5.

2.1.3   Random Variables

Even though probabilities are formally defined on sets, it is often more convenient to make use of random variables which take real numbers. Formally, a random variable is defined as a mapping from the sample space Ω into the real numbers, that is

$$Y : \Omega \rightarrow \mathbb{R}.$$

The probability is now defined with relation to the values of Y instead of on subsets of Ω. As a simple example, consider rolling two dice and define the random variable Y as the sum of the two dice. Then Ω is $\{1, 2, 3, 4, 5, 6\} \times \{1, 2, 3, 4, 5, 6\}$ and $Y_{2D}((a, b)) = a + b$. In general, random variables are used to describe random phenomena. The subset of possible values of $\mathbb{R}$ is called the support of Y. If the support is finite or countably infinite, the random variable is called discrete. For instance, the above random variable Y_{2D} has the finite support $\{2, \ldots, 12\}$ and can be characterised by the probability function $P(Y_{2D} = y)$. Random variables that have a support consisting of all real numbers (or an interval of them) are called continuous random variables, e.g. the height of a randomly chosen child or the lifetime of an electronic device. The stochastic behaviour of a continuous random

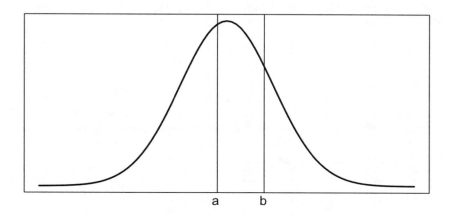

Fig. 2.1 Example of a density function

variable is described by its density function. The **density function** $f(.)$ is a positive function, such that

$$P(Y \in [a, b]) = \int_a^b f(y)dy,$$

as sketched in Fig. 2.1. From the second of Kolmogorov's axioms (2.1.2) we get $\int_{-\infty}^{\infty} f(t)dt = 1$.

For both discrete-valued and continuous random variables, the **distribution function** $F_Y(y) = P(Y \leq y)$ yields a unique description of the random variable Y, where y may take any possible real number, i.e. $y \in \mathbb{R}$. This is sometimes also called the **cumulative distribution function**. In the definition of $F_Y(y)$, we have introduced several notational concepts that are commonly used in statistics but are worth mentioning here explicitly. Firstly, random variables are commonly denoted with capital letters, while their realisations are denoted with lower case letters. Hence, whenever we write Y, it means that we do not know the variable's value, while $Y = y$ means that the random variable Y has taken the specific value y. In fact, this distinction is central to much of statistics, where models are generated with random variables, which are then put into practice by realising their values with real data. Secondly, when it is unclear, we put an index on the distribution or density function, meaning that this function (i.e. the probability model) belongs to the given random variable (e.g. F_Y). We will not be stringent with this index notation, because it is often obvious which random variable we are referring to. Note also that for

$y \to -\infty$ we have $F_Y(y) = 0$ and for $y \to \infty$ we get $F_Y(y) = 1$. Finally, we can relate $F_Y(.)$ to the density or probability function $f_Y(y)$ as follows:

$$F_Y(y) = \int_{-\infty}^{y} f_Y(t)dt. \tag{2.1.9}$$

It can be can notationally quite tedious, and not particularly informative, to thoroughly distinguish between discrete and continuous random variables. We therefore simplify the notation with the following convention, which will be used throughout the entire book. This allows us to have a general notation for both continuous and discrete random variables.

1. If Y is a discrete value random variable (e.g. with values $\{0, 1, 2, \ldots\}$), then the integral in Definition (2.1.9) refers to a sum, i.e.

$$\int_{-\infty}^{y} f_Y(\tilde{y})d\tilde{y} = \sum_{k:k \leq y} P(Y = k).$$

Function $f_Y(y)$ in this case is equivalent to the probability function $P(Y = y)$.
2. If Y is continuous, then the integral in (2.1.9) is an integral in the usual sense (assuming integrability), such that

$$f_Y(y) = F_Y'(y)$$

(at values y where $F_Y(y)$ is differentiable).
3. For mixed random variables that take discrete and continuous values, the integral in (2.1.9) is a mixture of sums and integrals.

The mathematically trained reader might not be satisfied with our "sloppyness" here. However, we do not want to get distracted from our main topic, which is statistics. For this reason, we content ourselves with being a little superficial with the notation. **Therefore, from now on, we denote** $f_Y(y)$ **as a density, even if** Y **is discrete valued**. This convention makes the notation much easier and we are convinced that the reader will quite easily understand when an integral refers to a sum or a real integral or a mixture of both. For a rigorous mathematical treatment of random variables, there are many textbooks on probability, such as Karr (1993).

Each random variable has some essential features, which we describe in the following definitions. The most prominent are the mean value, which we call the expectation, and the variability, which we call the variance.

Definition 2.5 (Expected Value, Variance and Moments)

1. The expected value or mean value of a random variable is defined as

$$E(Y) = \int y f_Y(y)dy.$$

2. The variance of a random variable is defined as

$$Var(Y) = E\left(\{Y - E(Y)\}^2\right) = \int \{y - E(Y)\}^2 f(y)dy,$$

which always results in $Var(Y) \geq 0$.
3. The k-th Moment is defined as $E(Y^k) = \int y^k f_Y(y)dy$ and the k-th central moment as $E(\{Y - E(Y)\}^k)$.

We often abbreviate the expectation with the parameter $\mu = E(Y)$ and the variance with the parameter $\sigma^2 = E(Y - E(Y)^2)$. Note also that the variance can be written in the following form:

$$\sigma^2 = E\left(\{Y - \mu\}^2\right) = E\left(Y^2 - 2Y\mu + \mu^2\right)$$
$$= E(Y^2) - 2\mu^2 + \mu^2 = E(Y^2) - \mu^2.$$
$$= E(Y^2) - \{E(Y)\}^2.$$

This follows from some properties of the expectation operator that we will define in the next section. We thereby assume that the corresponding integrals exist, i.e. are finite. We must also mention that for some exceptional distributions (e.g. the Cauchy-distribution) the expected value or the variance (or both) do not exist.

Property 2.2 (Expectation and Variance of a Sum of Random Variables) An important result is that the sum of the expectations of a set of random variables is equal to the expectation of their sum. Also important is that the variance of a sum of independent random variables is equal to the sum of their variances. For random variables $Y_1, \ldots, Y_n$ we get

$$E\left(\sum_{i=1}^{n} a_i Y_i\right) = \sum_{i=1}^{n} a_i E(Y_i), \tag{2.1.10}$$

and if $Y_1, \ldots, Y_n$ are independent, then

$$Var\left(\sum_{i=1}^{n} a_i Y_i\right) = \sum_{i=1}^{n} a_i^2 Var(Y_i), \tag{2.1.11}$$

for arbitrary known values $a_1, \ldots, a_n$.

2.1.4 Common Distributions

There are some distributions that appear so often in statistics that familiarity with their parameters and features would greatly benefit the reader. The distributions themselves depend on one or more parameters, which are conventionally denoted with Greek letters. We also follow the notational convention that the dependence of the distribution on the parameter is expressed with a semicolon, that is, the parameters of the distribution are listed after the semicolon. This convention is used throughout the rest of the book and is easily comprehended with the following standard distributions.

Definition 2.6 (Binomial Distribution $B(n, p)$)

- Application: Two outcomes: success | failure with a chance of success of $\pi \in$ [0, 1]. We count the number of successes in n independent trials.
- Probability function: $P(Y = k; \pi) = \binom{n}{k} \pi^k (1 - \pi)^{n-k}$, for $k = 0, \ldots, n$
- Expectation: $E(Y) = n\pi$
- Variance: $Var(Y) = n\pi(1 - \pi)$

Definition 2.7 (Poisson Distribution $Po(\lambda)$)

- Application: Count of events in a certain period of time with events occurring at an average rate of $\lambda > 0$.
- Probability function: $P(Y = k; \lambda) = \dfrac{\lambda^k}{k!} \exp(-\lambda)$ for $k = 0, 1, 2, \ldots$
- Expectation: $E(Y) = \lambda$
- Variance: $Var(Y) = \lambda$

Definition 2.8 (Normal Distribution $N(\mu, \sigma)$)

- Application: Metric symmetrically distributed random variables with mean $\mu \in$ $\mathbb{R}$ and variance $\sigma^2 > 0$.
- Density function: $f(y; \mu, \sigma) = \dfrac{1}{\sqrt{2\pi}\sigma} \exp\left(-\dfrac{(y - \mu)^2}{2\sigma^2}\right)$ for $y \in \mathbb{R}$
- Expectation: $E(Y) = \mu$
- Variance: $Var(Y) = \sigma^2$

Definition 2.9 (Exponential Distribution $Exp(\lambda)$)

- Application: Time between two events following a Poisson process, where the waiting time for the next event is on average $1/\lambda$, $\lambda > 0$.
- Density function: $f(y; \lambda) = \lambda \exp(-\lambda y)$ for $y \geq 0$
- Expectation: $E(Y) = \frac{1}{\lambda}$
- Variance: $Var(Y) = \frac{1}{\lambda^2}$

Definition 2.10 (t-Distribution $t(n)$)

- Application: Metric symmetrically distributed random variable with a considerable probability of extreme outcomes ("heavy tails"). Also used for statistical tests for the mean of normally distributed variables when the variance is unknown and estimated from the data and the degrees of freedom, n, is small.

- Density function: $f(y; d) = \dfrac{\Gamma(\frac{d+1}{2})}{\sqrt{d\pi}\,\Gamma(\frac{d}{2})}\left(1 + \dfrac{y^2}{d}\right)^{-\frac{d+1}{2}}$
  where $\Gamma(.)$ is the gamma function and $d = 1, 2, \ldots$ is the degree of freedom (which in principle can take any positive value but we only work with positive discrete values here)
- Expectation: $E(Y) = 0$ for $d \geq 1$
- Variance: $Var(Y) = \frac{d}{d-2}$ for $d \geq 3$

Definition 2.11 (Chi-Squared Distribution $\mathcal{X}^2(n)$)

- Application: Squared quantities like sample variances, sum of n independent squared normal distributed random variables.
- Density function: $f(y; n) = \dfrac{y^{\frac{n}{2}-1} e^{-\frac{y}{2}}}{2^{\frac{n}{2}} \Gamma(\frac{n}{2})}$ where $n = 1, 2, 3, \ldots$
- Expectation: $E(Y) = n$
- Variance: $Var(Y) = 2n$

2.1.5 Exponential Family Distributions

Several of the above distributions allow for a mathematical generalisation, that in turn allows for the development of theories and models that apply to the entire model class. One such class of distributions that is central to statistics is the exponential family of distributions. This class consists of many of the most common distributions in statistics, for example, the normal, the Poisson, the binomial distribution and many more. Although these distributions are very different, as can be seen above, how they are constructed is the same.

Definition 2.12 A class of distributions for a random variable Y is called an **exponential family**, if the density (or probability function) can be written in the form

$$f_Y(y; \theta) = \exp\{t^T(y)\theta - \kappa(\theta)\}h(y), \qquad (2.1.12)$$

where $h(y) \geq 0$ and $t(y) = (t_1(y), \ldots, t_p(y))^T$ is a vector of known functions and $\theta = (\theta_1, \ldots, \theta_p)^T$ is a parameter vector.

The density (or probability function) given in (2.1.12) is very general, and thus looks a little complicated, but we will see in the subsequent examples that the different terms simplify quite substantially in most real examples. Let us therefore consider the quantities in (2.1.12) in a little more depth. First of all, the function $t(y)$ is a function of the random variable and will later be called a statistic. The term θ is the parameter of the distribution, also called **natural parameter**, and quantity $\kappa(\theta)$ simply serves as normalisation constant, such that the density integrates out to

1. Hence, $\kappa(\theta)$ is defined by the following:

$$1 = \int \exp\{t^T(y)\theta\}h(y)dy\exp(-\kappa(\theta))$$

$$\Leftrightarrow \quad \kappa(\theta) = \log \int \exp\{t^T(y)\theta\}h(y)dy.$$

As a consequence, by differentiating $\kappa(\theta)$, we get

$$\frac{\partial\kappa(\theta)}{\partial\theta} = \frac{\int t^T(y)\exp\{t^T(y)\theta\}h(y)dy}{\int \exp\{t^T(y)\theta\}h(y)dy} = \int t^T(y)f_Y(y;\theta)dy = E\left(t^T(Y)\right).$$

Finally, the quantity $h(y)$ depends on the random variable and not on the parameter and we will see later that this quantity will not be of particular interest.

We will now demonstrate how some of the above introduced distributions can be written in the style of an exponential family. We thereby start with the normal distribution which can be rewritten as

$$f(y) = \frac{1}{\sqrt{2\pi\sigma^2}}\exp\left(-\frac{1}{2}\frac{(y-\mu)^2}{\sigma^2}\right)$$

$$= \exp\left(-\frac{1}{2}\frac{y^2-2y\mu+\mu^2}{\sigma^2} - \frac{1}{2}\log(\sigma^2)\right)\frac{1}{\sqrt{2\pi}}$$

$$= \exp\left(\underbrace{(-\frac{y^2}{2},y)}_{t^T(y)}\underbrace{\begin{pmatrix}\frac{1}{\sigma^2}\\\frac{\mu}{\sigma^2}\end{pmatrix}}_{\theta} - \underbrace{\frac{1}{2}(-\log\frac{1}{\sigma^2}+\frac{\mu^2}{\sigma^2})}_{\kappa(\theta)}\right)\underbrace{\frac{1}{\sqrt{2\pi}}}_{h(y)},$$

where $\theta_1 = \frac{1}{\sigma^2}$ and $\theta_2 = \frac{\mu}{\sigma^2}$ such that

$$\kappa(\theta) = \frac{1}{2}\left(-\log(\theta_1)+\frac{\theta_2^2}{\theta_1}\right) = -\frac{1}{2}\left(\log(\theta_1)-\frac{\theta_2^2}{\theta_1}\right).$$

Note that:

$$\frac{\partial\kappa(\theta)}{\partial\theta_1} = -\frac{1}{2}\left(\frac{1}{\theta_1}+\frac{\theta_2^2}{\theta_1^2}\right) = -\frac{1}{2}(\sigma^2+\mu^2) = E\left(-\frac{Y^2}{2}\right) = E\left(t_1(Y)\right)$$

$$\frac{\partial\kappa(\theta)}{\partial\theta_2} = \frac{\theta_2}{\theta_1} = \mu = E(Y) = E\left(t_2(Y)\right).$$

Hence, we see that the normal distribution is an exponential family distribution. The same holds for the Binomial distribution, because

$$f(y) = P(Y = y) = \binom{n}{y} \pi^y (1 - \pi)^{n-y}$$

$$= \left(\frac{\pi}{1 - \pi}\right)^y (1 - \pi)^n \binom{n}{y}$$

$$= \exp\left(\underbrace{y}_{t(y)} \underbrace{\log\left(\frac{\pi}{1 - \pi}\right)}_{\theta} - \underbrace{n \log\left(\frac{1}{1 - \pi}\right)}_{\kappa(\theta)}\right) \underbrace{\binom{n}{y}}_{h(y)},$$

where $\theta = \log(\frac{\pi}{(1-\pi)})$ is also known as log odds ratio and $\kappa(\theta)$ results in

$$\kappa(\theta) = n \log(1 + \exp(\theta)).$$

As $\pi = \exp(\theta)/(1 + \exp(\theta))$ one gets

$$\frac{\partial \kappa(\theta)}{\partial \theta} = n \frac{\exp(\theta)}{1 + \exp(\theta)} = n\pi = E(Y) = E\Big(t(Y)\Big).$$

Among many others the Poisson distribution is also part of the exponential family, the proof of which we leave as an exercise for the reader.

2.1.6   Random Vectors and Multivariate Distributions

We must also introduce the concept of multivariate random variables. A sequence of random variables $Y_1, Y_2, \ldots, Y_q$ is often represented as a random vector $(Y_1, Y_2, \ldots, Y_q)$. To allow for dependence between the elements of the random vector, we need to extend our previously introduced concept of the probability distribution towards a multivariate distribution function. The corresponding (multivariate) cumulative distribution function is given by

$$F(y_1, \ldots, y_q) = P(Y_1 \leq y_1, \ldots, Y_q \leq y_q).$$

If $F(.)$ is continuous and differentiable, we can relate the cumulative distribution function to the density function $f(y_1, \ldots y_q)$ with

$$P(a_1 \leq Y_1 \leq b_1, \ldots a_q \leq Y_q \leq b_q) = \int_{a_1}^{b_1} \cdots \int_{a_q}^{b_q} f(y_1, \ldots, y_q) dy_1 \ldots dy_q.$$

$$(2.1.13)$$

Integrating out all variables except for one gives the **marginal distribution**. We can also extend the concept of conditional probability to multivariate random variables.

Definition 2.13 (Marginal and Conditional Distribution) The marginal density with respect to Y_1 is given by

$$f_{Y_1}(y_1) = \int_{-\infty}^{\infty} \cdots \int_{-\infty}^{\infty} f(y_1, \ldots, y_q) dy_2 \ldots dy_q. \qquad (2.1.14)$$

The conditional density is analogue to the conditional probability. We give the definition for the two dimensional case as

$$f_{Y_1|Y_2}(y_1|y_2) = \frac{f(y_1, y_2)}{f(y_2)} \text{ for } f(y_2) > 0. \qquad (2.1.15)$$

From (2.1.15) we can calculate conditional expectation and variance, e.g.

$$E(Y_1|Y_2 = y_2) = \int y_1 f_{Y_1|Y_2}(y_1|y_2) dy_1$$

$$Var(Y_1|Y_2 = y_2) = \int \{y_1 - E(Y_1|Y_2 = y_2)\}^2 f_{Y_1|Y_2}(y_1|y_2) dy_1$$

$$= E(Y_1^2|Y_2 = y_2) - \{E(Y_1|Y_2 = y_2)\}^2.$$

Of particular interest is the dependence structure among the components of the random vector. This can be expressed in the covariance matrix, which is defined as follows:

$$Cov(Y) = \begin{pmatrix} Cov(Y_1, Y_1) & Cov(Y_1, Y_2) & \cdots & Cov(Y_1, Y_q) \\ Cov(Y_2, Y_1) & \ddots & & \vdots \\ \vdots & & \ddots & \\ Cov(Y_q, Y_1) & \cdots & & Cov(Y_q, Y_q) \end{pmatrix},$$

where

$$Cov(Y_j, Y_k) = E\left[\left\{Y_j - E(Y_j)\right\}\left\{Y_k - E(Y_k)\right\}\right] = E(Y_j Y_k) - E(Y_j)E(Y_k)$$

for $1 \leq j, k \leq q$. It is clear that $Cov(Y_j, Y_j) = Var(Y_j)$ and that if Y_j and Y_k are independent, then $Cov(Y_j, Y_k) = 0$. This follows easily, as under independence we have

$$f(y_j, y_k) = f_{Y_j}(y_j) f_{Y_k}(y_k),$$

such that $E(Y_j Y_k) = E(Y_j)E(Y_k)$. As well as the covariance, we will occasionally focus on the correlation, which is defined as follows:

$$Cor(Y_j, Y_k) = \frac{Cov(Y_j, Y_k)}{\sqrt{Var(Y_j)Var(Y_k)}}.$$

In Chap. 10 we will discuss more details and advanced statistical models for multivariate random vectors. For now, we will simply present most commonly encountered multivariate distribution, namely the multivariate normal distribution. It is defined as follows:

Definition 2.14 The random vector $(Y_1, \ldots, Y_q)^T$ follows a **multivariate normal distribution** if the density takes the form

$$f(y_1, \ldots, y_q) = \frac{1}{(2\pi)^{q/2}} |\Sigma|^{-1/2} \exp\left(-\frac{1}{2} \sum_{j=1}^{q} \sum_{k=1}^{q} (y_j - \mu_j)(y_k - \mu_k)\Omega_{jk}\right)$$

$$= \frac{1}{(2\pi)^{q/2}} |\Sigma|^{-1/2} \exp\left(-\frac{1}{2}(y - \mu)^T \Omega (y - \mu)\right),$$

where $\Omega = \Sigma^{-1}$ and $y = (y_1, \ldots, y_q)^T$. We denote this with $Y \sim N(\mu, \Sigma)$. The mean vector $\mu = (\mu_1, \ldots, \mu_q)^T$ is the vector of expected values, i.e. $\mu_j = E(Y_j)$ and the covariance matrix Σ has entries

$$\Sigma_{jk} = Cov(Y_j, Y_k).$$

A conditional normal distribution is derived as follows. Let vector $(Y_1, \ldots, Y_q)^T$ be divided into two subcomponents, labelled (Y_A^T, Y_B^T). Accordingly, let Σ be divided into the four submatrices

$$\Sigma = \begin{pmatrix} \Sigma_{AA} & \Sigma_{AB} \\ \Sigma_{BA} & \Sigma_{BB} \end{pmatrix},$$

where $A, B \subset \{1, \ldots, q\}$ with $A \cap B = \emptyset$ and $A \cup B = \{1, \ldots, q\}$. Then

$$Y_A | y_B \sim N\left(\mu_A - \Sigma_{AB}\Sigma_{BB}^{-1}(y_B - \mu_B), \Sigma_{AA} - \Sigma_{AB}\Sigma_{BB}^{-1}\Sigma_{BA}\right),$$

where μ_A and μ_B are the corresponding subvectors of μ. Hence, the conditional distribution of a multivariate normal distribution is again normal. This also holds for the marginal distribution, i.e.

$$Y_A \sim N(\mu_A, \Sigma_{AA}).$$

While the normal distribution is central for modelling multivariate continuous variables, the multinomial distribution is central for modelling discrete-valued random variables.

Definition 2.15 The random vector $Y = (Y_1, \ldots, Y_K)^T$ follows a multinomial distribution if

- $Y_k \in \mathbb{N}$ for all $k = 1, \ldots, K$
- $\sum_{k=1}^{K} Y_k = n$
- $P(Y_1 = y_1, \ldots, Y_K = y_K) = \frac{n!}{y_1! \cdots y_K!} \prod_{k=1}^{K} \pi_k^{y_k}$ where $\pi_k \in [0, 1]$ and $\sum_{k=1}^{K} \pi_k = 1$

We notate this as $Y \sim Multi(n, \boldsymbol{\pi})$ with $\boldsymbol{\pi} = (\pi_1, \ldots, \pi_K)^T$. The random vector $Y = (Y_1, \ldots, Y_K)^T$ has the expectation

$$E(Y) = \boldsymbol{\pi}$$

and the covariance matrix is given by

$$Var(Y) = diag(\boldsymbol{\pi}) - \boldsymbol{\pi} \boldsymbol{\pi}^T$$

$$= \begin{pmatrix} \pi_1(1 - \pi_1) & -\pi_1 \pi_2 & \ldots & -\pi_1 \pi_K \\ -\pi_2 \pi_1 & \pi_2(1 - \pi_2) & \ldots & \\ \vdots & & & \vdots \\ -\pi_K \pi_1 & & \ldots & \pi_K(1 - \pi_K) \end{pmatrix}.$$

Multivariate distributions can also be used to derive moments for subsets of the random vector. We exemplify this for a bivariate setting, where (Y, X) are two random variables. The mean and variance of Y can be derived from the conditional mean and variance of Y given $X = x$ as follows:

Property 2.3 (Iterated Expectation) For two random variables X and Y we find

$$E_Y(Y) = E_X(E_Y(Y|X))$$

$$Var_Y(Y) = E_X(Var_Y(Y|X)) + Var_X(E_Y(Y|X)),$$

where E_X and Var_X indicate the expectation and variance with respect to random variable X, while E_Y and Var_Y denote the corresponding conditional quantities for random variable Y.

We will need this statement later in the book and fortunately, the proof is rather simple as can be seen below.

Proof Note that

$$E_Y(Y) = \int yf(y)dy = \int \underbrace{\int yf(y|x)dy} \, f_X(x)dx$$

$$= \int E_Y(Y|x)f_X(x)dx = E_X\Big(E_Y(Y|X)\Big).$$

Moreover with $\mu = E_Y(Y)$ and $\mu(x) = E_Y(Y|x)$ we get

$$Var_Y(Y) = E_Y(Y^2) - \{E_Y(Y)\}^2$$

$$= E_X[\underbrace{E_Y(Y^2|X)}_{=}] - \{E_X[E_Y(Y|X)]\}^2$$

$$= E_X[\overbrace{Var_Y(Y|X) + \{E(Y|X)\}^2}] - \{E_X[E_Y(Y|X)]\}^2$$

$$= E_X[Var_Y(Y|X)] + \underbrace{E_X[E_Y(Y|X)^2] - \{E_X[E_Y(Y|X)]\}^2}_{=}$$

$$= E_X[Var_Y(Y|X)] + \overbrace{Var_X[E_Y(Y|X)]}.$$

$\square$

2.2 Limit Theorems

Numerous concepts in statistical reasoning rely on asymptotic properties. In its most simple case this is, for instance, the behaviour of the arithmetic mean of n independent observations. For a sample of n individuals, we assume that the observations from one individual do not depend on the observations from another individual, which we call independence. This concept of independence has important properties and justifies nearly all asymptotic arguments in statistical reasoning. "Asymptotic" thereby means that one explores how a quantity derived from a sample behaves if the sample size increases, i.e. that is the number of (independent) observations n tends to infinity. Asymptotic arguments will help to quantify uncertainty and we will see that asymptotic normality plays a fundamental role in statistical reasoning. Before motivating this in more depth, let us visualise this with a simple example. A random walk is recursively defined as $Y_t = Y_{t-1} + X_t$ for $t = 1, 2, \ldots$, where we assume

$$X_t = \begin{cases} 1 & \text{with probability } \pi \\ -1 & \text{with probability } 1 - \pi. \end{cases}$$

It is natural to start with $Y_0 = 0$, so that we can rewrite $Y_n = \sum_{i=1}^{n} X_i$. We assume
that the X_i are independent and set $\pi = 1/2$, which gives $E(X_i) = \pi - (1 -$
$\pi) = 0$ and $Var(X_i) = E(X_i^2) = 1$. We run 100.000 simulations of this random
walk and stop at $n = 10.000$. In Fig. 2.2 (top plot) we plot the resulting values
$y_1, y_2, \ldots, y_{10000}$ for 3 of these simulations. The shaded region of Fig. 2.2 shows the
range in which 90% of the observed values lie. The perceptive reader might note that
the boundary of the shaded area appears similar to $\sqrt{n}$. We therefore conclude that,
even though Y_n is a sum of n random variables, its spread does not grow with order
n but "only" with order $\sqrt{n}$. This is a fundamental property, which in fact justifies
most asymptotic reasoning in statistics. It implies that if we divide Y_n by $\sqrt{n}$, we
get the behaviour shown on the middle plot of Fig. 2.2. Because $Var(Y_n) = n$,
the variance of $\frac{Y_n}{\sqrt{n}}$ is 1. The reader can see that $Z_n := Y_n/\sqrt{n}$ shows a stabilised
behaviour and 90% of the simulations remain in the shaded region. We say that Y_n
has the asymptotic order of $\sqrt{n}$, meaning that Y_n is (stochastically) proportional to
$\sqrt{n}$. Without giving an exact definition here it means that the variance of $Y_n/\sqrt{n}$
does not depend on n (if n is large). In the bottom panel of Fig. 2.2, we show the
arithmetic mean $Y_n/n = \sum_{i=1}^{n} X_i/n$ and observe that the mean of the X-values
gets closer to 0 with increasing n. We say that $\frac{1}{n} \sum_{i=1}^{n} x_i$ converges in probability.
This property is known as the **law of large numbers**. Its central statement is that
the arithmetic mean of independent variables which have all the same distribution
converges to its expected value. An exact definition is given later.

Now let us explore whether we can say something about the distribution of
$Z_n = Y_n/\sqrt{n}$. In Fig. 2.3, we show the empirical distribution of the 100.000
simulations for $n = 10$ and $n = 100$ as a histogram. We overlay the plot
with a normal distribution. We see that for $n = 100$ a striking concordance
with a normal distribution is already apparent. Hence, if we divide the sum of
n independent random variables by $\sqrt{n}$ we obtain a stable behaviour, where the
distribution of $Z_n = Y_n/\sqrt{n}$ approximately follows a normal distribution. This
property is central to statistics and is expressed in the central limit theorem. All
in all, there are three theorems we want to highlight. All rely on the assumption
of independent and identically distributed random variables, which we explicitly
define in the next section. Formally, it means that in a sequence of random variables
$X_1, X_2, \ldots, X_n$

1. X_i and X_j are mutually independent,
2. X_i and X_j have the same distribution

for all $i \neq j$.

Property 2.4 (Central Limit Theorem)   Let $Y_1, Y_2, Y_3, \ldots$ be independent and
identically distributed random variables, with mean zero and variance σ^2. The
distribution of $Z_n = \sum_{i=1}^{n} Y_i/\sqrt{n}$ converges to a normal distribution with mean
zero and variance σ^2, which we denote as

$$Z_n \xrightarrow{d} N(0, \sigma^2).$$

Fig. 2.2 Top plot: Random
Walks with $n = 10.000$ steps.
Middle plot: Random Walks
divided by $\sqrt{n}$. Bottom plot:
divided by n

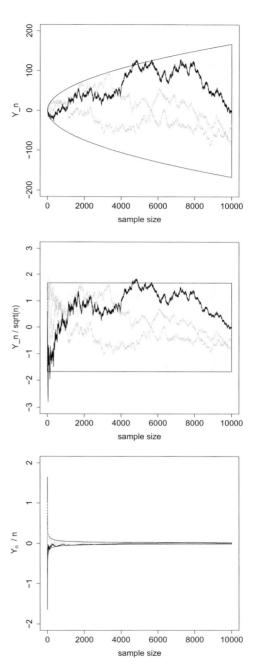

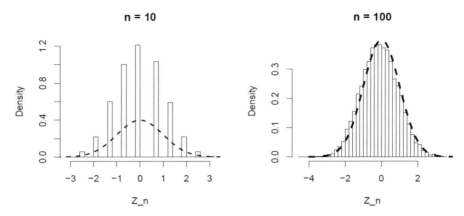

Fig. 2.3 Distribution of $Y_n/\sqrt{n}$ at $n = 10$ and $n = 100$ given 100.000 simulations

The word "central" has a double meaning here, as the central limit theorem truly is a core component of statistics. Note that the only assumption we make is that random variables $Y_1, \ldots, Y_n$ have a finite variance, but they can otherwise follow arbitrary distributions. Because the central limit theorem is so important, it seems worthwhile to sketch a proof.

Proof We show that the **moment generating function** of Z_n converges to the moment generating function of a normal distribution. The moment generating function of a random variable Y thereby is defined as

$$M_Y(t) = E\left(e^{tY}\right).$$

Note that

$$\left.\frac{\partial^k M_Y(t)}{(\partial t)^k}\right|_{t=0} = E(Y^k),$$

which justifies the name of the function as its k-th derivative generates the k-th moment of random variable Y, which, as a reminder, is $E(Y^k)$. A companion of the moment generating function is the cumulant generating function, defined as the logarithm of the moment generating function, i.e.

$$K_Y(t) = \log M_Y(t).$$

For a normally distributed random variable $Z \sim N(\mu, \sigma^2)$, the moment generating function is given by

$$M_Z(t) = \exp\left(\mu t + \frac{1}{2}\sigma^2 t^2\right),$$

which gives

$$K_Z(t) = \mu t + \frac{1}{2}\sigma^2 t^2.$$

This implies that the first two derivatives $\partial^k K_U(t)/\partial t\big|_{t=0}$ are equal to the mean μ and the variance σ^2 and all higher order derivatives are equal to zero.

Moreover, note that for any constant $c > 0$ one has

$$M_{cY}(t) = E\left(e^{tcY}\right) = M_Y(ct).$$

The last component needed for our proof is that a random variable is uniquely defined by its cumulant (or moment) generating function and vice versa (as long as the moments and cumulants are finite). Hence, if we prove convergence of the cumulant generating function, it implies convergence of the corresponding random variables as well. Assume now that $Y_1, \ldots, Y_n$ are independent and identically distributed random variables (but not necessarily normal), each drawn from the distribution $F_y()$ with mean value μ and variance σ^2. Then, the moment generating function of $Z_n = (Y_1 + \ldots + Y_n)/\sqrt{n}$ is given by

$$M_{Z_n}(t) = E\left(e^{t(Y_1+Y_2+\ldots+Y_n)/\sqrt{n}}\right)$$

$$= E\left(e^{tY_1/\sqrt{n}} \cdot e^{tY_2/\sqrt{n}} \cdot \ldots \cdot e^{tY_n/\sqrt{n}}\right)$$

$$= E\left(e^{tY_1/\sqrt{n}}\right) \cdot E\left(e^{tY_2/\sqrt{n}}\right) \cdot \ldots \cdot E\left(e^{tY_n/\sqrt{n}}\right)$$

$$= M_Y\left(t/\sqrt{n}\right)^n.$$

Correspondingly, the cumulant generating function for Z_n is given by

$$K_{Z_n}(t) = n K_Y\left(t/\sqrt{n}\right).$$

Taking derivatives gives

$$\frac{\partial K_{Z_n}(t)}{\partial t}\bigg|_{t=0} = \frac{n}{\sqrt{n}} \frac{\partial K_Y(t)}{\partial t}\bigg|_{t=0} = \sqrt{n}\mu$$

$$\frac{\partial^2 K_{Z_n}(t)}{(\partial t)^2}\bigg|_{t=0} = \frac{n}{n} \frac{\partial^2 K_Y(t)}{(\partial t)^2}\bigg|_{t=0} = \sigma^2$$

and all higher order derivatives tend to zero as n goes to infinity. Using Taylor series expansion we can derive $K_{Z_n}(t)$ around $K_Z(0)$ with

$$K_{Z_n}(t) = K_{Z_n}(0) + \frac{\partial K_{Z_n}(t)}{\partial t}\Big|_{t=0} t + \frac{1}{2}\frac{\partial^2 K_{Z_n}(t)}{(\partial t)^2} t^2 + \dots$$

$$= 0 + \sqrt{n}\mu t + \frac{1}{2}\sigma^2 t^2 + \dots,$$

where the terms collected in $\dots$ are of order $1/\sqrt{n}$ or smaller. Hence, for $n \to \infty$ one has

$$K_{Z_n}(t) \to K_Z(t),$$

where $Z \sim N(\sqrt{n}\mu, \sigma^2)$. This in turn proves the central limit theorem.   □

We should finally note that in fact the central limit theorem has been proven to apply under a much weaker set of assumptions, where the random variables need not be identically distributed or may be dependent.

The central limit theorem is visible in the middle plot of Fig. 2.2, but even clearer in Fig. 2.3. The bottom plot of Fig. 2.2, on the other hand, demonstrates what is known as the law of large numbers, which we define as follows.

Property 2.5 (Law of Large Numbers) . Let $Y_1, Y_2, Y_3, \dots$ be independent and identically distributed random variables with mean μ. Then for every $\varepsilon > 0$ we have

$$P(|\frac{1}{n}\sum_{i=1}^{n} Y_i - \mu| > \epsilon) \to 0$$

The law of large numbers states that the arithmetic mean (or any other quantity which is expressible as an arithmetic mean) converges to its mean value. In particular, the difference between the mean value μ and the arithmetic mean $\sum_i Y_i/n$ becomes infinitely small for the sample size n increasing. One important special case results when Y_i is a Bernoulli variable with $P(Y_i = 1) = p$ and $P(Y_i = 0) = 1 - p$ and $E(Y_i) = p$. Then $\frac{1}{n}\sum_{i=1}^{n} Y_i \to p$. Since $\frac{1}{n}\sum_{i=1}^{n} Y_i$ is the relative frequency of the outcome 1, this result can be interpreted as follows. When we independently repeat an experiment with a 0/1 outcome very often, then the relative frequency of successes converges to its success probability. This mirrors in an interesting way the Frequentist definition of probability (see Definition 2.2).

Let us go a step further and consider *i.i.d.* continuous random variables. As an example we use the exponential distribution with parameter $\lambda = 1$, i.e.

$$Y_i \sim Exp(\lambda), i = 1, \dots, n.$$

In Fig. 2.4, a histogram for sample sizes $n = 30$ and $n = 1000$ is given. The histogram comes to closely resemble the density function (solid line) as the sample

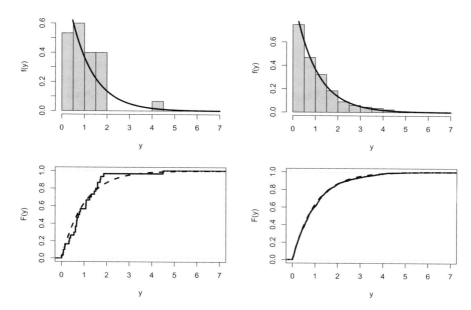

Fig. 2.4 Histograms (top row) and empirical distribution functions (bottom row) for n = 30 (left column) and n = 1000 (right column)

size increases. Alternatively, we can look at the empirical distribution function of the data. This is defined by

$$F_n(y) = \frac{1}{n} \sum_{i=1}^{n} 1_{\{Y_i \le y\}}$$

in other words, the proportion of data-points in the dataset with values less than y. For $n \to \infty$, the empirical distribution converges to the (theoretical) distribution $F(y)$, which is illustrated in Fig. 2.4.

Property 2.6 (Glivenko-Cantelli Theorem) Let $y_1, \ldots, y_n$ be a random sample from a distribution with distribution function $F(.)$. Then

$$lim_{n \to \infty} F_n(y) = F(y) \quad \text{for all } y \in \mathbb{R}. \tag{2.2.1}$$

$F_n(y)$ is the empirical distribution function for the sample $y_1, \ldots, y_n$.

The statement goes back to Glivenko (1933) and Cantelli (1933). The theorem is of central importance in data analysis as it states that the empirical distribution function is a good estimate for the true distribution function. We will extensively exploit the theorem in Chap. 9, when we discuss resampling, specifically with bootstrapping. For the proof we refer to the standard probability literature, e.g. Karr (1993), page 206.

2.3 Kullback–Leibler Divergence

In statistics, we are often faced with comparing two distributions. Assume that $f(y)$ is a density (or probability function) and so is $g(y)$. Our intention is to measure how far apart $f(.)$ and $g(.)$ are. Consider Fig. 2.5 (top plot), where we show two densities $f(.)$ and $g(.)$. One approach would be to look at the log ratio of the two distribution

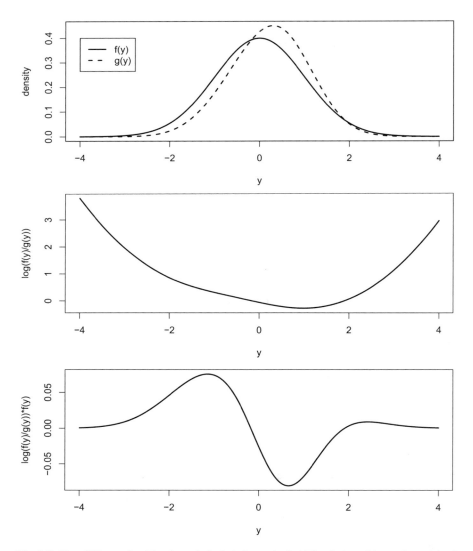

Fig. 2.5 Two different densities (top plot), their log ratio (middle plot), and log ratio weighted with $f(y)$ (bottom plot)

functions $\log(f(y)/g(y))$, which is visualised in the middle plot of Fig. 2.5. Clearly, this is largest in absolute terms at the boundary where both densities are small. To compensate for this, we can weight the ratio with respect to one of the two densities. This is shown in the bottom plot of Fig. 2.5, where we plot $\log(f(y)/g(y))f(y)$. After applying this change, the function is most pronounced around -1 and 1, where the difference (and hence distance) between the densities $f(y)$ and $g(y)$ matters most. We can now integrate this function, which leads us to the definition of the Kullback–Leibler divergence.

Definition 2.16 Let $f(y)$ and $g(y)$ be two densities or probability functions with the same support, i.e.$\{y : f(y) > 0\} = \{y : g(y) > 0\}$. The **Kullback–Leibler divergence** (KL divergence) is defined by

$$KL(f(.), g(.)) = \int_{-\infty}^{\infty} \log \frac{f(y)}{g(y)} f(y) dy = E_f\{\log \frac{f(Y)}{g(Y)}\}. \tag{2.3.1}$$

The Kullback–Leibler divergence will be used in many places throughout this book and thus it might be useful to discuss the measure in more depth. First of all, we need to emphasise that the Kullback–Leibler measure is a divergence and not a distance. The difference is that a distance measure needs to be symmetric, i.e. the distance from $f(.)$ to $g(.)$ is the same as the distance from $g(.)$ to $f(.)$. This simple property does not hold for the KL divergence and in fact $K(f(.), g(.)) \neq K(g(.), f(.))$ unless $g(.) = f(.)$. Nonetheless, it holds that:

$$KL(f(.), g(.)) = \begin{cases} 0 & \text{if } f(y) = g(y) \text{ for all } y \in \mathbb{R} \\ > 0 & \text{otherwise.} \end{cases}$$

This property can be easily seen as for the (natural) logarithm it holds $\log(x) \leq x - 1$ for all $x \geq 0$ and $\log(x) = x - 1$ if and only if $x = 1$. This implies for two densities $f(y)$ and $g(y)$ that

$$KL(f(.), g(.)) = \int \log \frac{f(y)}{g(y)} f(y) dy = -\int \log \frac{g(y)}{f(y)} f(y) dy$$

$$\geq \int \left(1 - \frac{g(y)}{f(y)}\right) f(y) dy = 1 - 1 = 0.$$

Note that the Kullback–Leibler divergence decomposes to

$$KL(f(.), g(.)) = \int \log(f(y)) f(y) dy - \int \log(g(y)) f(y) dy$$

$$= E_f\{\log f(Y)\} - E_f\{\log g(Y)\},$$

where the first component is also defined as the **entropy** of $f(y)$. The Kullback–Leibler divergence will be used at several places throughout the book. Indeed, it turns out that this quantity is of central importance in statistical reasoning.

2.4 Exercises

Exercise 1
A random variable Y with values in the interval $(0, 1)$ is sometimes modelled with a Beta distribution. The probability density function of a Beta(α, β) distributed random variable Y is given by

$$f(y) = \frac{1}{B(\alpha, \beta)} y^{\alpha-1}(1 - y)^{\beta-1} \, ,$$

where $B(\alpha, \beta)$ is the Beta function

$$B(\alpha, \beta) = \frac{\Gamma(\alpha)\Gamma(\beta)}{\Gamma(\alpha + \beta)} \, ,$$

and $\Gamma(\cdot)$ denotes the Gamma function. Show that the Beta distribution is a member of the exponential family and determine the (natural) parameters.

Exercise 2
A family of distributions with two parameters, the distribution function

$$F(y|a, b) = F_0 \left(\frac{y - a}{b} \right) , a \in \mathbb{R}, b > 0$$

and $F_0(y)$ defined for $a = 0, b = 1$ is called a location scale family. a is the location parameter and b is the scale parameter.

1. Show that the density f in the continuous case is

$$f(y|a, b) = \frac{1}{b} f_0 \left(\frac{y - a}{b} \right) .$$

2. The density of a generalised t-distributed random variable $Y \sim t_\nu(\mu, \sigma^2)$ with location parameter μ, scale parameter $\sigma \in \mathbb{R}_+$ and $\nu > 2$ degrees of freedom is

$$f(y) = \frac{\Gamma\left(\frac{\nu+1}{2}\right)}{\Gamma\left(\frac{\nu}{2}\right)\sqrt{\nu\pi}\sigma} \left(1 + \frac{(y - \mu)^2}{\nu\sigma^2} \right)^{-\frac{\nu+1}{2}} .$$

The moments are $E(Y) = \mu$ and $Var(Y) = \frac{\nu}{\nu-1}\sigma^2$. Show that the family of generalised t-distributions is an exponential family for fixed ν.

Exercise 3 (Use R Statistical Software)

Random numbers are often used to design simulation studies for evaluating the performance of statistical procedures. Generate a sample of $n = 200$ trivariate normal random numbers with mean $\mu = (3.0, 10.0, 50.0)$ and covariance matrix

$$\Sigma = \begin{pmatrix} 1.0 \ 0.6 \ 0.6 \\ 0.6 \ 1.0 \ 0.6 \\ 0.6 \ 0.6 \ 1.0 \end{pmatrix} .$$

1. As a first method, use the fact that, if X is a random vector of independent standard normal $N(0, 1)$ variables, then

$$Y = \mu + \Sigma^{1/2} X$$

has the desired mean and covariance matrix. Use e.g. the Cholesky decomposition in R to compute the matrix root $\Sigma^{1/2}$. By repeating the experiment $S = 1000$ times, check empirically that your random numbers have the desired expected values, variances and covariances.
2. A more comfortable way is to use the function rmvnorm in the package mvtnorm. Repeat your experiment using this function.
3. Calculate the (conditional) variance of the conditional distribution of $(Y_1|Y_2, Y_3)$.

Exercise 4 (Use R Statistical Software)

Write a simulation program using random numbers and plots to show that the central limit theorem holds for independent Poisson distributed random variables. Evaluate, whether the speed of convergence depends on the choice of the parameter $\lambda : \lambda = 0.1, \lambda = 0.5, \lambda = 1.0, \lambda = 5, \lambda = 10, \lambda = 50$. Discuss your results.

Chapter 3
Parametric Statistical Models

Now that we have introduced probability models, we have the tools we need to put our statisticians' hats on. We want to make a probabilistic model that best describes the world around us. How is it that we can best move from our set of observations to a good model—a model that not only describes our samples, but the process that generated them? In this chapter, we start by making the assumption that the observed data follow a probability model, whose properties are described by a set of parameters. Now that we have this data, the statistical question is: how can we draw information from the samples about the parameters of the distributions that generated them? One basic assumption that aids this process enormously is that of independence.

Typically, statistical reasoning is built upon the assumption of **independence** of the observed measurements. In fact, most of the theoretical results in the subsequent chapter rely on this concept in one way or another. Therefore, we often assume that our observations are (or could be) replicated and that the outcome of one measurement does not depend on the outcome of any other. In practice, however, this requirement is not always met and, in fact, one often needs to think very carefully about whether it applies at all. To make this clear we look again at a very (!) simple and artificial example. Assume we are interested in the height difference between men and women and measure the heights of a single man and a woman once per day. Our database grows larger and larger. However, the data carry very little information about our research question because independence is violated. The data are taken from only two individuals! From the data we could, for example, extract information about height variation of an individual over different days, i.e. their height conditional upon the individual chosen. Either way, we still could not draw a conclusion, even with thousands of data-points and means that differed substantially. This is because our question relies on the individual chosen for each sample to be independent, which is clearly not the case here. We hope that this little example demonstrates the importance of sample independence. For the moment we will continue with this common assumption.

© The Author(s), under exclusive license to Springer Nature Switzerland AG 2021   33
G. Kauermann et al., *Statistical Foundations, Reasoning and Inference*,
Springer Series in Statistics, https://doi.org/10.1007/978-3-030-69827-0_3

3.1 Likelihood and Bayes

Now is a good chance to introduce the concept of independence more formally. Let us begin with random variables $Y_1, \ldots, Y_n$ that come from the probability model

$$Y_i \sim F(y; \theta). \tag{3.1.1}$$

Here, $F(y; \theta)$ denotes a known distribution function, but the parameter θ is unknown. The parameter may contain multiple components, i.e. $\theta = (\theta_1, \ldots, \theta_p)$, but for simplicity's sake let us take θ to be univariate, i.e. $p = 1$. We let Θ be the set of possible parameter values, i.e. $\theta \in \Theta$. Note that the distribution function $F(y; \theta)$ is the same for all n observations. Finally, we postulate that the n observations are mutually independent, which gives

$$P(Y_1 \leq y_1, \ldots, Y_n \leq y_n) = P(Y_1 \leq y_1) \cdot \ldots \cdot P(Y_n \leq y_n) = \prod_{i=1}^{n} F(y_i; \theta).$$

This scenario is called the ***i.i.d.*** setting, where the abbreviation stands for *independently* and *identically distributed* random variables. We repeat that one should not accept this at face value, as in many real data situations, the *i.i.d.* assumption may be violated. For now, however, we rely on the *i.i.d.* assumption, defined as follows.

Definition 3.1 The data $y_1, \ldots, y_n$ are called **independently and identically distributed** (*i.i.d.*) if for $i = 1, \ldots, n$ we assume y_i to be the realisation of a random variable Y_i, with distribution $F(y; \theta)$ and that

$$P(Y_1 \leq y_1, \ldots, Y_n \leq y_n) = \prod_{i=1}^{n} F(y_i; \theta).$$

We now aim to use the observed data $y_1, \ldots, y_n$ to draw information about θ in the model. Given that the parameter θ is unknown, we may take the viewpoint of a subjective probability and assume that θ is random. Clearly this is a conceptional jump, but it has some attractive aspects that will become clear shortly. Bear in mind, a random variable is unknown but a realisation can be obtained by drawing and observing it.

Using the notion of subjective probability, we can formulate our uncertainty about θ in the form of a probability model. Following this path, we assume

$$\theta \sim F_\theta(\vartheta), \tag{3.1.2}$$

where $F_\theta(.)$ denotes a suitable distribution function for θ with $P(\theta \leq \vartheta) = F_\theta(\vartheta)$. This distribution is later called **prior distribution** (or sometimes *a priori* distribution) and it may depend on additional parameters, called hyperparameters, which are omitted for now. To distinguish between the random variable θ from its possible realisations, we denote possible concrete values of θ with ϑ. One way to interpret the prior distribution (3.1.2) is as a representation of our existing knowledge about the parameter before observing any data. For instance, if we are sure that θ takes a particular value θ_0, then F_θ can be taken as a step function such that:

$$F_\theta(\vartheta) = \begin{cases} 0 & \vartheta < \theta_0 \\ 1 & \text{otherwise.} \end{cases}$$

On the other hand, if we are not at all sure about θ, then $F_\theta()$ can also demonstrate our lack of knowledge before observing the data. Let us take the example of a Bernoulli distribution, i.e. an experiment with two outcomes $Y \in \{0, 1\}$, where $P(Y = 1) = \theta$ and $P(Y = 0) = 1 - \theta$. The unknown parameter θ can hold values between 0 and 1. One strategy to express total uncertainty is to assume a flat prior, i.e. with density $f_\theta(\vartheta) = 1$ if $0 \leq \vartheta \leq 1$ and $f_\theta(\vartheta) = 0$ otherwise. That is, we think that all valid values for θ are equally likely. There are other ways to express uncertainty in our prior that will be covered in more detail in Chap. 5.

Now we observe $y_1, \ldots, y_n$ as realisations of (3.1.1) and obtain information about θ which we need to quantify. So let us use $f(y; \theta)$ as the density or probability function derived from (3.1.1). We can now use **Bayes rule** and calculate the conditional density

$$f_\theta(\vartheta | y_1, \ldots, y_n) = \frac{\left[\prod_{i=1}^{n} f(y_i; \vartheta)\right] f_\theta(\vartheta)}{\int_\Theta \left[\prod_{i=1}^{n} f(y_i; \tilde{\vartheta})\right] f_\theta(\tilde{\vartheta}) d\tilde{\vartheta}}. \tag{3.1.3}$$

Hence $f_\theta(\vartheta | y_1, \ldots, y_n)$ denotes the conditional density or probability function corresponding to the distribution function $F_\theta(\vartheta | y_1, \ldots, y_n)$. The distribution (3.1.3) is also called the **posterior distribution** as it is calculated after observing the data. Note that the denominator in (3.1.3) is a normalisation constant that does not depend on ϑ.

Looking at Eq. (3.1.3), we see that our knowledge of θ changes with data $y_1, \ldots, y_n$ being observed. We multiply the prior distribution with the density function given our data to obtain the posterior. The first component in the numerator of the ratio (3.1.3) essentially tells us how likely we are to observe our data realisations, given a particular parameter value. The prior expresses our prior knowledge on how likely we are to observe that parameter value at all. Taken together, these functions give us an updated distribution of the parameter that favours parameter values that are more likely to have produced our data. The function that describes the likelihood of producing our data given a particular parameter is defined as the **likelihood** of the data.

Definition 3.2 The **likelihood function** $L(\theta; y_1, \ldots, y_n)$ for an *i.i.d.* sample $y_1, \ldots, y_n$ is defined by

$$L(\theta; y_1, \ldots, y_n) = \prod_{i=1}^{n} f(y_i; \theta) \tag{3.1.4}$$

with $f(y_i; \theta)$ as a density or probability function. Taking the logarithm defines the **log-likelihood function**

$$l(\theta; y_1, \ldots, y_n) = \sum_{i=1}^{n} \log f(y_i; \theta).$$

The likelihood function and likelihood theory play an essential role in statistics. This will be discussed in depth in Chap. 4. For now we conclude, because the denominator of (3.1.3) is simply a normalisation constant that does not depend upon ϑ, the posterior distribution is proportional to the product of the likelihood function and the prior, i.e.

$$f_\theta(\vartheta | y_1, \ldots, y_n) \propto L(\vartheta; y_1, \ldots, y_n) f_\theta(\vartheta),$$

where $\propto$ here means "is proportional to". This proportionality implies that the combination of the likelihood and the prior density provides the quantifiable information (under the given probability model) about θ that the data-points $y_1, \ldots, y_n$ provide. We will make use of this function in Chap. 5 when we discuss Bayesian inference. If there is no prior knowledge about θ, or if we do not want to include prior knowledge in our analysis, then a **non-informative prior** is assumed. If we use the previously mentioned flat prior for θ we find that all information about θ is contained in the likelihood function $L(\theta | y_1, \ldots, y_n)$, because our prior then gets absorbed into the denominator as a constant and we are left with

$$f(\vartheta | y_1, \ldots, y_n) \propto L(\vartheta; y_1, \ldots, y_n).$$

This in turn shows that the likelihood function is an essential tool for extracting information about θ.

3.2 Parameter Estimation

We have defined two central functions above for drawing information about the parameter θ in the model: the posterior distribution (3.1.3) and the likelihood function (3.1.4). In both cases, the parameter values that define models that better describe the data become themselves the more likely. The next step is to use these functions to find the best value for the unknown parameter given the data $y_1, \ldots, y_n$.

This is to say, we are interested in a single value $\hat{\theta}$, which is a *plausible guess* for the unknown parameter θ. We call this guess an **estimate**. In what follows, we tackle the question of how to construct good and useful estimates from our data and how to assess the quality of these estimates. Note that our estimate $\hat{\theta}$ depends on the observed data, i.e.

$$\hat{\theta} = t(y_1, \ldots, y_n), \tag{3.2.1}$$

where $t : \mathbb{R}^n \mapsto \mathbb{R}$. We call any function of the data (and not the model parameters) $t(\cdot)$ a **statistic**. It is often helpful to bear in mind that $t(\cdot)$ depends on the sample size n, which we sometimes denote as $t_{(n)}(\cdot)$. The reader should also note that $t(y_1, \ldots, y_n)$ depends on the realisations $y_1, \ldots, y_n$ of random variables $Y_1, \ldots, Y_n$. However, taking the random variables instead of the realised values, leaves the statistic $t(Y_1, \ldots, Y_n)$ itself as a random variable. This allows us to model the properties of the statistic on average, independently of the random sample that is taken. This view will be essential for statistical reasoning, as we evaluate an estimate, not based on the concrete value that results from the sample $t(y_1, \ldots, y_n)$, but based on its stochastic counterpart $t(Y_1, \ldots, Y_n)$.

In the coming section, we examine a number of different approaches to the parameter estimation process. Maximum Likelihood chooses the model parameterisation that is most likely to have generated the given data, while Bayes estimation also includes the effect of our prior knowledge of the parameter. This is done by treating it as a random variable and finding its posterior probability after observing the data. Method of moments estimation attempts to find the parameter by matching the theoretical moments to the observed moments of the distribution as a system of equations. A different perspective is to take a traditional optimisation approach and minimise a loss between predicted and actual values, or even minimise the difference between the two entire distributions using the Kullback–Leibler divergence. We begin by looking at Bayes estimation.

3.2.1 Bayes Estimation

From the Bayesian perspective, the posterior distribution (3.1.3) given the data $y_1, \ldots, y_n$ contains all of the information that we have about θ. Therefore, it seems that the posterior mean value would be a natural candidate for estimating θ.

Definition 3.3 The **posterior mean estimate** $\hat{\theta}_{postmean}$ is defined by

$$\hat{\theta}_{postmean} = E_\theta(\vartheta | y_1, \ldots, y_n) = \int_{\vartheta \in \Theta} \vartheta f_\theta(\vartheta | y_1, \ldots, y_n) d\vartheta. \tag{3.2.2}$$

An alternative is to use the mode of the distribution:

Definition 3.4 The **posterior mode estimate** $\hat{\theta}_{postmode}$ is defined by

$$\hat{\theta}_{postmode} = \arg\max_{\vartheta} f_\theta(\vartheta \mid y_1, \ldots, y_n).$$

Taking the mode has a couple of advantages compared to the mean. Firstly, in order to calculate the posterior mean (3.2.2), integration is required, which can be cumbersome and numerically demanding. In contrast, taking the mode simply requires finding the maximum of the function, which is usually numerically much easier to solve or can even be calculated analytically. Secondly, the maximum of the posterior density function is, loosely speaking, the value with the highest probability, which also seems like a more intuitive candidate for θ.

3.2.2   Maximum Likelihood Estimation

If we assume a uniform prior, that is that $f_\theta(\vartheta)$ is constant, we can directly maximise the likelihood function itself. This gives the **Maximum Likelihood estimate**, commonly called the **ML estimate**.

Definition 3.5 The **Maximum Likelihood estimate** (or ML estimate) is defined by

$$\hat{\theta}_{ML} = \arg\max_{\theta \in \Theta} L(\theta; y_1, \ldots, y_n).$$

The ML estimate can also be calculated by differentiating the log-likelihood function instead of the likelihood function. Not only is it simpler to differentiate a sum instead of a product, we will also learn in Chap. 4 that the log-likelihood has important properties. In fact, ML estimation is the most frequent estimation principle in statistics and we will cover it in the next chapter extensively. It is also important to note that, given that we are no longer considering our prior in this process, we can find the Maximum Likelihood estimate outside of the Bayesian framework. That is, we can estimate our parameter θ without assuming it to be random. The ML estimate and the posterior mode estimate obey the invariance property. This means that any transformation of the parameter directly yields the new estimate. Hence, if θ is the parameter and we are interested in the transformed parameter, $\gamma = g(\theta)$, where $g(.)$ is a bijective transformation function, i.e. there is exactly one g(x) for each x and vice versa, then $\hat{\gamma}_{ML} = g(\hat{\theta}_{ML})$ is the transformed ML estimate.

3.2.3   Method of Moments

We thought it remiss to introduce the concept of parameter estimation without introducing the first ever method for calculating parameter estimates. This method was introduced by Karl Pearson (1857–1936) and is based on relating the theoretical moments of random variables $Y_1, \ldots, Y_n$ to the empirical moments of the observed data $y_1, \ldots, y_n$. Assume that the expectation $E(Y^k) = \int y^k f(y; \theta) dy$ is a function of the unknown parameter θ. The empirical moments of the data are given by

$$m_k(y_1, \ldots, y_n) = \frac{1}{n} \sum_{i=1}^{n} y_i^k.$$

For instance, in the case of a normal distribution, the parameters μ and σ can be written as $\mu = E(Y)$ and $\sigma^2 = E(Y^2) - (E(Y))^2$. An obvious choice for estimating the mean parameter μ is to take the mean

$$\hat{\mu} = m_1 = \frac{1}{n} \sum_{i=1}^{n} y_i.$$

For the estimation of σ^2 one can use

$$\hat{\sigma}^2 = \frac{1}{n} \sum_{i=1}^{n} y_i^2 - \left(\frac{1}{n} \sum_{i=1}^{n} y_i \right)^2 = m_2 - m_1^2 = \frac{1}{n} \sum_{i=1}^{n} (y_i - \bar{y})^2.$$

The method of moments estimator relates the empirical and theoretical moments such that

$$E_{\hat{\theta}_{MM}}(Y^k) = m_k(y_1, \ldots, y_n),$$

where $\hat{\theta}_{MM}$ is the resulting method of moment estimator. Hence, the expectation derived from the probability model should match the observed value of the statistic. The calculation of $\hat{\theta}_{MM}$ is not always easy and sometimes requires numerical simulation routines. In exponential family distributions, however, the method of moments and Maximum Likelihood estimation coincide, which we show in the next chapter (see Example 9).

3.2.4   Loss Function Approach

So far, we have introduced probability models and then derived estimates for their parameters. An alternative approach is to make use of loss functions, which formalise a penalty for deviances of our estimate from the true value. To do this, we again make use of our data $y_1, \ldots, y_n$ to derive an estimate for the unknown parameter θ. This estimate is denoted as $\hat{\theta}$ and is calculated from the data with the statistic $\hat{\theta} = t(y_1, \ldots, y_n)$. With this in mind, we now want to find out how close our estimate $\hat{\theta}$ is to the true but unknown parameter θ. Clearly, if $\hat{\theta}$ is equal to θ, we have estimated the true parameter perfectly and if $\hat{\theta} \neq \theta$, we have made some sort of error. The idea is now to quantify the extent of this error with a loss function defined as follows:

Definition 3.6 Let $\Theta \subset \mathbb{R}$ be the parameter space and let $t : \mathbb{R}^n \to \mathbb{R}$ be a statistic which is intended to estimate the parameter θ. With $\mathcal{T}$ we define the set of possible outcomes of $t(Y_1, \ldots, Y_n)$. A **loss function** $\mathcal{L}$ is defined as

$$\mathcal{L} : \mathcal{T} \times \Theta \to \mathbb{R}^+,$$

where the minimum value is equal to zero and occurs if both elements are equal, i.e. $\mathcal{L}(\theta, \theta) = 0$.

One example of a very common loss function is the squared loss defined by

$$\mathcal{L}(t, \theta) = (t - \theta)^2.$$

Another example makes use of absolute distances, i.e. $\mathcal{L}(t, \theta) = |t - \theta|$. Both make sense to describe different objectives and their use is problem-dependent. For example, the squared loss clearly penalises estimates far away from the true value θ more than the absolute loss. Historically, the squared loss was used in many settings because it was easily differentiable, but the absolute loss is now widely available if appropriate, even for large models. For estimation, the intention is that the loss $\mathcal{L}(t, \theta)$ should be as small as possible. The reader should take note, however, that because θ is unknown, this loss cannot be calculated and thus $\mathcal{L}(t, \theta)$ is a purely theoretical quantity. Our loss also depends on our sample, and hence is a scalar that is dependent upon the data that we have gathered.

Let us now move on to the next question. What if we no longer want to evaluate a single estimate, but instead want to determine if the *estimator* $t(y_1, \ldots, y_n)$ itself is effective? That is, we want to determine if our method for estimating θ works consistently well, no matter what sample we get. As usual, we want to model the properties of this function in general, so let us take our sample, and by extension our statistic, to be random. We can now take the expectation of our loss to find out how it reacts on average. To deal with this notationally, we denote concrete realisations of t from a real dataset as $t = t(y_1, \ldots, y_n)$ and let $t(.)$ describe the function itself. So, instead of looking at the loss, which evaluates the performance of our estimate for a single realised sample, we evaluate a stochastic sample that takes the probability model of $Y_1, \ldots, Y_n$ into account. This takes us from loss to risk and allows us to define the related risk function.

Definition 3.7 For a given loss function $\mathcal{L}(t, \theta)$ we define the **risk function** with

$$R(t(.), \theta) = E\Big(\mathcal{L}(t(Y_1, \ldots, Y_n), \theta)\Big) = \int_{-\infty}^{\infty} \cdots \int_{-\infty}^{\infty} \mathcal{L}(t(y_1, \ldots, y_n), \theta) \prod_{i=1}^{n} f(y_i; \theta) dy_i.$$

Note that this is an n-dimensional integral and due to independence the density results as product of marginal densities.

The parameter θ in the risk is the true unknown parameter. Taking the squared loss allows us to split the risk as calculated above into bias and variance components, an idea that is central to both statistics and machine learning. Let $\mathcal{L}(t, \theta) = (t - \theta)^2$ such that for $Y = (Y_1, \ldots, Y_n)$ we get

$$\begin{aligned}
R(t(.), \theta) &= E\Big(\{t(Y) - \theta\}^2\Big) \\
&= E\Big(\{t(Y) \underbrace{-E(t(Y)) + E(t(Y))}_{=0} -\theta\}^2\Big) \\
&= \underbrace{E\Big(\{t(Y) - E(t(Y))\}^2\Big)}_{1} + \underbrace{E\Big(\{E(t(Y)) - \theta\}^2\Big)}_{2} + \\
&\quad \underbrace{2E\Big(\{t(Y) - E(t(Y))\}\{E(t(Y)) - \theta\}\Big)}_{3}.
\end{aligned}$$

Note that Component 1 is given by the variance of $t(Y) = t(Y_1, \ldots, Y_n)$ while Component 2 defines the systematic error, which we will define as bias below. Finally, component 3 is by definition equal to 0, i.e. $E_\theta\Big(t(Y) - E_\theta(t(Y))\Big) \equiv 0$. So we can see that the squared loss decomposes the risk into the sum of the variance (Component 1) and the squared bias (Component 2). The variance describes how much variation we have in estimates sampled from the distribution with the given parameters, i.e. for a given θ, while the bias captures how much we systematically over or underestimate the correct θ. Together, these components define the **mean squared error** (MSE).

Definition 3.8 Given an estimate $t = t(Y_1, \ldots, Y_n)$, the **mean squared error** (MSE) is defined as

$$MSE(t(.), \theta) = E\Big(\{t(Y) - \theta\}^2\Big) = \text{Var}_\theta\Big(t(Y_1, \ldots, Y_n)\Big) + \text{Bias}(t(.); \theta)^2,$$

where $\text{Var}_\theta\Big(t(Y_1, \ldots, Y_n)\Big)$ is the variance of estimate $t = t(Y_1, \ldots, Y_n)$ and $\text{Bias}(t(.), \theta)$ is the systematic error, called **bias**, defined by

$$\text{Bias}(t(.), \theta) = E\Big(t(Y_1, \ldots, Y_n)\Big) - \theta.$$

An estimate with vanishing bias, i.e. Bias$(t(.), \theta) = 0$ for all $\theta \in \Theta$, is also called an **unbiased** estimate.

If we now take an estimator that minimises the risk we can be sure that we have a good estimator, but only for our unknown true parameter θ. For example, if our estimator was simply a constant, i.e. $t(.) = c$ and our true parameter just so happened to be equal to be c, the risk would be 0. Of course, this would be a terrible estimator if θ took any other value, so somehow our best estimator is still dependent upon our true θ. Therefore, a useful strategy would be to estimate θ with the function $t(.)$ such that the risk is minimised for all $\theta \in \Theta$. In special cases, this distinction does not matter. With the mean squared error this happens if the estimate is unbiased and the variance does not depend on θ. In this case, we can simply select the estimator $t(.)$ such that the risk is minimal.

This approach will give the best estimator no matter what the true θ value is. In general, however, we need to be more precise when minimising the risk and need to take θ into account. A "cautious" strategy would be to apply the **minimax** approach. First we choose θ such that the risk is highest (maximal), then we select $t(.)$ such that the highest risk is smallest (minimal). In mathematical notation this means that we are looking for

$$\hat{\theta}_{minimax} = \arg\min_{t(.)} \left(\max_{\theta \in \Theta} R(t(.), \theta) \right), \tag{3.2.3}$$

where $t(.)$ is selected from all possible statistics, i.e. functions from $\mathbb{R}^n$ to $\mathbb{R}$. While the minimax approach may appear sound, it is certainly a very cautious strategy, as we aim to minimise the worst case error for our estimation and is therefore often not very practical.

The dependence of the risk $R(t(.), \theta)$ on the parameter θ can also be circumvented by following a Bayesian approach. In this case, we assume that the parameter θ has a prior probability $f_\theta(.)$, which allows us to find the expected value over all possible values of θ. This leads to the **Bayes risk**

$$R_{Bayes}\Big(t(.)\Big) = E_\theta\Big(R(t(.), \theta)\Big) = \int_\Theta R(t(.), \vartheta) f_\theta(\vartheta) d\vartheta,$$

where the risk of course depends on the prior distribution. The **Bayes-optimal** estimation is then found by minimising the Bayes risk, that is

$$\hat{\theta}_{Bayes} = \arg\min_{t(.)} R_{Bayes}\Big(t(.)\Big), \tag{3.2.4}$$

where again the minimum is taken over all possible statistics $t(.)$. Note that the risk is calculated with respect to the prior distribution. It seems, however, strategically more appropriate to use the posterior distribution instead of the prior. A more useful

alternative to the Bayes risk is therefore to use the posterior distribution yielding the
posterior Bayes risk.

$$R_{post.Bayes}\left(t(.) \mid y_1, \ldots, y_n\right) = \int_{\Theta} \mathcal{L}\left(t(y_1, \ldots, y_n), \vartheta\right) f_\theta(\vartheta \mid y_1, \ldots, y_n) d\vartheta$$

$$= E_{\theta \mid y}\left(\mathcal{L}(t(y), \theta) \mid y\right).$$

We may again minimise $R_{post.Bayes}(.)$ with respect to $t(.)$ yielding a posterior Bayes
risk estimate.

$$\hat{\theta}_{post.Bayes.risk} = \arg\min_{t(.)} R_{post.Bayes}\left(t(.) \mid y_1, \ldots, y_n\right).$$

3.2.5   Kullback–Leibler Loss

A loss function evaluates an estimate $\hat{\theta}$ by evaluating how far it is from the unknown
parameter θ. An alternative would be to compare the two distributions directly, one
parameterised by our estimate and the other by the true parameter. We therefore
assume that θ is estimated by $\hat{\theta} = t(y_1, \ldots, y_n)$, such that $f(\tilde{y}; \hat{\theta})$ is the estimated
probability function at some arbitrary value $\tilde{y} \in \mathbb{R}$. We now look at the log ratio

$$\log \frac{f(\tilde{y}; \theta)}{f(\tilde{y}; \hat{\theta})}, \tag{3.2.5}$$

which indicates the discrepancy between the true density $f(\tilde{y}; \theta)$ and the estimated
density $f(\tilde{y}; \hat{\theta})$. Clearly, if $\hat{\theta} = \theta$ the ratio equals 1 and taking the log makes the
term vanish. The log ratio (3.2.5) depends on the particular value of $\tilde{y}$. We can
consider all possible values for $\tilde{y}$ by taking the expectation with respect to the
true distribution. This suggests making use of the Kullback–Leibler divergence as
introduced in Chap. 2.3.

With a slight change of notation we define

$$KL(\theta, \hat{\theta}) = KL(f(;\theta), f(;\hat{\theta})) = \int \log \frac{f(y; \theta)}{f(y; \hat{\theta})} f(y; \theta) dy.$$

Note that $KL(\theta, \hat{\theta}) > 0$ unless $\theta = \hat{\theta}$.

Note that the KL divergence is in fact a (non-symmetric) loss function, which
may be explicitly defined as $\mathcal{L}_{KL}(t, \theta) = KL(\theta, t)$. With this in mind, we can
construct a risk measure from the KL loss. Taking the statistic $t(.)$ this would be
defined by

$$R_{KL}(t(.), \theta) = \int_{-\infty}^{\infty} \cdots \int_{-\infty}^{\infty} \mathcal{L}_{KL}\left(t(y_1, \ldots, y_n), \theta\right) \prod_{i=1}^{n} f(y_i; \theta) dy_i. \tag{3.2.6}$$

Note again that the above integral is n-dimensional, and due to independence, the density ends up as a product of marginal univariate densities. It seems like things are getting a bit complicated here, because we are integrating over the possible sample values $y_1, \ldots, y_n$, while in the Kullback–Leibler divergence itself we integrate over a possible observation $\tilde{y}$. This will be essential in Chap. 9, when we introduce model selection based on the Kullback–Leibler divergence. To simplify the calculations, we rely on a simple second-order Taylor series approximation and write

$$\log f(\tilde{y}, \hat{\theta}) \approx \log f(\tilde{y}, \theta) + \frac{\partial \log f(\tilde{y}, \theta)}{\partial \theta}(\hat{\theta} - \theta) + \frac{1}{2}\frac{\partial^2 \log f(\tilde{y}, \theta)}{(\partial \theta)^2}(\hat{\theta} - \theta)^2.$$

Writing $\hat{\theta} = t(y_1, \ldots, y_n)$ allows us to approximate the Kullback–Leibler risk (3.2.6) as

$$\int \cdots \int \left[\int \log \frac{f(\tilde{y}; \theta)}{f(\tilde{y}; t(y_1, \ldots, y_n))} f(\tilde{y}; \theta) d\tilde{y} \right] \prod_{i=1}^{n} f(y_i; \theta) dy_i$$

$$= \int \cdots \int \left[\int \log f(\tilde{y}; \theta) f(\tilde{y}, \theta) d\tilde{y} - \int \log f(\tilde{y}; t(y_1, \ldots, y_n)) f(\tilde{y}; \theta) d\tilde{y} \right] \prod_{i=1}^{n} f(y_i; \theta) dy_i$$

$$\approx - \int \cdots \int \underbrace{\left(\int \frac{\partial \log f(\tilde{y}; \theta)}{\partial \theta} f(\tilde{y}; \theta) d\tilde{y} \right)}_{1} \left(t(y_1, \ldots, y_n) - \theta \right) \prod_{i=1}^{n} f(y_i; \theta) dy_i$$

$$+ \frac{1}{2} \int \cdots \int \underbrace{\left(-\int \frac{\partial^2 \log f(\tilde{y}; \theta)}{(\partial \theta)^2} f(\tilde{y}; \theta) d\tilde{y} \right)}_{2} \left(t(y_1, \ldots, y_n) - \theta \right)^2 \prod_{i=1}^{n} f(y_i; \theta) dy_i.$$

We will see in Chap. 4 that component 1 is equal to zero and component 2 will later be defined as the Fisher information. Given that the Fisher information does not depend on the sample $y_1, \ldots, y_n$, we see that the Kullback–Leibler risk is roughly proportional to

$$\int \cdots \int \left(t(y_1, \ldots, y_n) - \theta \right)^2 \prod_{i=1}^{n} f(y_i; \theta) dy_i = E\left((\hat{\theta} - \theta)^2 \right),$$

which we defined as mean squared error risk above. The reverse also holds, meaning that if we are able to choose an estimate such that the mean squared error is small, then the estimated density will be close to the true density in the Kullback–Leibler sense. Minimising the mean squared error is therefore a useful strategy.

3.3 Sufficiency and Consistency, Efficiency

Now we have a number of different ways to construct estimators $t(.)$ for a sample $y_1, \ldots, y_n$, such as Maximum Likelihood estimation and posterior mean/mode estimation.

We also have a number of different ways of analysing an estimator's overall performance with various risk measures for a range of true parameter values, such as minimax and Bayesian risk. Let us now look at some further properties of our estimators that might also guide our choice, which will also introduce us to a number of important mathematical concepts. Instead of going too deep into theory here, we hope to simply motivate the key building blocks of and remain on a somewhat informal level.

3.3.1 Sufficiency

The first property we want to introduce is sufficiency. If we have a sample $y_1, \ldots, y_n$, we can try to condense all of the relevant information contained within this sample to the quantity $t(y_1, \ldots, y_n)$. Hence, from originally n different values we calculate a single value $t(y_1, \ldots, y_n)$, that should represent everything that we need to know. At first, this sounds like a tremendous loss of information. The question is now whether the information that we lose when taking our statistic instead of the entire sample is relevant or not. If $t(y_1, \ldots, y_n)$ has sufficient information about θ, it would mean that all information we get about θ from our sample is contained in $t(y_1, \ldots, y_n)$. This leads us to the concept of sufficiency. In fact, we can completely erase our data if we have calculated a sufficient estimate. This can be rather useful, particularly in the age of big data, as it states that one only needs to calculate and store $t(y_1, \ldots, y_n)$ in order to get all information from our sample relevant to our parameter of interest θ. To proceed, let us now define sufficiency more formally.

Definition 3.9 A statistic $t(Y_1, \ldots, Y_n)$ is called **sufficient** for θ if the conditional distribution $P(Y_1 = y_1, \ldots, Y_n = y_n | t(Y_1, \ldots, Y_n) = t_0; \theta)$ does not depend on θ.

This states, once again, that the distribution of the single values $y_1, \ldots, y_n$ is non-informative, if we know the value of the statistic $t(y_1, \ldots, y_n)$. This idea can be a little confusing at first glance, so an example might help to clarify.

Example 1 Let $Y_1, \ldots, Y_n$ be Bernoulli variables with values 0 or 1 and $P(Y_i = 1) = \pi$. Let our statistic be $t(.) = t(Y_1, \ldots, Y_n) = \sum_{i=1}^{n} Y_i / n$, i.e. the mean. It holds that statistic $t(.)$ is sufficient for π. We can see that the statistic takes values in set $\{0, \frac{1}{n}, \frac{2}{n}, \ldots, \frac{n}{n}\}$ and we denote the resulting value with t_0. Correspondingly, $n_0 = nt_0$ is the sum of all of our variables. Then, using Bayes rule,

$$P(Y_1 = y_1, \ldots, Y_n = y_n | t(Y_1, \ldots, Y_n) = t_0; \pi)$$

$$= \frac{P\left(Y_1 = y_1, \ldots, Y_n = y_n, \sum_{i=1}^{n} Y_i = n_0; \pi\right)}{P\left(\sum_{i=1}^{n} Y_i = n_0; \pi\right)}$$

$$= \begin{cases} \dfrac{\prod_{i=1}^{n} \pi^{y_i}(1 - \pi)^{1-y_i}}{\binom{n}{n_0} \pi^{n_0}(1 - \pi)^{(n-n_0)}} & \text{for } \sum_{i=1}^{n} y_i = n_0 \\ 0 & \text{otherwise} \end{cases}$$

$$= \begin{cases} \dfrac{1}{\binom{n}{n_0}} & \text{for } \sum_{i=1}^{n} y_i = n_0 \\ 0 & \text{otherwise} \end{cases}$$

Apparently, this distribution does not depend on π.   ▷

It can be difficult to prove sufficiency in the above defined form, but the **Neyman-factorisation** makes it simple to find a sufficient statistic.

Property 3.1 A statistic $t(Y_1, \ldots, Y_n)$ is sufficient for θ if and only if the density or probability function decomposes to

$$f(y_1, \ldots, y_n; \theta) = h(y_1, \ldots, y_n) g\left(t(y_1, \ldots, y_n); \theta\right), \tag{3.3.1}$$

where $h(.)$ does not depend on θ and $g(.)$ depends on the data only through the statistic $t(y_1, \ldots, y_n)$.

The proof of this statement is rather simple and given at the end of this section. Note that sufficiency itself is a rather weak statement, as the original sample $(y_1, \ldots, y_n)$ itself is already sufficient. We therefore also require some concept of minimality defined as follows:

Definition 3.10 The statistic $t(y_1, \ldots, y_n)$ is minimal sufficient for θ if $t(.)$ is sufficient and for any other sufficient statistic $\tilde{t}(y_1, \ldots, y_n)$ there exists a function $m(.)$ such that $t(y_1, \ldots, y_n) = m(\tilde{t}(y_1, \ldots, y_n))$.

The definition states that if there exists a minimal sufficient statistic, then it can be calculated directly from any other sufficient statistic. Hence, we may reduce the data $y_1, \ldots, y_n$ to the value of the minimal sufficient statistic, but we may not reduce it further without losing information about the parameter θ.

Sufficient statistics are also closely related to the exponential family distributions that we described in Sect. 2.1.5. We have written an exponential family distribution in the form

$$f(y; \boldsymbol{\theta}) = \exp \left(t^T(y)\boldsymbol{\theta} - \kappa(\boldsymbol{\theta}) \right) h(y).$$

This directly shows that, for a sample $y_1, \ldots, y_n$, one obtains $\sum_{i=1}^n t(y_i)$ as a minimal sufficient statistic.

Proof We here prove Neyman-factorisation. Assume that $t(.)$ is sufficient, then $f(y_1, \ldots, y_n | t(y) = t, \theta)$ does not depend on θ. Because $t(.)$ is calculated from $y_1, \ldots, y_n$, with the basic definition of conditional probabilities, we get

$$f(y_1, \ldots, y_n; \theta) = \underbrace{f(y_1, \ldots, y_n | t(y_1, \ldots, y_n) = t; \theta)}_{h(y_1, \ldots, y_n)} \underbrace{f_t(t | y_1, \ldots, y_n; \theta)}_{g(t(y_1, \ldots, y_n); \theta)},$$

where the first component does not depend on θ and the second component is the distribution of $t(.)$ which by construction depends only on $t(y_1, \ldots, y_n)$ and θ. Let us assume now that the density $f(y_1, \ldots, y_n)$ is factorised as in (3.3.1). The marginal density for $t(y_1, \ldots, y_n)$ is

$$f_t(t; \theta) = \int_{t(y_1, \ldots, y_n)=t} f(y_1, \ldots, y_n; \theta) dy_1 \ldots dy_n$$

$$= \int_{t(y_1, \ldots, y_n)=t} h(y_1, \ldots, y_n) g(t; \theta) dy_1 \ldots dy_n$$

$$= g(t; \theta) \int_{t(y_1, \ldots, y_n)=t} h(y_1, \ldots, y_n) dy_1 \ldots dy_n.$$

The conditional distribution can then be written as

$$f(y_1, \ldots, y_n | t(y_1, \ldots, y_n) = t; \theta) = \frac{f(y_1, \ldots, y_n, t(y_1, \ldots, y_n) = t; \theta)}{f_t(t; \theta)}$$

$$= \begin{cases} \dfrac{h(y_1, \ldots, y_n) g(t; \theta)}{g(t, \theta)} & \text{for } t(y_1, \ldots, y_n) = t \\ 0 & \text{otherwise,} \end{cases}$$

which does not depend on θ because $g(t, \theta)$ cancels out.   □

3.3.2 Consistency

The principle of sufficiency describes how data can be condensed without losing information about the quantity of interest. The next question is how to quantify the gain of information for increasing sample size. We have seen that by taking the squared loss function, the risk equals the mean squared error (MSE). We have also seen that the MSE as loss measure approximates the Kullback–Leibler divergence. Let us look at this statement again, but explicitly focus on the role of the sample size n. In principle, we know that information increases with sample size, which means the estimate should approach the true value. This property can be formulated in mathematical terms, by once again making use of the mean squared error.

Definition 3.11 An estimate $\hat{\theta} = t(y_1, \ldots, y_n)$ is called **consistent** if

$$MSE(\hat{\theta}, \theta) = \mathrm{Var}_\theta(\hat{\theta}) + \mathrm{Bias}(\hat{\theta}, \theta)^2 \longrightarrow 0 \text{ as } n \to \infty.$$

Hence, if $\hat{\theta}$ is consistent, it approaches the unknown value θ with increasing sample size n. Using Chebyshev's inequality (Dudewicz and Mishra 1988) we can show that for any $\delta > 0$,

$$P\left(|\hat{\theta} - E(\hat{\theta})| \geq \delta\right) \leq \frac{\mathrm{Var}_\theta(\hat{\theta})}{\delta^2}.$$

Hence, if both, $\mathrm{Bias}(\hat{\theta}, \theta)^2 \to 0$ and $\mathrm{Var}_\theta(\hat{\theta}) \to 0$ for increasing sample size n, it follows that $P(|\hat{\theta} - \theta| \geq \delta) \to 0$. In other words, the estimate is getting closer to and more peaked around the true unknown parameter θ as n grows. This property is usually called weak consistency. There are many other consistency properties in use in statistics, see e.g. Dudewicz and Mishra (1988) or Lehmann and Casella (1998). For the purpose of this book, however, we will stick with the (MSE) consistency defined above. Consistency is thereby an essential property which any reasonable estimate should fulfil. It simply means that the larger the data, the more precise the conclusions are that can be drawn about the parameter of interest.

3.3.3 Cramer-Rao Bound

As the MSE is a key property of an estimator, the intention should therefore be to have the MSE as small as possible. This poses the question whether there is a lower threshold for the MSE. That is to say, for a given sample size n, the mean squared error cannot be smaller than a lower limit. Logically, it follows that some error will be introduced by the sampling process that cannot be overcome, no matter how good our estimator is. In fact, such a lower limit exists and is known as the **Cramer-Rao bound**. This bound holds for a wide class of distributions which are **Fisher-regular**.

Definition 3.12 The distribution $f(y; \theta)$ is **Fisher-regular** if the following properties hold:

1. The support of Y is not dependent upon θ, i.e. the set $\{y : f(y; \theta) > 0\}$ does not
   depend on θ.
2. The possible parameter space Θ is open, i.e. if θ is univariate, it has the form
   $\Theta = (a, b)$ with $a < b$.
3. The probability function $f(y; \theta)$ can be differentiated twice with respect to θ.
4. Integration and differentiation are exchangeable, i.e.

$$\int \frac{\partial}{\partial \theta} f(y; \theta) dy = \frac{\partial}{\partial \theta} \int f(y; \theta) dy.$$

It should be noted that Fisher-regularity is not a strong assumption. Still, it can be
violated. One standard counterexample is the uniform distribution on $[0, \theta]$, i.e.

$$f(y; \theta) = \begin{cases} \frac{1}{\theta} & \text{for } y \in [0, \theta] \\ 0 & \text{otherwise} \end{cases}.$$

Here, requirement 1 is violated as the support of Y depends on θ.

We will deal with Fisher-regular distributions in depth in Chap. 4. For now we
use the idea of Fisher-regularity to find the lower bound for the mean squared error.
This requires a further definition, namely that of the Fisher information. The Fisher
information can be understood as the information available about θ in the data
$y_1, \ldots, y_n$. It plays a central role in statistics and will also be treated in depth in
Chap. 4. We will briefly provide a definition here as we need the Fisher information
to describe the lower bound of the MSE.

Definition 3.13 Assume that $f(y; \theta)$ is Fisher-regular. We define with

$$I(\theta) = E\left[-\frac{\partial^2 \log f(Y; \theta)}{\partial \theta \partial \theta} \right] \tag{3.3.2}$$

the **Fisher information**. The component

$$J(\theta) = -\frac{\partial^2 \log f(y; \theta)}{\partial \theta \partial \theta} \tag{3.3.3}$$

is also known as **observed Fisher information**. If θ is multivariate, we define
$I(\theta) = E\left[\frac{\partial^2 \log f(Y;\theta)}{\partial \theta \partial \theta^T} \right]$ as the Fisher information matrix and similarly $J(\theta)$ as
the observed Fisher information matrix.

We will show in Chap. 4 that the Fisher-matrix can equivalently be written as

$$I(\theta) = E\left[\left(\frac{\partial \log f(Y; \theta)}{\partial \theta} \right) \left(\frac{\partial \log f(Y; \theta)}{\partial \theta} \right)^T \right]. \tag{3.3.4}$$

With these definitions in place, we now have all the tools we need to derive the Cramer-Rao bound.

Property 3.2 Let $\hat{\theta} = t(Y_1, \ldots, Y_n)$ be an estimate for θ, $Y_1, \ldots, Y_n$ *i.i.d.* and $Y_i \sim f(y; \theta)$ is drawn from a Fisher-regular distribution. The mean squared error of $t(.)$ is always larger than the lower **Cramer-Rao bound**

$$MSE(\hat{\theta}, \theta) \geq \text{Bias}^2(\hat{\theta}, \theta) + \frac{\left(1 + \frac{\partial \text{Bias}(\hat{\theta}, \theta)}{\partial \theta}\right)^2}{I(\theta)}.$$

In particular, if the estimate is unbiased one has

$$MSE(\hat{\theta}, \theta) \geq \frac{1}{I(\theta)}. \tag{3.3.5}$$

The Cramer-Rao bound demonstrates why the Fisher information plays a central role. In fact, if we are able to find an unbiased estimate with variance equal to the inverse Fisher information, then we have found one of the best possible estimates. We will show in the next chapter that such a class of estimators can be calculated with the Maximum Likelihood approach.

Example 2 Let us demonstrate the Cramer-Rao bound with a simple example. Assume we have normally distributed random variables $Y_i \sim N(\mu, \sigma^2)$, which are *i.i.d.* We are interested in estimating the mean and propose the estimate

$$t(y) = \sum_{i=1}^{n} w_i y_i$$

for some weights w_i. To obtain unbiasedness it is easy to see that we need to postulate $\sum_{i=1}^{n} w_i = 1$, because

$$E(t(Y)) = \sum_{i=1}^{n} w_i \mu.$$

We now question how to choose the weights such that the variance of $t(y)$ is minimised. Note that

$$Var(t(Y)) = \sum_{i=1}^{n} w_i^2 \sigma^2.$$

If we set $w_i = 1/n + d_i$ we get $\sum_{i=1}^{n} d_i = 0$ and hence

$$Var(t(Y)) = \sum_{i=1}^{n} (\frac{1}{n} + d_i)^2 \sigma^2 = \sum_{i=1}^{n} (\frac{1}{n^2} + d_i^2) \sigma^2$$

$$= \sigma^2/n + \sum_{i=1}^{n} d_i^2 \sigma^2.$$

Because $\sum_{i=1}^{n} d_i^2 \geq 0$, unless $d_i = 0$, we see that the arithmetic mean, which results when we set the weights w_i equal to $1/n$, in fact has the smallest variance, just as the Cramer-Rao bound states.   ▷

Proof Here we prove the Cramer-Rao bound. We emphasise that the proof requires properties which will be derived in the next chapter. It is given here for completeness and may be skipped on a first reading. Note that for unbiased estimates we have

$$\theta = E(\hat{\theta}) = \int t(y) f(y; \theta) dy.$$

Differentiating both sides with respect to θ yields

$$1 = \int t(y) \frac{\partial f(y; \theta)}{\partial \theta} dy$$

$$= \int t(y) \frac{\partial \log f(y; \theta)}{\partial \theta} f(y; \theta) dy$$

$$= \int t(y) s(y; \theta) f(y; \theta) dy,$$

where $s(\theta; y) = \partial \log f(y; \theta)/\partial \theta$. We will later call $s(\theta; y)$ the score function and prove in Chap. 4 that

$$E\left(s(\theta; y)\right) = \int s(\theta; y) f(y; \theta) dy = 0.$$

This implies that

$$1 = \int t(y) s(\theta; y) f(y; \theta) dy = \int \left(t(y) - \theta\right)\left(s(\theta; y) - 0\right) f(y; \theta) dy$$

$$= Cov\left(t(Y); s(\theta; Y)\right).$$

With the **Cauchy-Schwarz** inequality one obtains

$$1 = Cov\left(t(Y); s(\theta; Y)\right) \leq \sqrt{Var_\theta(t(Y))} \sqrt{Var_\theta(s(\theta; Y))}.$$

Note that

$$\text{Var}_\theta \Big(s(\theta; Y) \Big) = \int \Big(s(\theta; y) - 0 \Big)^2 f(y, \theta) dy$$

$$= \int s^2(\theta; y) f(y; \theta) dy$$

$$= \int \frac{\partial \log f(y; \theta)}{\partial \theta} \frac{\partial \log f(y; \theta)}{\partial \theta} f(y; \theta) dy$$

$$= I(\theta)$$

using (3.3.4). As for unbiased estimates we have the equality $\text{Var}_\theta(t(Y)) = MSE(t(Y), \theta)$, we obtain

$$1 \leq \sqrt{\text{Var}_\theta(t(Y))} \sqrt{\text{Var}_\theta(s(\theta; Y))}$$

$$\frac{1}{\sqrt{\text{Var}_\theta(s(\theta; Y))}} \leq \sqrt{\text{Var}_\theta(t(Y))}$$

$$\frac{1}{I(\theta)} \leq MSE(t(y), \theta).$$

$\square$

3.4   Interval Estimates

3.4.1   Confidence Intervals

We have now defined and discussed a few properties relevant to the evaluation of point estimates. Our estimate $\hat{\theta} = t(Y_1, \ldots, Y_n)$ itself, however, is just a single value and no information is given yet about how close our estimate is to the true parameter θ. Even though we know that for consistent estimates we have

$$P\left(|\hat{\theta} - \theta| > \delta \right) \to 0 \tag{3.4.1}$$

for $n \to \infty$ and for any $\delta > 0$. While these results are useful and informative, they do not tell us anything about what occurs in real life. In real life, despite our best efforts, we can never take infinitely sized samples. As a consequence, some knowledge about the range of likely values would definitely be of value. Let us therefore move to constructing **interval estimates** instead of single point estimates. Such intervals will show us how close our estimate is to the true value. Interval estimates are defined by a left boundary $t_l(Y_1, \ldots, Y_n)$ and a right boundary $t_r(Y_1, \ldots, Y_n)$, which give an interval of $\big[t_l(Y_1, \ldots, Y_n), t_r(Y_1, \ldots, Y_n) \big]$. Clearly, to achieve a useful interval $t_l(Y_1, \ldots, Y_n) < t_r(Y_1, \ldots, Y_n)$ for all values of $Y_1, \ldots, Y_n$. These statistics can be defined as follows:

Definition 3.14 The interval $CI = [t_l(Y_1, \ldots, Y_n), t_r(Y_1, \ldots, Y_n)]$ is called a **confidence interval** for θ with confidence level $1 - \alpha$ if

$$P_\theta\left(t_l(Y_1, \ldots, Y_n) \le \theta \le t_r(Y_1, \ldots, Y_n)\right) \ge 1 - \alpha \qquad (3.4.2)$$

for all θ. The value $(1 - \alpha)$ is called the confidence level and α is chosen as a small value, e.g. $\alpha = 0.01$ or $\alpha = 0.05$.

The probability statement (3.4.2) can be reformulated as

$$P_\theta\left(\theta \in [t_l(Y_1, \ldots, Y_n), t_r(Y_1, \ldots, Y_n)]\right) \ge 1 - \alpha.$$

It is important to bear in mind that $t_l(Y_1, \ldots, Y_n)$ and $t_r(Y_1, \ldots, Y_n)$ are both random variables. However, it is also clear that defining $t_l(Y_1, \ldots, Y_n) = -\infty$ and $t_r(Y_1, \ldots, Y_n) = \infty$ gives a $1-\alpha$ confidence interval, as $P(-\infty \le \theta \le \infty) = 1$ for any $\theta \in \mathbb{R}$. This is, of course, a useless interval. However, it demonstrates that our intention should be to choose the smallest possible interval and that means choosing "$\ge 1 - \alpha$" in (3.4.2) in fact means "$= 1 - \alpha$".

The construction of such a confidence interval is easy if a pivotal statistic exists.

Definition 3.15 A quantity $g(Y_1, \ldots, Y_n; \theta)$ is called a **pivotal statistic** if its distribution does not depend on θ. The distribution of $g(Y_1, \ldots, Y_n; \theta)$ is also called a pivotal distribution.

Exact pivotal quantities are rare. However, approximate pivotal distributions are quite common due to the central limit theorem. In fact, if the sample size n is large, the estimate $\hat\theta = t(Y_1, \ldots, Y_n)$ in many cases follows approximately a normal distribution. We denote this as

$$\hat\theta = t(Y_1, \ldots, Y_n) \overset{a}{\sim} N\left(\theta, Var(\hat\theta)\right), \qquad (3.4.3)$$

where the letter a in the formula above stands for asymptotic. We then can construct an approximate pivotal statistic with:

$$g\left(t((Y_1, \ldots, Y_n); \theta)\right) = \frac{t(Y_1, \ldots, Y_n) - \theta}{\sqrt{Var(\hat\theta)}} = \frac{\hat\theta - \theta}{\sqrt{Var(\hat\theta)}} \overset{a}{\sim} N(0, 1), \qquad (3.4.4)$$

which asymptotically follows a standard normal $N(0, 1)$ distribution. As $N(0, 1)$ does not depend on θ, the statistic is pivotal. The pivotal distribution can now be used to construct a confidence interval in the following way. With (3.4.4) we have

$$1 - \alpha \approx P\left(z_{\alpha/2} \leq \frac{\hat{\theta} - \theta}{\sqrt{Var(\hat{\theta})}} \leq z_{1-\alpha/2}\right)$$

$$\Leftrightarrow 1 - \alpha \approx P\left(\hat{\theta} + z_{\alpha/2}\sqrt{Var(\hat{\theta})} \leq \theta \leq \hat{\theta} + z_{1-\alpha/2}\sqrt{Var(\hat{\theta})}\right),$$

where $z_{\alpha/2}$ is the $\alpha/2$ quantile of a $N(0, 1)$ distribution and accordingly, $z_{1-\alpha/2}$ the $1 - \alpha/2$ quantile. Because $z_{\alpha/2} = -z_{1-\alpha/2}$, we obtain the confidence interval as

$$CI = \left[\underbrace{\hat{\theta} - z_{1-\alpha/2}\sqrt{Var(\hat{\theta})}}_{t_l(Y_1,\ldots,Y_n)}, \underbrace{\hat{\theta} + z_{1-\alpha/2}\sqrt{Var(\hat{\theta})}}_{t_r(Y_1,\ldots,Y_n)}\right]. \qquad (3.4.5)$$

One should note that $Var(\hat{\theta})$ may itself depend on θ, the unknown parameter. It may also depend on some other parameters, which are unknown. In the first case it is reasonable to replace θ with its estimate $\hat{\theta}$. If the variance $Var(\hat{\theta})$ depends on other parameters, one needs to estimate them. In fact, by doing so we estimate the variance $Var(\hat{\theta})$ and denote this with $\widehat{Var(\hat{\theta})}$. If the estimator $\widehat{Var(\hat{\theta})}$ is consistent, then the confidence interval is still asymptotically valid and (3.4.5) now changes to

$$CI = \left[\underbrace{\hat{\theta} - z_{1-\frac{\alpha}{2}}\sqrt{\widehat{Var(\hat{\theta})}}}_{t_l(Y_1,\ldots,Y_n)}, \underbrace{\hat{\theta} + z_{1-\frac{\alpha}{2}}\sqrt{\widehat{Var(\hat{\theta})}}}_{t_r(Y_1,\ldots,Y_n)}\right]. \qquad (3.4.6)$$

The construction of a confidence interval based on asymptotic normality (3.4.3) is easy and convenient and works in a large number of practical applications. It fails, however, if the observed data are extreme in the sense of the underlying distribution. A typical and quite common situation occurs if we are interested in rare events. Assume, for instance, we are interested in the failure of a technical component and the available database consists of n observations (e.g. trials) out of which in Y cases a technical component failed. Assuming that Y is the realisation of the binomial distribution

$$Y \sim B(n, \pi),$$

we are interested in a confidence interval for the failure probability π. If π is small then Y is small and arguments relying on the asymptotic normality are questionable. In this case, we need to look more closely into the distributional model. Bear in mind that if we observe $y = 0$, for instance, it does not mean that $\pi = 0$. Technical systems do fail and just because we have not observed a failure yet does not mean that we can be certain that we will never observe a failure in the future. We therefore need to construct a confidence interval for π which also includes values with $\pi > 0$. In this case, we can use (3.4.2) and restructure it by allowing the error probability α to be allocated in equal size on both sides of the interval. That is, we construct the confidence interval $\left[t_l(Y), t_r(Y) \right]$ with $t_l(Y) < t_r(Y)$ as

$$P\left(t_l(Y) > \pi; \pi \right) \leq \alpha/2 \text{ and } P\left(t_r(Y) < \pi; \pi \right) \leq \alpha/2,$$

such that

$$P\left(t_l(Y) \leq \pi \leq t_r(Y); \pi \right) \geq 1 - \alpha.$$

Let us first look at the left-hand side. Note that $Y \in \{0, 1, 2, \ldots, n\}$ and thus $t_l :$ $\{0, \ldots, n\} \to [0, 1]$. If we could "invert" t_l we could derive the following:

$$P\left(t_l(Y) > \pi; \pi \right) \leq \alpha/2 \tag{3.4.7}$$

$$\Leftrightarrow P\left(Y > t_l^{-1}(\pi); \pi \right) \leq \alpha/2$$

$$\Leftrightarrow P\left(Y \leq t_l^{-1}(\pi); \pi \right) \geq 1 - \alpha/2.$$

Let $t_l^{-1}(\pi) \in \{0, \ldots, n\}$, where clearly for value $t_l^{-1}(\pi) = n$ the inequality is fulfilled for all values of π. We therefore look at values $k \in 0, \ldots, n-1$ and consider

$$P(Y \leq k; \pi)$$

as a function of π. For every k there exists a unique left sided π_{lk}, such that

$$P\left(Y \leq k; \pi_{lk} \right) = \alpha/2,$$

where index l refers to the left side. Setting $\pi_{ln} = 1$, we have thereby defined the inverse function $t_l^{-1} : \{\pi_{l0}, \ldots, \pi_{ln}\} \to \{0, \ldots, n\}$ that fulfils (3.4.7). Accordingly, for $Y \in \{0, \ldots, n\}$, we can now set

$$t_l(y) = \pi_{ly}.$$

In the same way, we can define the right-hand statistic $t_r(y) = \pi_{ry}$. This construction yields confidence intervals $[t_l(y), t_r(y)]$ which are shown for various values of n and y/n in Fig. 3.1. Note that for $y \equiv 0$ and hence $y/n = 0$, we always

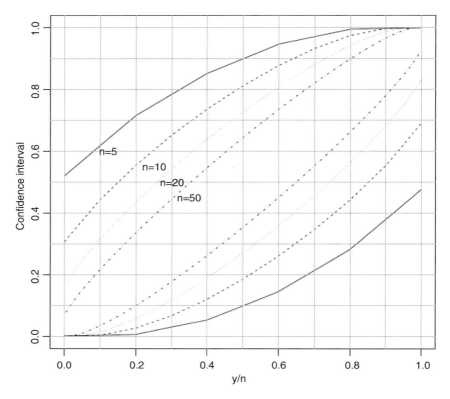

Fig. 3.1 Confidence intervals for π for binomial distribution

obtain $t_l(0) = 0$. For instance, if $n = 5$ and $y = 0$ then $t_r(0)$ is even larger than 0.5. The same holds symmetrically for $t_l(n)$.

This construction for confidence intervals was devised by Clopper and Pearson (1934) and is therefore sometimes called the Clopper-Pearson interval. Another name for it is the exact binomial confidence interval, to distinguish it from the asymptotic confidence interval, which is only adequate for large n.

Example 3 When the nuclear disaster happened in Fukushima in 2011, it was the second serious nuclear accident after Chernobyl in 1986. These are just two events, but it raised doubts as to whether the safety of nuclear power plants corresponds to the stated risk of one accident per 250,000 years. In Kauermann and Kuechenhoff (2011) the question is tackled with statistical tools. We give the main reasoning and conclusion here again. Setting the number of nuclear power plants to 442 (which was the number of actively running nuclear power plants in 2011), we consider this number as valid for the years 1981–2011 (which is a good proxy, as there is little variation in this number over the 30 years). We consider the binary events $Y_{it} \in \{0, 1\}$, where Y_{it} indicates whether the i-th power plant had a serious disaster in year t. Assuming independence between the years and between the power plants we have

$$Y = \sum_i \sum_t Y_{it}$$

as Binomial distributed variable with $n = 30 \cdot 442$ and π being the yearly risk for a nuclear disaster of a power plant. We have two accidents observed, so our data to fit the model is $Y = 2$. The Maximum Likelihood estimate results as

$$\hat{\pi} = \frac{2}{30 \cdot 442} \approx \frac{1.5}{10000} \approx \frac{1}{6607}.$$

In other words, given the data we estimate the probability of an accident in the order of every 6667 years for each reactor. This is apparently much larger than the one every 250,000 years, which is the reported safety risk. More important, however, is to assess the confidence in our estimate. This can be done with the methods just described leading to the following exact confidence interval for π:

$$\left[\frac{1}{54000}, \frac{1}{1800} \right].$$

We see that the safety level $1/250,000$ is not within the confidence interval meaning that the fact that two nuclear accidents have been observed indicates that the proposed safety level is not valid. We do not question the model here nor do we want to interpret the result in depth. The example is given to demonstrate that statistical reasoning can also be used if events are rare. ▷

In practice, confidence intervals are a very important tool to handle uncertainty of estimators. However, the correct interpretation of confidence intervals is very important. Given a concrete sample $y_1, \ldots, y_n$ one can estimate a corresponding confidence interval with:

$$\left[t_l(y_1, \ldots, y_n), t_r(y_1, \ldots, y_n) \right].$$

Note that if $y_1, \ldots, y_n$ are realised values, then the interval boundaries $t_l(y_1, \ldots, y_n)$ and $t_r(y_1, \ldots, y_n)$ are realised values as well—they are concrete numbers. This implies that probability statements like "What is the probability that parameter θ lies within the interval $\left[t_l(y_1, \ldots, y_n), t_r(y_1, \ldots, y_n) \right]$?" are in principle not valid using the probability model for Y. As the values $y_1, \ldots, y_n$ are observed, there is no randomness with respect to random variables. As a consequence, our only technical answer would be that we simply do not know whether θ lies in the interval or not. This is not a very satisfactory conclusion. A way out of this problem could, of course, be to use a Bayesian viewpoint and formulate our uncertainty about θ with a probability function. We will demonstrate this later. For now we still want to consider θ as a fixed but unknown parameter for which a (realised) confidence interval is given. Even though we cannot quantify with a formal probability statement whether θ lies in the interval, we may formulate this with subjective probabilities using DeFinetti's approach (see Definition 2.3) leading to a confidence statement. In fact, our *confidence* that θ lies in the interval $\left[t_l(y_1, \ldots, y_n), t_r(y_1, \ldots, y_n)\right]$ can be quantified to be $1 - \alpha$. That is also to say that in the long run, if we repeated the data collection process, we expect in $(1 - \alpha) \cdot 100$ % of the cases that the parameter lies in the interval.

3.4.2 Credibility Interval

Let us conclude this chapter by exploring confidence intervals from a Bayesian perspective. Our knowledge about the parameter θ is given by the posterior distribution $f_\theta(\vartheta \mid y_1, \ldots, y_n)$. The posterior distribution can directly be used to construct an interval $\left[t_l(y_1, \ldots, y_n), t_r(y_1, \ldots, y_n)\right]$ such that

$$P_\theta\left(\vartheta \in \left[t_l(y_1, \ldots, y_n), t_r(y_1, \ldots, y_n)\right] \mid (y_1, \ldots, y_n)\right) = \int_{t_l(y_1, \ldots, y_n)}^{t_r(y_1, \ldots, y_n)} f_\theta(\vartheta \mid y_1, \ldots, y_n) d\vartheta \geq 1 - \alpha.$$

In the Bayesian terminology, such an interval is called a **credibility interval**. A natural choice is to set $t_l()$ and $t_r()$ such that

$$\int_{-\infty}^{t_l(y_1, \ldots, y_n)} f_\theta(\vartheta \mid y_1, \ldots, y_n) d\vartheta = \int_{t_r(y_1, \ldots, y_n)}^{\infty} f_\theta(\vartheta \mid y_1, \ldots, y_n) d\vartheta = \frac{\alpha}{2},$$

that is, we cut off a probability mass of $\alpha/2$ on the left and right of the posterior probability. This choice is not optimal, as it may occur that for $\vartheta_1 \notin [t_l(y_1, \ldots, y_n), t_r(y_1, \ldots, y_n)]$ and for $\vartheta_2 \in [t_l(y_1, \ldots, y_n), t_r(y_1, \ldots, y_n)]$ one has

$$f_\theta(\vartheta_1 \mid y_1, \ldots, y_n) > f_\theta(\vartheta_2 \mid y_1, \ldots, y_n).$$

Hence, the density may be larger for values outside of the credibility region compared to values within the credibility region. This drawback can be avoided by using a **highest posteriori density** credibility interval or, in short, the highest density interval

$$HDI(y_1, \ldots, y_n) = \{\theta; f_\theta(\theta \mid y_1, \ldots, y_n) \geq c\},$$

where c is chosen such that

$$\int_{\vartheta \in HDI(y_1,...,y_n)} f_\theta\big(\vartheta \mid (y_1, \ldots, y_n)\big)d\vartheta = 1 - \alpha.$$

That is, we choose an interval where values are greater than a threshold density c, such that the region integrates to $1 - \alpha$. To demonstrate the idea let us use the binomial example from above again. Assume that $Y_1, .., Y_n \in \{0, 1\}$ are independent Bernoulli variables, such that $Y = \sum_{i=1}^{n} Y_i$ is binomial with parameter n and π. We assume a flat prior for π that the posterior is

$$f_\pi(\pi \mid y) \propto \pi^y(1 - \pi)^{n-y},$$

where $\propto$ means "is proportional to". As will be shown in Chap. 5, this gives a beta distribution, such that

$$\pi \mid y \sim Beta(1 + y, 1 + n - y).$$

The resulting highest density credibility intervals are shown in Fig. 3.2 for the setting $n = 5$ and $y = 0$ (left plot) and $n = 5$ and $y = 2$ (right plot). The generic credibility intervals for different values of n and y/n are shown in Fig. 3.3.

Clearly, the plot resembles the one in Fig. 3.1, although they are built upon different reasoning. While Fig. 3.1 gives what is called exact confidence intervals, Fig. 3.3 relies on Bayesian reasoning and provides credibility intervals. Interestingly, the coincidence of the confidence intervals and credibility intervals occurs in many places and we will see it later in the book.

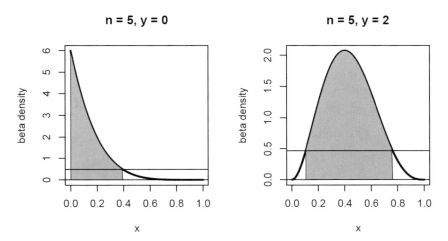

Fig. 3.2 Highest density credibility interval for binomial data with $n = 5$ and $y = 0$ (left) and $n = 5$ and $y = 2$ (right)

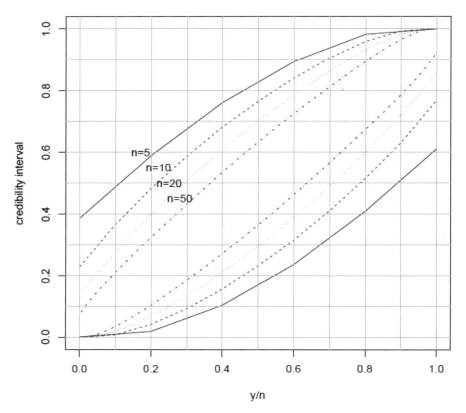

Fig. 3.3 Highest posterior density credibility intervals (given a flat prior) for different values of n and y/n

Example 4 Let us illustrate the important role of confidence and credibility intervals in some real applications. In surveys on voting behaviour, randomly sampled people are asked about their potential vote. In order to determine the share of population that votes for party A, we can define the random variable:

$$Y_i^{(A)} \text{ with } Y_i^{(A)} = \begin{cases} 1 & \text{if the i-th person votes for A} \\ 0 & \text{otherwise.} \end{cases}$$

If we define the unknown true share as π_A, the distribution of Y_i is $B(1, \pi_A)$ assuming a simple random sample from a large population. Furthermore, independence of the Y_i can be assumed and $Y_1, \ldots, Y_n$ are *i.i.d.*. Then $\sum_{i=1}^{n} Y_i \sim B(n, \pi_A)$ and the estimator is $\hat{\pi}_A = \frac{1}{n} \sum_{i=1}^{n} Y_i$, which is the relative frequency of voters for party A in our sample. For the construction of a confidence interval, we can

Table 3.1 95% Confidence intervals for the proportion of voting intention for different parties based on a survey conducted by Infratest dimap on May, 2nd, 2019

Party	No. of votes in sample	$\hat{p}$ in %	Asymptotic CI in %	Exact CI
CDU/CSU	421	28	[25.7; 30.2]	[25.7;30.0]
SPD	271	18	[16.1; 19.9]	[16.1; 20.0]
AFD	181	12	[10.4; 13.7]	[10.4; 13.8]
FDP	120	8	[6.6; 9.3]	[6.7; 9.5]
Linke	135	9	[7.5; 10.4]	[7.6; 10.5]
Gruene	301	20	[18; 22]	[18.0; 22.1]

use (3.4.3) for a large sample size n applying the asymptotic normality of $\hat{\pi}_A$. As $Var(\hat{\pi}_A) = \frac{\pi_A(1-\pi_A)}{n}$, the interval is

$$\left[\hat{\pi}_A - z_{1-\frac{\alpha}{2}} \cdot \sqrt{\frac{\hat{\pi}_A(1-\hat{\pi}_A)}{n}}, \hat{\pi}_A + z_{1-\frac{\alpha}{2}} \cdot \sqrt{\frac{\hat{\pi}_A(1-\hat{\pi}_A)}{n}} \right].$$

For a survey conducted by Infratest dimap on May 2nd, 2019, $n = 1505$ were sampled and asked for their voting intention. The results including asymptotic and Clopper-Pearson intervals are given in Table 3.1.

Note that the intervals are relatively large in spite of the rather high sample size.

▷

Example 5 (Validation of Machine Learning Algorithms) Let us examine how confidence intervals can be calculated for machine learning methods. In an analysis of professional soccer players, Rossi et al. (2018) developed an algorithm to predict injuries of players with GPS training data. The authors use a test sample to check the predictive power of their algorithm. In their approach, 9 out of 14 injuries could be predicted, while in the second best approach 6 out of 14 injuries could be successfully predicted. The rate is 64% (43%). The respective confidence intervals for the rate of successful prediction are [0.39, 0.83] (Method 1) and [0.17, 0.71] (second method). The intervals show a high uncertainty in the prediction performance of the model, which is due to the small sample size of the validation data.

▷

3.5   Exercises

Exercise 1

Let $Y_i \in \{0, 1\}$ be independent Bernoulli variables, such that $Y = \sum_{i=1}^{n} Y_i \sim B(n, \pi)$. Given the data y we want to estimate π.

1. Derive the ML estimate and the method of moment estimate.

2. We now look at estimates of the form

$$t(y) = \frac{y + a}{a + b + n},$$

   where a and b need to be chosen appropriately. Derive the $\mathrm{MSE}(t, \pi)$.
3. Taking the squared risk $\mathcal{L}(t, \pi) = (t - \pi)^2$, we obtain (with differentiation) the maximum risk given a and b. Plot the risk for different values of a and b, including $a = 0$ and $b = 0$. Given your results choose the minimax estimate.

Exercise 2 (Use R Statistical Software)

We consider a sample $Y_1, \ldots, Y_n$ from a uniform distribution on the interval $[0, \theta]$ with density function $f(y|\theta) = 1/\theta$ for $y \in [0, \theta]$ and $f(y|\theta) = 0$ otherwise. We estimate θ with the maximum value in the sample, i.e. $\hat{\theta} = Y_{(n)}$.

1. Illustrate why $\hat{\theta}$ is a biased estimate.
2. Show that $\hat{\theta}$ is the Maximum Likelihood estimate.
3. Show that $\theta^* = 2\bar{Y} = \frac{2}{n} \sum_{i=1}^{n} Y_i$ is an unbiased estimate for θ.
4. Check your results empirically by generating uniform random numbers in the interval $[0, 5]$. Try different sample sizes n: $n = 5$, $n = 10$, $n = 50$, $n = 100$, $n = 500$ and discuss your findings.

Exercise 3 (Use R Statistical Software)

In the data file `injured.csv`, the weekly numbers of pedestrians severely injured in traffic in a city are recorded for 10 years ($n = 520$). We assume that the numbers are independent realisations of a Poisson distributed random variable Y with constant intensity parameter λ.

1. Derive and calculate the Maximum Likelihood estimate $\hat{\lambda}_{\mathrm{ML}}$ of λ given the available data.
2. Calculate the method of moments estimate for λ.
3. Calculate a 95% confidence interval, assuming asymptotic normality of the Maximum Likelihood estimate.

Chapter 4
Maximum Likelihood Inference

In the last chapter, we explored Maximum Likelihood as a common approach for estimating model parameters given a sample from the population. In this chapter, we examine in more detail this estimate as well as the likelihood and its related functions: the log-likelihood, score function and Fisher information. We also derive properties of their asymptotic distributions, which tells us how much the functions themselves vary between samples. This is useful for deriving confidence intervals and statistical tests. We begin this chapter with a peek into the controversial history of the likelihood function. Not only does this give an insight into the often fraught process of discovery, but also makes clear the sometimes confusing difference between likelihood and probability.

Maximum Likelihood estimation was first proposed by Fisher (1922) and, although likelihood reasoning is absolutely essential to modern-day statistics, it was met with considerable resistance upon publication. In fact, as a graduate student (!), Fisher proposed the central ideas of likelihood reasoning in 1912 (see Fisher 1912), describing it as a method of maximum probability. Following the parametric distribution models of the previous chapter, he assumed that the parameter θ is unknown but follows a distribution. Therefore, the uncertainty about θ can also be represented by a distribution, whose maximum can then be found. Although this resembles the Bayesian approach, Fisher did not specify a prior distribution for the parameter, but instead suggested using only the likelihood function. That is, he considered the distribution of the data y but treated parameter θ as argument of the function, which needs to be set in relation to the data. This concept caused confusion and outrage, in part because Fisher referred to the outcome as a probability, which it is not. The discussion was settled 10 years later when Fisher finally defined the likelihood, not as a probability distribution, but as the "likelihood" function (see Fisher 1922). The name was not the only important factor. The entire concept he proposed in 1912 also had no solid theoretical justification. This was developed by Fisher over the next decade, leading to his seminal work in 1922. The major development was that, while the likelihood function itself may have no deep

© The Author(s), under exclusive license to Springer Nature Switzerland AG 2021
G. Kauermann et al., *Statistical Foundations, Reasoning and Inference*,
Springer Series in Statistics, https://doi.org/10.1007/978-3-030-69827-0_4

meaning, its maximum most certainly does. Fisher investigated the likelihood at its maximum and showed that the resulting estimate has useful properties. We refer the curious reader to Stigler (2007), Aldrich (1997) and Edwards (1974) for a comprehensive history.

4.1   Score Function and Fisher Information

In the last chapter, we introduced the likelihood function as a representation of how much we believe that a given sample $y_1, \ldots, y_n$ was generated by a model parameterised by θ. As before, we are now interested in how this function varies with the random sampling process. Therefore, instead of assuming a concrete sample, we treat our sample as a series of n identically distributed random variables

$$Y_i \sim f(y; \theta) \quad i.i.d.$$

Here, the distribution $f(.; \theta)$ is assumed to be Fisher-regular as given in Definition 3.12. To remind the reader, Fisher-regularity primarily means that the set of possible values of Y does not depend on θ, and that the order of integration with respect to y and differentiation with respect to θ does not matter. The parameter θ can be multidimensional with $\theta \in \mathbb{R}^P$, although for notational simplicity we continue to assume that $p = 1$. Following Definition 3.2 we define the likelihood function as

$$L(\theta; y_1, \ldots, y_n) = \prod_{i=1}^{n} f(y_i; \theta).$$

The log-likelihood is given by

$$l(\theta; y_1, \ldots, y_n) = \log L(\theta; y_1, \ldots, y_n) = \sum_{i=1}^{n} \log f(y_i; \theta).$$

Definition 4.1 The first derivative of the log-likelihood function is called the **score function**

$$s(\theta; y_1, \ldots, y_n) = \frac{\partial l(\theta, y_1, \ldots, y_n)}{\partial \theta}. \tag{4.1.1}$$

Differentiating the score function and taking negative expectation gives the **Fisher information** already defined in (3.3.4), that is

$$I(\theta) = -E\left(\frac{\partial s(\theta; Y_1, \ldots, Y_n)}{\partial \theta}\right) = -E\left(\frac{\partial^2 l(\theta; Y_1, \ldots, Y_n)}{\partial \theta \partial \theta}\right).$$

The score function is the slope of the likelihood function and our intention to maximise the likelihood function corresponds to finding the root of the score function. That is, we find $s(\hat{\theta}) = 0$, assuming that the likelihood function is differentiable. Moreover, the Fisher information is the expected second order derivative of the likelihood function and it quantifies the curvature of the likelihood. In particular, at the maximum of the likelihood function, where the slope and thus the score function are zero, the second order derivative expresses the "peakiness" of the maximum. We will see that this in turn gives the amount of information in the data about the parameter.

We will now investigate the score function. Let us stick with a single random variable $Y \sim f(y; \theta)$ for notational simplicity, but later show the same results for a random sample. The reader should note that the outputs of the score function and Fisher information are now random as well. Let us now show that the score function has mean zero, for all possible true parameter values. Note that for each parameter value θ, we obtain that $f(\cdot; \theta)$ is a density function (or a probability function), such that the integral over all possible values of a random variable given the parameter θ is one, i.e.

$$1 = \int f(y; \theta) dy. \tag{4.1.2}$$

Differentiating both sides of Eq. (4.1.2) with respect to θ and making use of the fact that integration and differentiation are exchangeable gives

$$0 = \frac{\partial 1}{\partial \theta} = \int \frac{\partial f(y; \theta)}{\partial \theta} dy = \int \frac{\partial f(y; \theta)}{\partial \theta} \frac{f(y; \theta)}{f(y; \theta)} dy$$

$$= \int \frac{\partial \log f(y; \theta)}{\partial \theta} f(y; \theta) dy = \int s(\theta; y) f(y; \theta) dy,$$

where $s(\theta; y) = \frac{\partial}{\partial \theta} \log f(y; \theta)$. This shows that the score function has mean zero, i.e.

$$E\left(s(\theta; Y)\right) = 0. \tag{4.1.3}$$

Property (4.1.3) is called the first **Bartlett identity**. Equation 4.1.3 shows now that, although its position may vary, on average there is a peak in the likelihood function at θ. Let us further differentiate both sides of Eq. (4.1.3) with respect to θ, which gives

$$0 = \frac{\partial 0}{\partial \theta} = \frac{\partial}{\partial \theta} \int \frac{\partial}{\partial \theta} \log f(y; \theta) f(y; \theta) dy$$

$$= \int \left(\frac{\partial^2}{\partial \theta \partial \theta} \log f(y; \theta)\right) f(y; \theta) dy + \int \frac{\partial \log f(y; \theta)}{\partial \theta} \frac{\partial f(y; \theta)}{\partial \theta} dy$$

$$= E \left(\frac{\partial^2}{\partial \theta \partial \theta} \log f(Y; \theta) \right) + \int \frac{\partial \log f(y; \theta)}{\partial \theta} \frac{\partial \log f(y; \theta)}{\partial \theta} f(y; \theta) dy$$

$$\Leftrightarrow E \Big(s(\theta; Y) s(\theta; Y) \Big) = E \left(-\frac{\partial^2}{\partial \theta \partial \theta} \log f(Y; \theta) \right).$$

Because $E \big(s(\theta; Y) \big) = 0$, we obtain the **second order Bartlett identity**

$$Var\,(s(Y; \theta)) = E \left(-\frac{\partial^2 \log f(Y; \theta)}{\partial \theta \partial \theta} \right). \qquad (4.1.4)$$

Hence, the variance of the score function is equal to the Fisher information. The reader should also note that if $\theta \in \mathbb{R}^p$, then the score function is vector valued and second order differentiation leads to a matrix. In this case $I(\theta)$ is called the Fisher information matrix or simply the Fisher matrix. It is defined as

$$I(\theta) = -E \left(\frac{\partial^2 l(\theta; Y)}{\partial \theta \partial \theta^T} \right).$$

Given the importance of the above properties, we formulate identical results for an *i.i.d.* sample $Y_1, \ldots, Y_n$.

Property 4.1 Given the *i.i.d.* sample $Y_1, \ldots, Y_n$ from a Fisher-regular distribution, the score function

$$s\,(\theta; Y_1, \ldots, Y_n) = \sum_{i=1}^{n} \frac{\partial \log f(Y_i; \theta)}{\partial \theta}$$

has zero mean; i.e. $E \big(s(\theta; Y_1, \ldots, Y_n) \big) = 0$, and its variance is given by the Fisher information (3.3.4), that is

$$I(\theta) = E \left(-\frac{\partial s(\theta; Y_1, \ldots, Y_n)}{\partial \theta} \right) = Var \big(s(\theta; Y_1, \ldots, Y_n) \big).$$

Example 6 Here we show the log-likelihood, score function and Fisher information for the normal distribution. Let $Y_i \sim N(\mu, \sigma^2)$ *i.i.d.* for $i = 1, \ldots, n$ where for simplicity σ^2 is assumed to be known. The log-likelihood function is given by (up to constants not depending on μ)

$$l(\mu; y_1, \ldots, y_n) = -\frac{1}{2} \sum_{i=1}^{n} \frac{(y_i - \mu)^2}{\sigma^2}.$$

Taking the derivative with respect to μ gives the score function

$$s(\mu; y_1, \ldots, y_n) = \sum_{i=1}^{n} \frac{y_i - \mu}{\sigma^2} = \frac{\sum_{i=1}^{n} y_i}{\sigma^2} - \frac{n}{\sigma^2}\mu,$$

which clearly gives $E(s(\mu; Y_1, \ldots, Y_n)) = 0$, as $E(Y_i) = \mu$. The Fisher information is

$$I(\mu) = \frac{n}{\sigma^2}$$

and with a bit of calculation we can see that $Var(s(\mu; Y_1, \ldots, Y_n)) = \frac{n}{\sigma^2} = I(\mu)$.

▷

The Maximum Likelihood estimate is obtained by maximising the likelihood function. Because the score function has a non-negative variance, it follows from (4.1.4) that the second order derivative is negative in expectation. This in turn guarantees a well defined optimisation problem as the likelihood function is concave and hence the maximum is unique. This property holds in expectation only and for a concrete sample we might experience non-unique maxima, i.e. local maxima. Nonetheless, we see that the Fisher information plays a fundamental role as it mirrors whether the maximisation problem is well defined or not. With the above definitions, we also define the Maximum Likelihood estimate with reference to the score function:

Definition 4.2 For a random sample $Y_1, \ldots Y_n$, the **Maximum Likelihood** (ML) estimate is defined by

$$\hat{\theta}_{ML} = \arg\max l(\theta; Y_1, \ldots, Y_n),$$

which for Fisher-regular distributions occurs when

$$s\left(\hat{\theta}_{ML}; y_1, \ldots, y_n\right) = 0.$$

For simplicity of notation and if there is no ambiguity, we sometimes drop the index ML and just write $\hat{\theta}$ for the ML estimate. Let us look at the case of the binary variable $Y_i \sim B(1, \pi)$ $i.i.d., i = 1, \ldots, n$ with unknown parameter $\pi = P(Y_i = 1)$. As the number of successes $Y = \sum_{i=1}^{n} Y_i$ is a sufficient statistic, we look at Y, which has a Binomial distribution with $Y \sim B(n, \pi)$. Assume we have observed $y = 10$ with $n = 30$. The log-likelihood function is plotted in Fig. 4.1. The maximum is at $\bar{y} = \frac{10}{30}$, indicated as dashed vertical line. Now assume that we have generated the data using $\pi = 0.4$, shown as solid vertical line. Then $y = 10$ occurs with probability $P(y = 10) = \binom{30}{10}0.4^{10}0.6^{20} = 0.115$. In other words, if we were to draw Y again, we would get a different value of Y with a rather large probability. Consequently, the resulting log-likelihood function would be different.

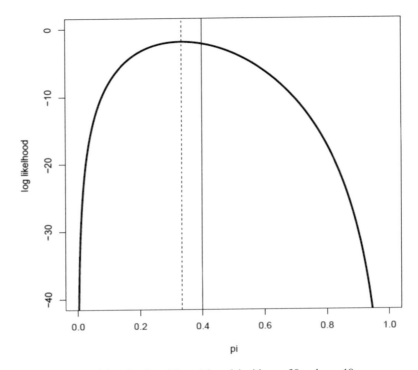

Fig. 4.1 Log-likelihood function for a Binomial model with $n = 30$ and $y = 10$

This is to say that the log-likelihood function itself must be considered as random variable. In Fig. 4.2, we plot the resulting random log-likelihood functions $l(\pi, Y)$, where the width of the lines is proportional to the probability that this log-likelihood function results. The functions are normed to take maximum value 0. The true parameter value $\pi = 0.4$ is shown with a vertical line. We see that even though the log-likelihood functions are random, their maxima are centred around the true parameter. We may also visualise the effect of increasing sample sizes. In Fig. 4.3 we plot the log-likelihood function for the binomial model for different values of n, assuming always that the arithmetic mean $\bar{y}$ is 0.4. Apparently, the larger the sample size, the more exposed is the maximum.

The next step is to quantify the variation around the true value, meaning that we aim to assess the variability of the maximum of the log-likelihood function. Note that the maximum is the root of the score function $s(\pi; Y)$ and we have already derived the variance of the score. We will make use of this result shortly.

Example 7 To demonstrate a two dimensional likelihood function, let us look once again at the normal distribution and assume that $Y_i \sim N(\mu, \sigma^2)$ *i.i.d.*, $i = 1, \ldots, n$, where we set $\mu = 0$ and $\sigma^2 = 1$. For a given sample of size $n = 10$ we obtain an arithmetic mean of $\bar{y} = -0.41$ and a standard deviation of 0.86. We plot the resulting likelihood function for both the mean value μ and the standard

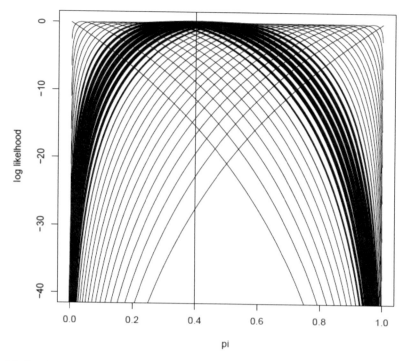

Fig. 4.2 Randomness of log-likelihood functions. Thickness is proportional to the probability that this log-likelihood function occurs

deviation σ in Fig. 4.4. This demonstrates the use of the likelihood in the case of multidimensional parameters. A horizontal line indicates the true value of μ and a vertical line that of σ. We should again bear in mind that we consider the likelihood function as random if it takes a random sample $Y_1, \ldots, Y_n$.

▷

4.2 Asymptotic Normality

Now that we understand how the score function and Fisher information vary with a sample $Y_1, \ldots, Y_n$, we can turn our attention to the Maximum Likelihood estimate itself. The ML estimate has asymptotic properties, which we will derive in the following section. These properties will be very helpful in predicting how much our estimate varies and for providing confidence intervals for the parameters.

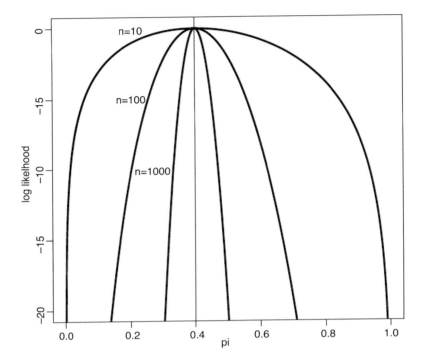

Fig. 4.3 Log-likelihood function for different sample sizes

Property 4.2 Assuming a Fisher-regular distribution with parameter θ for an *i.i.d.* sample $Y_1, \ldots, Y_n$, the ML estimate is asymptotically normally distributed with

$$\hat{\theta}_{ML} \overset{a}{\sim} N\left(\theta, I^{-1}(\theta)\right). \tag{4.2.1}$$

In particular, this means that the ML estimate has asymptotically the smallest possible variance: the Cramer-Rao bound given in Eq. (3.3.5). Note that we also obtain that $\hat{\theta}$.

The proof of this statement is a bit lengthy and the mathematically less experienced reader may omit it. We emphasise, however, that (4.2.1) is *the* central result in Maximum Likelihood theory and proves very useful in inferring the properties of θ.

Proof In order to motivate and prove asymptotic properties of the Maximum Likelihood estimate $\hat{\theta}_{ML}$, we need to modify the notation slightly. Firstly, we take the score function $s\left(\theta; Y_1, \ldots, Y_n\right)$ as a random quantity. Secondly, we decompose the score function into the sum of the scores of each individual sample

$$s(\theta; Y_1, \ldots, Y_n) = \frac{\partial}{\partial \theta} \sum_{i=1}^{n} \log f(Y_i; \theta) = \sum_{i=1}^{n} \frac{\partial}{\partial \theta} \log f(Y_i; \theta) =: \sum_{i=1}^{n} s_i(\theta).$$

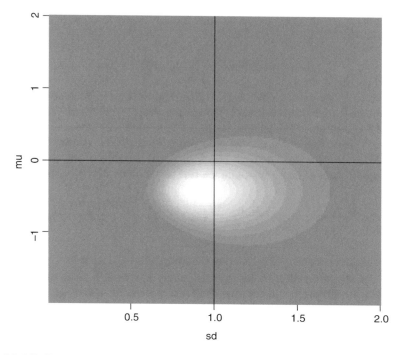

Fig. 4.4 Likelihood function for a sample of standard normally distributed variables. Lighter indicates an increased log-likelihood

Finally, to emphasise the role of the sample size, we include n as subscript and denote the score function as

$$s_{(n)}(\theta) = \sum_{i=1}^{n} s_i(\theta).$$

Note that $E(s_i(\theta)) = 0$. The Fisher information is also labelled with subscript n and similarly the sum of the Fisher Information of each individual sample

$$I_{(n)}(\theta) = \text{Var}\left(s_{(n)}(\theta)\right) = \sum_{i=1}^{n} \text{Var}(s_i(\theta)) =: \sum_{i=1}^{n} I_i(\theta).$$

The index n for the sample size is also used for the log-likelihood, i.e. we explicitly write the log-likelihood as

$$l_{(n)}(\theta) = \sum_{i=1}^{n} l_i(\theta) = \sum_{i=1}^{n} \log f(Y_i; \theta).$$

As both $s_i(\theta)$ and Y_i are $i.i.d.$, we may directly apply the central limit theorem to show that

$$s_{(n)}(\theta) \overset{a}{\sim} N\left(0, I_{(n)}(\theta)\right). \tag{4.2.2}$$

This follows easily with the two Bartlett Identities derived in Property 4.1. Finally, we also make explicit the influence of the sample size n on the Maximum Likelihood estimate and denote the ML estimate as $\hat{\theta}_{(n)}$. Property (4.2.2) is central to Maximum Likelihood estimation and it allows us to generally derive an asymptotic normality for the ML estimate. Note that the ML estimate is the root of the score function. Using Taylor series expansion around the true parameter value θ_0 leads to

$$0 = s_{(n)}(\hat{\theta}_{(n)}) = s_{(n)}(\theta_0) + \frac{\partial s_{(n)}(\theta_0)}{\partial \theta}\left(\hat{\theta}_{(n)} - \theta_0\right) + \frac{1}{2}\frac{\partial^2 s_{(n)}(\tilde{\theta})}{(\partial\theta)^2}\left(\hat{\theta}_{(n)} - \theta_0\right)^2, \tag{4.2.3}$$

where $\tilde{\theta}$ lies between θ_0 and $\hat{\theta}_{(n)}$, using the mean value theorem for differentiation. In fact, the above equation holds approximately if we set $\tilde{\theta}$ to θ_0 in the second derivative. To simplify notation we drop the parameter when we evaluate the functions at the true parameter value θ_0, i.e.

$$s_{(n)} = s_{(n)}(\theta_0), \ s'_{(n)} = \frac{\partial s_{(n)}(\theta_0)}{\partial \theta} \text{ and } s''_{(n)} = \frac{\partial^2 s_{(n)}(\theta_0)}{(\partial\theta)^2}.$$

Equation (4.2.3) reads now in approximate form as

$$0 \approx s_{(n)} + s'_{(n)}(\hat{\theta}_{(n)} - \theta_0) + \frac{1}{2}s''_{(n)}(\hat{\theta}_{(n)} - \theta_0)^2. \tag{4.2.4}$$

Because (4.2.4) is an approximate quadratic equation, we can solve it for $\hat{\theta}_{(n)} - \theta_0$ using standard calculus to get

$$\left(\hat{\theta}_{(n)} - \theta_0\right) \approx -\frac{1}{s''_{(n)}}\left(s'_{(n)} \pm \sqrt{s'^2_{(n)} - 2s_{(n)}s''_{(n)}}\right). \tag{4.2.5}$$

We will now look at the components on the right-hand side of Eq. (4.2.5) and try to simplify them with respect to their asymptotic behaviour, i.e. we investigate what happens if n tends to infinity. Taylor expansion of the square root component in (4.2.5) leads to

$$\left(s'^2_{(n)} - 2s_{(n)}s''_{(n)}\right)^{1/2} = s'_{(n)} - \frac{s_{(n)}s''_{(n)}}{s'_{(n)}} + \dots, \tag{4.2.6}$$

where we can ignore the remaining components collected in ... as they are of negligible asymptotic order for increasing n. Taking (4.2.6) and inserting it in (4.2.5) yields

$$\hat{\theta}_{(n)} - \theta_0 = -\frac{s_{(n)}}{s'_{(n)}} \tag{4.2.7}$$

and a second solution which is not of any further interest to us here. Looking at (4.2.7), we can simplify this further by decomposing the expectation of $s'_{(n)}$ into its mean value and the remainder, where its mean is the negative Fisher information. Hence

$$s'_{(n)} = -I_{(n)} + u_{(n)},$$

where $u_{(n)} = s'_{(n)} + I_{(n)}$ is a sum of independent zero mean variables. With the results discussed in Sect. 2.2, where we looked at the sum of independent zero mean random variables, we can see that $u_{(n)}$ is of asymptotic order $\sqrt{n}$. Moreover $I_{(n)}$ is of order n, i.e. $I_{(n)}$ is proportional to n. Using Taylor series again we get

$$\left(s'_{(n)}\right)^{-1} = \left(I_{(n)} + u_{(n)}\right)^{-1} = I_{(n)}^{-1} + I_{(n)}^{-2} u_{(n)} + \dots,$$

where components included in ... are of ignorable size as n increases. Looking at the order of the terms, we find $I_{(n)}^{-1}$ to be proportional to n^{-1} while the second component is asymptotically proportional to $n^{-2}\sqrt{n} = n^{-3/2}$ and can therefore also be ignored. Consequently, we can simplify (4.2.7) to

$$\hat{\theta}_{(n)} - \theta_0 = I_{(n)}^{-1} s_{(n)} + \dots, \tag{4.2.8}$$

where, as above, the components collected in ... are of ignorable size. As $s_{(n)}$ is a sum of zero mean variables, the central limit theorem (4.2.2) applies which proves (4.2.1).   □

Occasionally, we need or want to transform a parameter so that $\gamma = g(\theta)$ is the transformed value and $g(.)$ is an invertible and differentiable transformation, such that $\theta = g^{-1}(\gamma)$. The ML estimate $\hat{\gamma}_{ML}$ is **transformation invariant**, which means that it can simply be calculated as $\hat{\gamma}_{ML} = g(\hat{\theta}_{ML})$. This is easily seen, because the log-likelihood with parameterisation γ is defined by $l(g^{-1}(\gamma))$. Hence

$$\frac{\partial l(g^{-1}(\gamma))}{\partial \gamma} = \frac{\partial g^{-1}(\gamma)}{\partial \gamma} \frac{\partial l(\theta)}{\partial \theta} = \frac{\partial \theta}{\partial \gamma} \frac{\partial l}{\partial \theta}$$

such that for $\hat{\gamma}_{ML} = g(\hat{\theta}_{ML}) \Leftrightarrow \hat{\theta}_{ML} = g^{-1}(\hat{\gamma}_{ML})$ it follows

$$\frac{\partial l\big(g^{-1}(\hat{\gamma}_{ML})\big)}{\partial \gamma} = \frac{\partial g^{-1}(\gamma)}{\partial \gamma} \underbrace{\frac{\partial l(\hat{\theta}_{ML})}{\partial \theta}}_{=0} = 0.$$

The root of the original score function is also the root with the new transformed parameter. Transformation of parameters often requires the calculation of the Fisher information for the transformed parameter. This is fortunately straightforward. If we use the parameterisation with γ instead of θ, we get for the second order derivative

$$\frac{\partial^2 l\big(g^{-1}(\gamma)\big)}{\partial \gamma \partial \gamma} = \frac{\partial}{\partial \gamma} \left(\frac{\partial g^{-1}(\gamma)}{\partial \gamma} \frac{\partial l(\theta)}{\partial \theta} \right)$$

$$= \frac{\partial^2 g^{-1}(\gamma)}{(\partial \gamma)^2} \frac{\partial l(\theta)}{\partial \theta} + \frac{\partial g^{-1}(\gamma)}{\partial \gamma} \frac{\partial^2 l(\theta)}{\partial \theta \partial \theta} \frac{\partial g^{-1}(\gamma)}{\partial \gamma}.$$

The first component has mean zero when calculated at the true parameter θ_0, because the expectation of the score function $\partial l(\theta_0)/\partial \theta$ vanishes. Hence for the Fisher information we get

$$I_\gamma(\gamma) = \frac{\partial g^{-1}(\gamma)}{\partial \gamma} I(\theta) \frac{\partial g^{-1}(\gamma)}{\partial \gamma} = \frac{\partial \theta}{\partial \gamma} I(\theta) \frac{\partial \theta}{\partial \gamma}, \qquad (4.2.9)$$

where $I_\gamma(.)$ refers to the Fisher information using parameter γ. This result allows us to derive asymptotic normality for transformed ML estimates, a property sometimes called the **delta rule**.

Property 4.3 Let $\gamma = g(\theta)$ be a transformed parameter with an invertible and differentiable function g and let $\hat{\theta}$ be the Maximum Likelihood estimate with Fisher information $I(\theta)$. Then

$$\hat{\gamma}_{ML} - \gamma_0 \overset{a}{\sim} N\left(0, \frac{\partial \gamma}{\partial \theta} I^{-1}(\theta_0) \frac{\partial \gamma}{\partial \theta}\right).$$

In other words, transformation of parameters does not require the recalculation of the ML estimate or its variance. Instead, one can simply make use of the estimate of the untransformed parameter. However, this result is asymptotic and the quality of the approximation could be distorted by the transformation. We finish off this section with a number of examples that demonstrate the usefulness of this property.

Example 8 This example demonstrates the use of the delta rule to transform the parameter of a binomial distribution. Assume that Y is binomially distributed with $Y \sim B(n, \pi)$. The ML estimate is given by

$$\hat{\pi}_{ML} = Y/n \overset{a}{\sim} N(\pi, \underbrace{\frac{\pi(1-\pi)}{n}}_{I^{-1}(\pi)}).$$

If we now use the log odds as the transformed parameter we get

$$\vartheta = g(\pi) = \log\left(\frac{\pi}{1-\pi}\right)$$

$$\frac{\partial g(\pi)}{\partial \pi} = \frac{1-\pi}{\pi} \cdot \frac{(1-\pi)+\pi}{(1-\pi)^2} = \frac{1}{\pi(1-\pi)}$$

$$\hat{\vartheta}_{ML} = \log\left(\frac{Y/n}{1-Y/n}\right) \overset{a}{\sim} N\left(\log\left(\frac{\pi}{1-\pi}\right), \frac{1}{n(\pi(1-\pi))}\right).$$

▷

Example 9 We are now able to prove an interesting result in parameter estimation, namely that the Maximum Likelihood and method of moments estimates are the same for exponential family distributions. Assume an exponential family distribution

$$f(y; \theta) = \exp\{t^T(y)\theta - \kappa(\theta)\}.$$

The log-likelihood function then is given by

$$l(\theta; y_1, \ldots, y_n) = \sum_{i=1}^{n} t^T(y_i)\theta - n\kappa(\theta)$$

and the score function is vector valued and is given by

$$s(\theta; y_1, \ldots, y_n) = \sum_{i=1}^{n} t(y_i) - nE(t(Y))$$

$$\Rightarrow I(\theta) = Var(s(\theta, Y_1, \ldots, Y_n)) = n \cdot Var(t(Y_i)).$$

where we used the fact that $\partial \kappa(\theta) = E(t(Y))$. Therefore the Fisher information given by

$$I(\theta) = Var(s(\theta, Y_1, \ldots, Y_n)) = n \cdot Var(t(Y_i)),$$

Note also that

$$\partial^2 \kappa(\theta)/\partial\theta\partial\theta^T = Var(t(Y)).$$

Hence, the ML estimate $\hat{\theta}$ fulfils

$$\sum_{i=1}^{n} t(y_i) = n E_{\hat{\theta}}(t(Y)),$$

where the expectation is calculated with parameter $\hat{\theta}$. Consequently, we can interpret the ML estimate as a method of moments estimate.

$\triangleright$

Example 10 This example demonstrates the calculation of the asymptotic distribution of the ML estimate for the mean and variance of normally distributed random variables. Let $Y_i \sim N(\mu, \sigma^2)$ *i.i.d.*. We denote with $y = (y_1, \ldots, y_n)$ the sample, such that the log-likelihood is given by

$$l(\mu, \sigma^2; y) = -\frac{n}{2}\log\sigma^2 - \frac{1}{2\sigma^2}\sum(y_i - \mu)^2.$$

This leads to the two dimensional score function

$$\frac{\partial l(\mu, \sigma^2; y)}{\partial \mu} = \frac{1}{\sigma^2}\sum_{i=1}^{n}(y_i - \mu) \stackrel{!}{=} 0 \qquad (4.2.10)$$

$$\frac{\partial l(\mu, \sigma^2; x)}{\partial \sigma} = -\frac{n}{\sigma} + \frac{1}{\sigma^3}\sum_{i=1}^{n}(y_i - \mu)^2 \stackrel{!}{=} 0 \qquad (4.2.11)$$

$$\Leftrightarrow -\frac{n}{2\sigma^2} + \frac{1}{2\sigma^4}\sum_{i=1}^{n}(y_i - \mu)^2 \stackrel{!}{=} 0.$$

The system of equations can be solved by solving (4.2.10), which gives $\hat{\mu} = \bar{y} = \sum_{i=1}^{n} y_i/n$. This can then be inserted into (4.2.11), giving

$$\frac{n}{\sigma} = \frac{1}{\sigma^3}\sum(y_i - \bar{y})^2 \qquad \Leftrightarrow \qquad \hat{\sigma}^2 = \frac{1}{n}\sum(y_i - \bar{y})^2.$$

Note that the estimation of σ^2 is biased, but for n going to infinity we have asymptotic unbiasedness. With the asymptotic results derived for the ML estimate, we get

$$\sqrt{n}\left(\frac{\bar{Y} - \mu}{\hat{\sigma}^2 - \sigma^2}\right) \stackrel{a}{\sim} N(0, \Sigma)$$

with

$$\Sigma = I^{-1}(\mu, \sigma^2) = \begin{pmatrix} \frac{n}{\sigma^2} & 0 \\ 0 & \frac{n}{2\sigma^4} \end{pmatrix}^{-1} = \frac{1}{n} \begin{pmatrix} \sigma^2 & 0 \\ 0 & 2\sigma^4 \end{pmatrix}.$$

Note that the covariance of the two estimators $\hat{\mu}$ and $\hat{\sigma}^2$ is 0, which means that the estimators of the parameters μ and σ^2 are independent.

▷

4.3 Numerical Calculation of ML Estimate

The calculation of the ML estimate corresponds to finding the root of the score function $s(\theta; y) \overset{!}{=} 0$. Apart from very simple cases, an analytical solution is not available and numerical methods are needed to solve the score function. A convenient and commonly used method in this respect is Fisher scoring. This is a statistical version of the classical Newton-Raphson procedure. Note that in first order approximation we have

$$0 = s(\hat{\theta}_{ML}; y) \approx s(\theta_0; y) + \frac{\partial s(\theta_0; y)}{\partial \theta}(\hat{\theta}_{ML} - \theta_0).$$

Rewriting this gives

$$\hat{\theta}_{ML} = \theta_0 - \left(\frac{\partial s(\theta_0; y)}{\partial \theta} \right)^{-1} s(\theta_0; y).$$

In many models, the derivative of the score is rather complicated and in order to simplify it, we replace $\partial s(\theta; y)/\partial \theta$ with its expectation:

$$\hat{\theta}_{ML} \approx \theta_0 + I^{-1}(\theta_0)s(\theta_0; y).$$

A simple iteration scheme follows naturally from this formula. Starting with an initial estimate $\theta_{(0)}$ and setting $t = 0$ we iterate as follows:

1. Calculate $\theta_{(t+1)} := \theta_{(t)} + I^{-1}(\theta_{(t)})s(\theta_{(t)}; y)$
2. Repeat step 1 until $\|\theta_{(t+1)} - \theta_{(t)}\| < d$
3. Set $\hat{\theta}_{ML} = \theta_{(t+1)}$

As with any Newton-Raphson procedure, the process may fail if the starting value $\theta_{(0)}$ is too far away from the target value $\hat{\theta}_{ML}$. In this case, it can help to work with a reduced step size. Hence, one can add some $0 < \delta < 1$:

$$\theta_{(t+1)} = \theta_{(t)} + \delta I^{-1}(\theta_{(t)})s(\theta_{(t)}; y).$$

This step size can even be chosen adaptively based on the current step t, i.e. $\delta(t)$, which can even be traced as far back as Robbins and Monro (1951).

In recent years, models have become more complex and sometimes the calculation of neither the score function nor the Fisher matrix is possible analytically. In this case, simulation based methods can help. This can be applied to exponential family distributions as follows. Let

$$f(y; \theta) = \exp\{t^T(y)\theta - \kappa(\theta)\}$$

such that

$$s(\theta; y_1, \ldots, y_n) = \sum_{i=1}^{n} t(y_i) - nE\big(t(Y)\big).$$

Hence, the Maximum Likelihood estimate is defined by

$$\frac{1}{n}\sum_{i=1}^{n} t(y_i) = E_{\hat{\theta}_{ML}}\big(t(Y)\big).$$

The idea is now to simulate Y from $f(y; \theta)$. If this is possible, we may simulate for a given $\theta_{(t)}$

$$Y_j^* \sim f(y; \theta_{(t)}) \qquad j = 1, \ldots, N.$$

With these simulations, we can now estimate the mean and the variance of $t(Y)$ with

$$\widehat{E_{\theta_{(t)}}(t(Y))} = \frac{1}{n}\sum_{j=1}^{N} t(y_j^*)$$

$$\widehat{Var_{\theta_{(t)}}(t(Y))} = \frac{1}{N}\sum_{j=1}^{N} \Big(t(y_j^*) - \widehat{E_{\theta_{(t)}}(t(Y))}\Big)\Big(t(y_j^*) - \widehat{E_{\theta_{(t)}}(t(Y))}\Big).$$

This allows now to replace the iteration step during Fisher scoring with

$$\theta_{(t+1)} = \theta_{(t)} + \widehat{Var_{\theta_{(t)}}(t(Y))}^{-1} \widehat{E_{\theta_{(t)}}(t(Y))}.$$

Although it sounds like a formidable effort to simulate the score function and the Fisher information, with increasing computing power these operations are becoming bearable. More details regarding simulation based calculation of the ML estimate are provided in Geyer (1992).

4.4 Likelihood-Ratio

So far we have looked at the asymptotic properties of the ML estimate but the properties of the log-likelihood function itself are also very useful. We therefore define the likelihood-ratio as

$$lr(\hat{\theta}; \theta) = l(\hat{\theta}) - l(\theta) = \log \frac{L(\hat{\theta})}{L(\theta)} \geq 0.$$

The likelihood-ratio is positive and it takes value 0 if $\theta = \hat{\theta}$. If we consider the estimate $\hat{\theta}$ as a random variable, the likelihood-ratio itself becomes a random variable, for which we can derive some asymptotic properties. With Taylor series expansion we get

$$l(\theta) \approx l(\hat{\theta}) + \underbrace{\frac{\partial l(\hat{\theta})}{\partial \theta}}_{=0}(\theta - \hat{\theta}) + \frac{1}{2} \frac{\partial^2 l(\hat{\theta})}{\partial \theta \partial \theta}(\theta - \hat{\theta})^2. \tag{4.4.1}$$

Using Eq. (4.2.7), or similarly (4.2.8), we can approximate $\hat{\theta} - \theta$ with $I^{-1}(\theta)s(\theta; Y_1, \ldots, Y_n)$. We can similarly approximate the second order derivative in (4.4.1) with $-I(\theta)$, such that (4.4.1) simplifies to

$$l(\theta_0) \approx l(\hat{\theta}) - \frac{1}{2} \frac{s^2(\theta_0; Y_1, \ldots, Y_n)}{I(\theta_0)}. \tag{4.4.2}$$

We have shown the asymptotic normality of $s(\theta_0; Y_1, \ldots, Y_n)$ in (4.2.2). Note that $Var(s(\theta_0; Y_1, \ldots, Y_n)) = I(\theta_0)$, such that the latter component in (4.4.2) is equal to the square of the score divided by its standard deviation. In other words, the last component in (4.4.2) is asymptotically equal to Z^2 with $Z \sim N(0, 1)$ being standard normally distributed. Hence, we asymptotically get a chi-squared distribution for the likelihood-ratio.

Property 4.4 (Likelihood-Ratio) The likelihood-ratio for a Fisher-regular distribution converges for sample size n increasing to a chi-squared distribution $\mathcal{X}_1^2$, that is

$$2\{l(\hat{\theta}_0) - l(\theta)\} \overset{a}{\sim} \mathcal{X}_1^2. \tag{4.4.3}$$

Subscript 1 in (4.4.3) refers to the degrees of freedom of the chi-squared distribution and, in fact, if $\theta \in \mathbb{R}^p$, we find that the likelihood-ratio converges to a chi-squared distribution with p degrees of freedom. The likelihood-ratio has proven itself to be quite powerful in statistical testing, which will be discussed in the next chapter.

4.5   Exercises

Exercise 1
In a clinical study for a certain disease, n patients are treated with a new drug while another n patients are treated with a placebo. Let $Y_1 \sim Bin(n, p_1)$ the number of diseased patients in the drug group and $Y_0 \sim Bin(n, p_0)$ the number of diseased patients in the placebo group. We assume that the groups are independent. An interesting measure is the *relative risk*

$$RR = \frac{p_1}{p_0} \in \mathbb{R}_+ .$$

Consider a family of estimates

$$\widehat{RR}_\theta = \frac{\hat{p}_1 + \theta}{\hat{p}_0 + \theta} , \theta > 0 \text{ and } \hat{p}_1 = Y_1/n, \ \hat{p}_0 = Y_0/n.$$

1. Derive the asymptotic distribution of $\log(\widehat{RR})_\theta$. (Note: assume that $\hat{p}_0$ and $\hat{p}_1$ are independent and asymptotically normally distributed and apply the delta rule.)
2. Calculate the asymptotic mean value and the variance of the estimate $\log(\widehat{RR})_\theta$.
3. Derive an asymptotic 95% confidence interval for RR.

Exercise 2 (Use R Statistical Software)
The toxicity of a chemical substance is tested by exposing it to beetles in different concentrations. The data are given in the file ch4exerc2.csv. The following table shows the results:

Experiment	Concentration	NumberExposed	NumberDied
1	1.70	60	5
2	1.72	60	12
3	1.75	62	19
4	1.79	60	29
5	1.80	60	51
6	1.84	60	54
7	1.86	65	62
8	1.89	65	65

numberExposed is the number of beetles exposed to the corresponding concentration, numberDied is the number of beetles that died at that concentration.

1. Three different models are used to estimate the influence of toxicity on the probability that a beetle dies given a certain concentration x. Let Y be the binary response with $Y = 1$ if a beetle dies at concentration x with $\pi(x) = P(Y = 1|x)$ and $Y = 0$ otherwise.

   (a) The probability is linked to the concentration x through the logistic function

   $$\pi(x) = \frac{\exp(\alpha_1 + \beta_1 x)}{1 + \exp(\alpha_1 + \beta_1 x)}.$$

   (b) The probability is linked through the probit function

   $$\pi(x) = \Phi(\alpha_2 + \beta_2 x),$$

   where $\Phi(\cdot)$ is the distribution function of the normal distribution.

   (c) The probability is linked through the complementary log-log function

   $$\pi(x) = 1 - \exp[-\exp(\alpha_3 + \beta_3 x)].$$

   For all three models, determine the likelihood and log-likelihood given the above data. Find the Maximum Likelihood estimates of the parameters (α_j, β_j), $j = 1, 2, 3$ using a generic optimisation function in R, e.g. the function optim.

2. Alternatively, a Fisher Scoring or Newton algorithm can be used for maximising the likelihood. Develop a suitable algorithm for the three models and derive the score function as well as the expected and observed Fisher information.

3. Using the Maximum Likelihood estimates $(\hat{\alpha}_j, \hat{\beta}_j)$, $j = 1, 2, 3$, calculate the expected proportion of dead beetles for each concentration. Compare the results with the raw proportions (numberDied/numberExposed) of the data and visualise the results in an appropriate plot.

4. Think about how to determine which of the three models has the best fit to the observed proportions.

Chapter 5
Bayesian Statistics

We already briefly introduced the Bayesian approach to statistical inference in Chap. 3 and in this chapter we will dive deeper into this methodology. Whole books have been written about the different techniques in Bayesian statistics, which is a huge and very well developed field that we could not hope to cover in a single chapter. For this reason we will focus on major principles and will provide a list of references for deeper exploration.

A comprehensive history of Bayes reasoning can be found in McGrayne (2011), in which one can find many interesting ways it has been applied in the last centuries. The field was named after the reverend Thomas Bayes (1701–1761), who solved the "inverse probability" problem. His solution is nowadays known as Bayes' rule. This work was published posthumously in his name by Richard Price in 1763. It was also developed independently by Laplace (1774) but then was somewhat forgotten. When the field of Bayesian reasoning emerged in the 1950s, it was given his name as the process was largely based on manipulating conditional distributions.

5.1 Bayesian Principles

The fundamental principle in Bayesian reasoning is that uncertainty about a parameter θ is expressed in terms of a probability. This implies that the parameter θ is considered a random variable, with some prior probability distribution $f_\theta(\vartheta)$, where $\vartheta \in \Theta$ and Θ is the parameter space containing all possible (or reasonable) values for the parameter θ. If we assume a distributional model for the random variables $Y_1, \ldots Y_n$, this gives us, after observing $y_1, \ldots, y_n$, the posterior distribution

$$f_\theta(\vartheta | y_1, \ldots y_n) = \frac{f(y_1, \ldots, y_n; \vartheta) f_\theta(\vartheta)}{f(y_1, \ldots y_n)}.$$

© The Author(s), under exclusive license to Springer Nature Switzerland AG 2021
G. Kauermann et al., *Statistical Foundations, Reasoning and Inference*,
Springer Series in Statistics, https://doi.org/10.1007/978-3-030-69827-0_5

The denominator

$$f(y) = f(y_1, \ldots, y_n) = \int_\Theta f(y_1, \ldots, y_n; \vartheta) f_\theta(\vartheta) d\vartheta$$

can be seen from multiple perspectives. It can be seen as the marginal density for the observations after integrating out the uncertainty about the parameter. It can also be seen as the probability of the data based on our prior understanding of likely values for θ. A further way to view $f(y)$, as it does not depend on θ, is as the normalisation constant of the posterior distribution. The posterior distribution itself is proportional to the product of the prior and the likelihood function, that is

$$f_\theta(\vartheta|y_1, \ldots, y_n) \propto L(\vartheta; y_1, \ldots, y_n) f_\theta(\vartheta),$$

where $\propto$ stands for "is proportional to". We will see that in general the marginal distribution $f(y)$ is difficult to derive analytically and numerical methods are absolutely necessary to make Bayesian statistics work in practice. An exception can be found in **conjugate priors**.

Definition 5.1 Assume that $Y \sim f(y; \theta)$ follows a given probability model coming from a family of distributions $\mathcal{F}_y = \{f(y; \theta), \theta \in \Theta\}$ and that the prior distribution $f_\theta(\vartheta)$ comes from a family of distributions $\mathcal{F}_\theta = \{f_\theta(\vartheta; \gamma), \gamma \in \Gamma\}$. $\mathcal{F}_\theta$ is **conjugate** to $\mathcal{F}_y$, if for the posterior distribution it holds that $f_\theta(\vartheta|y) \in \mathcal{F}_\theta$.

This definition requires some explanation. First, the reader should note that the distribution of the parameter θ also depends on some further parameters, defined as γ above. These parameters are called **hyperparameters** and need to be specified in advance. Moreover, we can see that the families $\mathcal{F}_y$ and $\mathcal{F}_\theta$ are linked, meaning that the conjugate prior depends on our model of Y. To better understand this definition and the role of conjugate priors in Bayesian reasoning, let us look at a few examples.

Example 11 Assume $Y_i \sim N(\mu, \sigma^2)$ i.i.d. and that for the moment σ^2 is known. For the mean μ we postulate the prior

$$\mu \sim N(\gamma, \tau^2),$$

where both hyperparameters γ and τ^2 are assumed to be known. Let $y = (y_1, \ldots, y_n)$ be the data. We find that the posterior of μ is proportional to the prior distribution multiplied by the likelihood, i.e.

$$f_\mu(\mu|y) \propto f_y(y; \mu, \sigma^2) f_\mu(\mu; \gamma, \tau^2)$$

$$\propto \prod_{i=1}^{n} \exp\left(-\frac{1}{2}\frac{(y_i - \mu)^2}{\sigma^2}\right) \exp\left(-\frac{1}{2}\frac{(\mu - \gamma)^2}{\tau^2}\right)$$

$$\propto \exp\left(-\frac{1}{2}\frac{\sum_{i=1}^{n}y_i^2 - 2\sum_{i=1}^{n}y_i\mu + n\mu^2}{\sigma^2} - \frac{1}{2}\frac{\mu^2 - 2\mu\gamma + \gamma^2}{\tau^2}\right)$$

$$\propto \exp\left(-\frac{1}{2}\left\{\mu^2\left[\frac{n}{\sigma^2} + \frac{1}{\tau^2}\right] - 2\mu\left[\frac{n\bar{y}}{\sigma^2} + \frac{\gamma}{\tau^2}\right]\right\}\right). \tag{5.1.1}$$

Note that the normal distribution is a member of the exponential family of distributions, as shown in Sect. 2.1.5. In fact, we can derive from (5.1.1) that $f(\mu|y)$ has the form of a normal distribution with parameters

$$\mu|y \sim N\left(\left(\frac{n}{\sigma^2} + \frac{1}{\tau^2}\right)^{-1}\left(\frac{n\bar{y}}{\sigma^2} + \frac{\gamma}{\tau^2}\right), \left(\frac{n}{\sigma^2} + \frac{1}{\tau^2}\right)^{-1}\right).$$

Note that terms or factors that do not depend on μ can be ignored on the right-hand side of (5.1.1). With this, we can conclude that, when estimating the mean of a normal distribution, the normal prior is also a conjugate prior. Furthermore, the posterior mean is a weighted average of the sample mean $\bar{y}$ and the prior mean γ.

If we look at the parameters of the posterior, we can see an asymptotic property of Bayesian statistics. With appropriate prior parameters, i.e. γ bounded and τ^2 bounded away from zero, and increasing sample size n the posterior distribution of μ given y converges to a normal $N(\bar{y}, \sigma^2/n)$ distribution. That is,

$$\mu|y \qquad \overrightarrow{n \to \infty} \qquad N\left(\bar{y}, \frac{\sigma^2}{n}\right).$$

Hence, with increasing sample size we find that the posterior distribution of μ becomes centred around $\bar{y}$ with variance $\frac{\sigma^2}{n}$. Note that this is exactly the "opposite" probability formulation as for the Maximum Likelihood estimate $\hat{\mu} = \bar{y}$. For the ML estimate we found asymptotic normality of $\bar{y}$ around θ, while for Bayes inference we find posterior normality of θ around $\bar{y}$. ▷

Example 12 Let us again look at the normal distribution, but this time focus on the variance σ^2 and, for simplicity, fix the mean μ to a known value. We assume that σ^2 comes from an inverse gamma distribution with hyperparameters α and β, such that

$$f_{\sigma^2}(\sigma^2; \alpha, \beta) = \frac{\beta^\alpha}{\Gamma(\alpha)}(\sigma^2)^{-\alpha-1}\exp\left(-\frac{\beta}{\sigma^2}\right),$$

where $\alpha > 0, \beta > 0$ and $\Gamma(.)$ is the Gamma function (which is the continuous extension of the factorial operation, such that $\Gamma(k) = (k - 1)!$ for $k \in \mathbb{N}$). We denote the prior model with $\sigma^2 \sim IG(\alpha, \beta)$. The posterior is then obtained with

$$f_{\sigma^2}(\sigma^2|y) \propto (\sigma^2)^{-n/2} \exp\left(-\frac{1}{2}\frac{\sum_{i=1}^{n}(y_i - \mu)^2}{\sigma^2}\right)(\sigma^2)^{-\alpha-1} \exp\left(-\frac{\beta}{\sigma^2}\right)$$

$$\propto (\sigma^2)^{-(\alpha+n/2)-1} \exp\left(-\frac{(\beta + \frac{1}{2}\sum_{i=1}^{n}(y_i - \mu)^2)}{\sigma^2}\right),$$

such that $\sigma^2|y \sim IG(\alpha + n/2, \beta + \frac{1}{2}\sum_{i=1}^{n}(y_i - \mu)^2)$.   ▷

We have already seen a further example of a conjugate prior in Sect. 3.4.2, with the beta distribution for the parameter π of a Binomial distribution. The last example relates to the Poisson distribution.

Example 13 Let $Y \sim \text{Poisson}(\lambda)$ and assume for λ a Gamma prior of the form

$$f_\lambda(\lambda|\alpha, \beta) = \frac{\lambda^{\alpha-1} \exp(-\lambda\beta)}{\beta^{-\alpha}\Gamma(\alpha)}.$$

Then the posterior of λ given y is again a Gamma distribution with parameters $(\alpha + y)$ and $(\beta + 1)$ because

$$f_\lambda(\lambda|y; \alpha, \beta) \propto \lambda^y \exp(-\lambda)\lambda^{\alpha-1} \exp(-\lambda\beta)$$

$$\propto \lambda^{\alpha+y-1} \exp(-(\beta + 1)\lambda).$$

We continue with this example and visualise the effect of the sample size. Let us take a flat prior with $\alpha = 1$ and $\beta = 0$ and calculate the posterior as given above. Note that the resulting prior is not a proper distribution, but this turns out not to be a problem, since we are interested in the posterior. Note that for Y_i sampled $i.i.d.$ we get

$$f_\lambda(\lambda|y_1, \ldots, y_n; \alpha, \beta) \propto \lambda^{\alpha+n\bar{y}-1} \exp(-(\beta + n)\lambda).$$

We plot this density in Fig. 5.1 for three different sample sizes, namely $n = 10$, $n = 100$ and $n = 1000$, where $\bar{y} = 2$ in all three cases. The increasing amount of information becomes obvious, as with increasing sample size the posterior becomes more centred around $\bar{y}$.

                                                                                            ▷

We can conclude that with conjugate priors Bayesian reasoning becomes rather simple. Unfortunately, conjugate priors are often not a reasonable way to model and quantify our prior knowledge. Moreover, beyond the classical cases discussed above it can be difficult, or even impossible, to find a conjugate prior for a more

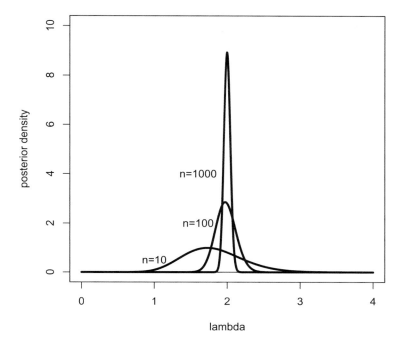

Fig. 5.1 Posterior distribution for Poisson distributed variables with Gamma prior and different sample sizes

complicated model, particularly if θ is multidimensional. Therefore, it is essential that we discuss different prior structures and numerical routines for calculating posterior distributions under more general conditions. We will begin with the choice of the prior.

Example 14 We extend our analysis from Chapter 3 on evaluating opinion polls for election preferences (see Bauer et al. 2018). In a multi-party system, survey participants are asked for their preferences between various political parties. The answers of the participants can be modelled by a multinomial distribution, which is a generalisation of the binomial distribution. Having the choice between K parties, let $Y_1, \ldots, Y_K$ be the number of participants that decided in favour of party $k = 1, \ldots, K$. The total number of participants is denoted by n. Then, the probability distribution is given by

$$P(Y_1 = y_1, \ldots, Y_K = y_K) = \frac{n!}{y_1! \ldots y_K!} \prod_{k=1}^{K} \theta_k^{y_k},$$

with $\sum_{k=1}^{K} y_k = n$. The parameters θ_k are the probabilities for choosing each party k. Assuming a simple random sample, θ_k is the (unknown) proportion of voters for each party k in the population. This is exactly the parameter of interest and we can approach inference on the parameter vector $\boldsymbol{\theta} = (\theta_1, \ldots, \theta_K)$ from a Bayesian perspective.

To this end, we first need to specify a prior distribution for $\boldsymbol{\theta}$. The conjugate prior for the multinomial distribution is the Dirichlet distribution, which is given by

$$f(\theta_1, \ldots, \theta_K) = \frac{1}{B(\boldsymbol{\alpha})} \prod_{k=1}^{K} \theta_k^{\alpha_i - 1} \text{ for all } 0 \le \theta_k \le 1 \text{ and } \sum_{k=1}^{K} \theta_k = 1. \qquad (5.1.2)$$

Here, $\boldsymbol{\alpha} = (\alpha_1, \ldots, \alpha_K)$ is the parameter vector and $B(\boldsymbol{\alpha})$ the normalisation constant, such that (5.1.2) is a well defined density. A common choice for the prior distribution is $\boldsymbol{\alpha} = (\frac{1}{2}, \ldots, \frac{1}{2})$. The calculation of the posterior distribution is straightforward in this case, because

$$f_{post}(\theta|y) = P(Y = y|\theta) \cdot f_\theta(\theta)$$

$$\propto \prod_{k=1}^{K} \theta_j^{y_k} \cdot \prod_{k=1}^{K} \theta_k^{-\frac{1}{2}} = \prod_{k=1}^{K} \theta_k^{y_k - 1/2}.$$

Therefore, the posterior is again a Dirichlet distribution with parameter $\boldsymbol{\alpha} = (y_1 + \frac{1}{2}, \ldots, y_K + \frac{1}{2})$:

$$(\theta|y) \sim Dirichlet(y_1 + \frac{1}{2}, \ldots, y_K + \frac{1}{2}). \qquad (5.1.3)$$

The resulting expectation is given by

$$E(\boldsymbol{\theta}|y) = \left(\frac{y_1 + \frac{1}{2}}{\sum_{k=1}^{K}(y_k + \frac{1}{2})}, \ldots, \frac{y_K + \frac{1}{2}}{\sum_{k=1}^{K}(y_k + \frac{1}{2})} \right).$$

Furthermore, we can calculate probabilities of relevant events. For example, assume that we are interested whether Party 1 has the most votes or whether Party 2 has more than 5% of the votes, which is the limit for entering parliament in Germany. For concrete calculations of these probabilities, a Monte Carlo approach can be applied, where a sample of sufficient size (e.g. 10^4) is drawn from the posterior distribution. Given the sample we can now calculate

$$P(\theta_1 = max(\theta_1, \ldots, \theta_k)|y) = \frac{\#(\text{samples with } \theta_1 = max(\theta_1, \ldots, \theta_K))}{10^4},$$

$$P(\theta_2 \ge 0.05|y) = \frac{\#(\text{samples with } \theta_2 \ge 0.05)}{10^4}.$$

With this method it is possible to calculate posterior probabilities for all relevant events concerning the multivariate distribution of $(\theta_1, \ldots, \theta_K)$, which is one of the key advantages of the Bayesian approach, see Bauer et al. (2018).

5.2 Selecting a Prior Distribution

The necessity of choosing a prior distribution is the Achilles heel of Bayesian statistics. It can be problematic if the result of a statistical evaluation depends too much on the choice of the prior distribution. To perform an objective analysis based on the data, the prior information inserted into the model needs to be as small as possible. Therefore, one tries to apply a **non-informative** prior. This will be explored in the coming sections.

5.2.1 Jeffrey's Prior

Let us assume $f_\theta(.)$ as the prior distribution of our parameter θ, of which we have no prior knowledge. Because we lack any prior knowledge, we decide to set $f_\theta(.)$ to be constant. This is commonly called a **flat prior**. If we now transform the parameter θ to $\gamma = g(\theta)$, where $g(.)$ is an invertible transformation function, then the prior for γ is given by

$$f_\gamma(\gamma) = f_\theta(g^{-1}(\gamma)) \left| \frac{\partial g^{-1}(\gamma)}{\partial \gamma} \right|.$$

We assumed that no prior knowledge exists for θ, and hence that $f_\theta(.)$ is constant. However, we can clearly see that $f_\gamma(.)$ is not constant in expressing knowledge about the *transformed* parameter γ. Hence, if we assume no prior knowledge of θ, we implicitly have a non-uniform prior for γ. This may or may not be reasonable, depending upon the situation.

Example 15 Let us run a simple Binomial experiment with $Y \sim B(n, \pi)$. We assume a non-informative prior on π, which is given by $\pi \sim Beta(1, 1)$, such that $f_\pi(\pi) = 1$ for $\pi \in [0, 1]$. Hence, the prior is constant. Instead of π we now reparameterise the model with the log odds $\theta = \log(\pi/(1 - \pi))$. Then $\pi = g^{-1}(\theta) = \exp(\theta)/(1 + \exp(\theta))$ and the resulting prior for θ is given by

$$f_\theta(\theta) = f_\pi(g^{-1}(\theta)) \left| \frac{\partial g^{-1}(\theta)}{\partial \theta} \right| = \frac{\exp(\theta)}{(1 + \exp(\theta))^2}$$

which is clearly not constant. This is visualised in the top row of Fig. 5.2. ▷

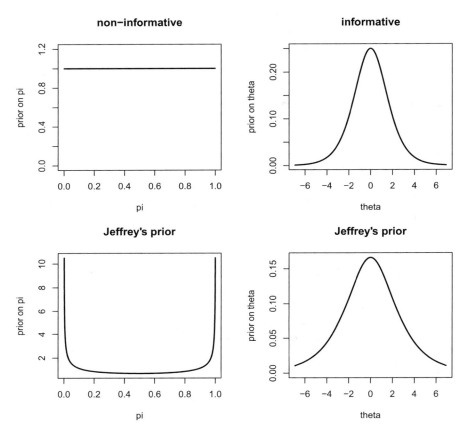

Fig. 5.2 Different priors for the binomial distribution. Top row: flat prior for π (left) and resulting informative prior for the log odds ratio θ. Bottom row: Jeffrey's prior for π (left) and θ (right)

Transformations of a parameter induce a structure on the prior of the transformed parameter. If we consider this unreasonable, we might want to formulate a prior distribution which remains unchanged, even under transformation. Such a transformation-invariant prior is called the Jeffery's prior. Transformation-invariance means that if we make use of Jeffrey's prior for one parameterisation, then a transformation of the parameter gives the Jeffrey's prior in this alternative parameterisation.

Definition 5.2 (Jeffrey's Prior) For Fisher-regular distributions $f(y; \theta)$ Jeffrey's prior is proportional to

$$f_\theta(\theta) \propto \sqrt{I_\theta(\theta)},$$

where $I_\theta(\theta)$ is the Fisher information.

To see that Jeffrey's prior is transformation invariant, we can make use of our previous result in Eq. (4.2.9) about the response of the Fisher matrix to parameter transformation. If $\gamma = g(\theta)$, then the Fisher information for γ is given by

$$I_\gamma(\gamma) = \frac{\partial g^{-1}(\gamma)}{\partial \gamma} I_\theta(g^{-1}(\gamma)) \frac{\partial g^{-1}(\gamma)}{\partial \gamma}.$$

If $f_\theta(\theta) \propto \sqrt{I_\theta(\theta)}$, then

$$f_\gamma(\gamma) \propto f_\theta(g^{-1}(\gamma)) \left| \frac{\partial g^{-1}(\gamma)}{\partial \gamma} \right| \propto \sqrt{\frac{\partial g^{-1}(\gamma)}{\partial \gamma} I_\theta(g^{-1}(\gamma)) \frac{\partial g^{-1}(\gamma)}{\partial \gamma}} = \sqrt{I_\gamma(\gamma)}$$

which is the Jeffrey's prior for the transformed parameter γ. For the binomial case, see the bottom row of Fig. 5.2.

An interesting property of the Jeffrey's prior is that it maximises the information gain from the data. This means that the distributional change from prior to posterior is as large as possible. In order to continue, we first need to clarify a few details. Firstly, we need to quantify the information difference between the prior and the posterior distribution. In Sect. 3.2.5, we introduced the Kullback–Leibler divergence to quantify the difference between two distributions (i.e. densities). This leads us to the following divergence measure between the prior and the posterior:

$$KL(f_\theta(.|y), f_\theta(.)) = \int_\Theta \log\left(\frac{f_\theta(\vartheta|y)}{f_\theta(\vartheta)}\right) f_\theta(\vartheta|y) \partial\vartheta.$$

The intention is now to choose the prior $f_\theta(.)$, such that the Kullback–Leibler divergence is maximised. Note that the Kullback–Leibler divergence depends on the data y, but the prior distribution does not. Hence, we also need to integrate out the data, which leaves us with the expected information (see Berger et al. 2009)

$$I(f_\theta) = \int KL(f_\theta(.|y), f_\theta(.)) f(y) dy,$$

where $f_y(y) = \int f(y; \vartheta) f_\theta(\vartheta) d\vartheta$. Clarke and Barron (1994) showed that, under appropriate regularly conditions, the information is maximised when $f_\theta(.)$ is chosen as the Jeffrey's prior. Thus, Jeffrey's prior has the additional property that the difference between the prior and posterior distribution is maximal and the maximum possible information is drawn from the data. Even though this supports the use of Jeffrey's prior, one should be aware that Jeffrey's prior still imposes prior knowledge on the parameter, as seen in Fig. 5.2. In fact, looking at Jeffrey's prior for the parameter π of a binomial distribution gives the impression that, before seeing any data, we favour values around 0 and 1 and a priori find values around 0.5 less plausible. If, however, we really have no prior knowledge about π, why should we choose this clearly informative prior?

For this reason, the use of Jeffrey's prior is contested and it is, in fact, not common to use non-informative priors, even if no prior information on the parameter is available. As Bernardo and Smith (1994) state: "Put bluntly: data cannot even speak entirely for themselves; every prior specification has *some* informative posterior (...) implications. (...) There is no objective prior that represents ignorance". We certainly do not want to engage in an ideological discussion of this issue here, but nevertheless emphasise that the choice of the prior may always be criticised and may have an influence on the outcome of Bayesian analyses. Nevertheless, Bayesian reasoning is generally useful and for increasing sample size, the influence of the prior vanishes. Therefore, a reasonable prior still needs to be chosen to make the Bayesian machinery run, which is why we will now discuss common alternatives.

5.2.2 Empirical Bayes

In this section, we discuss the derivation of priors from data with the "empirical Bayes" approach. Let us begin by noting that, as parameter θ is considered a random variable, we may integrate it out to get the distribution of the data, i.e.

$$f(y) = \int f(y; \vartheta) f_\theta(\vartheta) d\vartheta.$$

We can also assume that the prior distribution $f_\theta(.)$ depends on further hyperparameters, i.e. $f_\theta(\theta; \gamma)$ with γ as the (possibly multidimensional) hyperparameter. By extension, it should be clear that the marginal distribution $f(y)$ also depends on γ, because

$$f(y; \gamma) = \int f(y; \vartheta) f_\theta(\vartheta; \gamma) d\vartheta.$$

This is exactly the type of model that we dealt with in previous chapters and what we have learnt thus far would suggest choosing γ such that the likelihood $L(\gamma) = f(y; \gamma)$ is maximal. This approach is called **empirical Bayes**. In principle, however, empirical Bayes completely contradicts the Bayesian thinking. The prior is supposed to express the prior knowledge of the parameter *before* seeing data. As such, its hyperparameters should technically not be chosen *based on* the data. Counterintuitive as it may sound, empirical Bayes routines have become quite fashionable in some areas of statistics, such as penalised smoothing, and we will come back to the approach in Chap. 7. For now, however, we must stress that empirical Bayes technically contradicts the fundamentals of Bayesian reasoning.

Example 16 To clarify the relation between empirical Bayes and classical likelihood theory, let us look at a simple example. Let us take $Y \sim \text{Binomial}(n, \pi)$ and $\pi \sim \text{Beta}(\alpha, \beta)$. Then, integrating out π, we get

$$f(y|(\alpha, \beta)) = \binom{n}{y} \frac{1}{\text{beta}(\alpha, \beta)} \int \pi^{y+\alpha-1}(1-\pi)^{n-y+\beta-1} d\pi$$

$$= \binom{n}{y} \frac{\text{beta}(\alpha + y, \beta + n - y)}{\text{beta}(\alpha, \beta)},$$

where beta(,) is the beta function. The resulting distribution is known as Beta-Binomial model. Hence by integrating out π we obtain a "new" distribution which we can now use to estimate our hyperparameters α and β with Maximum Likelihood theory.

▷

5.2.3 Hierarchical Prior

A common strategy for the specification of the prior is to shift the problem of quantifying prior knowledge (or prior uncertainty) to a "higher level". This becomes an option if the form or the family of the prior distribution is fixed, but the prior distribution again depends on some hyperparameter(s) γ, which needs to be specified. Hence, the prior distribution takes the form $f_\theta(\vartheta; \gamma)$. The choice of the prior now simplifies to the question how to choose the hyperparameter γ. We can again take this from a Bayesian perspective and specify our knowledge of γ with a hyper-prior, i.e. γ itself has a distribution $f_\gamma(\gamma)$. In this case, we are interested in the posterior distribution of γ, which in turn will lead to a posterior distribution on θ, the parameter of interest. A model following this pattern takes the form

$$y|\theta \sim f_y(y; \theta); \quad \theta|\gamma \sim f_\theta(\theta; \gamma); \quad \gamma \sim f_\gamma(\gamma).$$

Clearly, the hyper-prior $f_\gamma(\gamma)$ may again depend on unknown parameters (i.e. hyper-hyperparameters), which would again need to be specified. That is, we have shifted the problem of quantifying our prior uncertainty to the "next level" and it may even seem that specifying our prior is now even more complicated than before! In fact, this is not necessarily the case. The intuition behind putting an extra layer of distributions on top of the hyperparameter is that it may be easier to select a distribution for the hyperparameters instead of fixing them to a specific value. This approach is commonly known as **hierarchical Bayes** and has proven to be quite powerful and applicable to a wide field of problems. We will see, however, that it will no longer be possible to calculate the posterior analytically.

Example 17 Assume $Y \sim B(n, \pi)$ and let $\pi \sim \text{Beta}(\alpha, \beta)$ be the conjugate prior. Instead of specifying the hyperparameters α, and β, we assume that

$$(\alpha, \beta) \sim f_{\alpha, \beta}((\alpha, \beta)|\varsigma),$$

where ς is a, possibly multidimensional, hyper-hyperparameter specifying the prior distribution of the Beta distribution parameters. Robert and Casella (2010), for instance, propose the hyper-prior

$$f_{\alpha, \beta}((\alpha, \beta); \varsigma) \propto \text{beta}(\alpha, \beta)^{\varsigma_0} \varsigma_1^{\alpha} \varsigma_2^{\beta}, \tag{5.2.1}$$

where $\varsigma = (\varsigma_0, \varsigma_1, \varsigma_2)$ are the hyper-hyperparameters. Clearly, the hyper-prior distribution does not have a common form, but is conjugate with respect to the Beta distribution. We are interested in the parameter π and therefore intend to find the posterior distribution $f_\pi(\pi|y)$. Note that,

$$f(\pi, (\alpha, \beta)|y) \propto f(y; \pi) f_\pi(\pi; (\alpha, \beta)) f_{(\alpha, \beta)}((\alpha, \beta); \varsigma),$$

such that

$$f(\pi|y) \propto f(y; \pi) \underbrace{\int f_\pi(\pi; (\alpha, \beta)) f_{(\alpha, \beta)}((\alpha, \beta); \varsigma) d\alpha d\beta}_{:= f(\pi|\varsigma)},$$

which requires numerical integration. We can see that this extra layer in principle just defines a more complex prior distribution $f(\pi; \varsigma)$, which now depends on hyper-hyperparameters ς instead of hyperparameters α and β. Figure 5.3 (top row) shows the hyper-prior for α and β on the left, with $\varsigma_0 = 0$, $\varsigma_1 = \varsigma_2 = 0.9$ and the resulting prior for π on the right. This prior for π is clearly more complex than one coming from a simple Beta distribution. We get a similar shape if we replace the rather artificial prior (5.2.1) with a more pragmatic prior, assuming that $\log(\alpha)$ and $\log(\beta)$ are independently normally distributed, i.e. that α and β are log-normal. This is shown in Fig. 5.3 (bottom row) where we set the mean value of $\log(\alpha)$ and $\log(\beta)$ to 1.   ▷

We can conclude that with hierarchical priors we get a more flexible parameter distribution, at the cost of extra computation. This naturally leads us to the next section, where we discuss numerical approaches for calculating the posterior distribution for arbitrarily chosen priors.

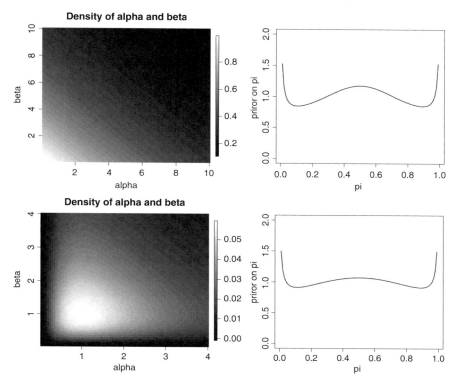

Fig. 5.3 Prior distribution for α and β of a beta distribution (left column) and resulting prior structure for π (right column) by integrating out α and β. Top row is for conjugate priors, bottom row is for independent normal priors for $\log(\alpha)$ and $\log(\beta)$

5.3   Integration Methods for the Posterior

5.3.1   Numerical Integration

The distribution that we are ultimately interested in is the posterior distribution of the parameter θ given data y, that is

$$f_\theta(\theta|y) = \frac{f(y;\theta)f_\theta(\theta)}{\int f(y;\vartheta)f_\theta(\vartheta)d\vartheta}. \tag{5.3.1}$$

In this setting we ignore (for the moment) that the prior $f_\theta(.)$ might depend on hyperparameters which themselves might have their own distributions. Instead, we assume that $f_\theta(.)$ is known and can be evaluated, that is, for every value of θ we can (easily) calculate $f_\theta(\theta)$. In this case, the unknown quantity in (5.3.1) is the denominator, which, for an arbitrary prior distribution $f_\theta(.)$, requires numerical integration. This could simply be carried out by standard numerical integration algorithms, such as rectangular, trapezoid or Simpson approximation. For example, with trapezoid approximation we use a grid $a = \theta_0 < \theta_1 < \ldots < \theta_k = b$. When Θ is an interval, i.e. $\Theta = [a, b]$, a and b are the boundary values of the parameter space Θ. If Θ is unbounded, a and b are chosen such that the integral components beyond a and b are of ignorable size, i.e. $\int_{-\infty}^{a} f(y; \vartheta) f_\theta(\vartheta) d\vartheta \approx 0$ and $\int_{b}^{\infty} f(y; \vartheta) f_\theta(\vartheta) d\vartheta \approx 0$. Trapezoid approximation of the denominator in (5.3.1) gives us

$$\int_{\Theta} f(y; \vartheta) f_\theta(\vartheta) d\vartheta \approx \sum_{k=1}^{K} \frac{f(y; \theta_k) f_\theta(\theta_k) + f(y; \theta_{k-1}) f_\theta(\theta_{k-1})}{2} (\theta_k - \theta_{k-1}).$$

Other routines for numerical integration, like Simpson approximation or rectangular approximation, are described in Stoer and Bulirsch (2002). While numerical integration is perfectly reasonable for one-dimensional integrals, it becomes problematic when applied to higher-dimensional problems. Even though we have thus far made the simplifying assumption that θ is univariate, we are in trouble when faced with the common task of approximating the integral of multivariate (often high dimensional) parameter vectors θ. There are, however, a number of alternative techniques for the approximation of this integral.

5.3.2   Laplace Approximation

Note that we typically draw n independent samples from $f(y|\theta)$. Hence, the data at hand are given by $y_1, \ldots, y_n$ and the posterior distribution takes the form

$$f(\theta|y_1, \ldots, y_n) = \frac{\prod_{i=1}^{n} f(y_i|\theta) f_\theta(\theta)}{\int \prod_{i=1}^{n} f(y_i|\theta) f_\theta(\theta) d\theta}.$$

Looking again at the denominator, we can rewrite the integral component as

$$\int \exp\left\{ \sum_{i=1}^{n} \log f(y_i|\theta) + \log f_\theta(\theta) \right\} d\theta = \int \exp\left\{ l_{(n)}(\theta; y_1, \ldots, y_n) + \log f_\theta(\theta) \right\} d\theta,$$

(5.3.2)

where $l_{(n)}(.)$ is the log-likelihood as defined in Chap. 4. Note that $l_{(n)}(.)$ is increasing in n, as discussed in Sect. 4.2. Hence, with increasing sample size n, the first component in the exp(.) in (5.3.2) becomes dominant. The idea is now to approximate the inner part of the exp(.) term with a second order Taylor approximation. We therefore twice differentiate the inner component of the exp(.) and get

$$
s_{P,(n)}(\theta) := \frac{\partial l_{(n)}(\theta; y_1, \ldots, y_n)}{\partial \theta} + \frac{\partial \log f_\theta(\theta)}{\partial \theta}
$$

$$
= s_{(n)}(\theta; y_1, \ldots, y_n) + \frac{\partial \log f_\theta(\theta)}{\partial \theta},
$$

$$
J_{P,(n)}(\theta) := -\frac{\partial^2 l_{(n)}(\theta; y_1, \ldots, y_n)}{\partial \theta \partial \theta} - \frac{\partial^2 \log f_\theta(\theta)}{\partial \theta \partial \theta}, \tag{5.3.3}
$$

where the subscript n, similar to that of the score and Fisher information in Chap. 4, makes clear the influence of the sample size. The additional index P indicates that the prior distribution is also being considered in the calculations. Analogously, we define with $l_{P,(n)}(\theta)$ the component in the exp(.) in (5.3.2). Now we let $\hat{\theta}_P$ be the posterior mode estimate such that $s_{P,(n)}(\hat{\theta}_P) = 0$. With second order Taylor approximation we get

$$
\int \exp(l_{P,(n)}(\vartheta)) d\vartheta \approx \int \exp\left(l_{P,(n)}(\hat{\theta}_P) - \frac{1}{2} J_{P,(n)}(\hat{\theta}_P)(\vartheta - \hat{\theta}_P)^2\right) d\vartheta.
$$

A formal proof shows that with increasing sample size n the approximation error vanishes, which makes use of similar arguments as those applied to proving asymptotic normality of the ML estimate in Sect. 4.2. We refer the curious reader to Severini (2000) for a detailed discussion. Note that the integral now simplifies to

$$
\exp(l_{P,(n)}(\hat{\theta}_P)) \int \exp\left(-\frac{1}{2} J_{P,(n)}(\hat{\theta}_P)(\vartheta - \hat{\theta}_P)^2\right) d\vartheta.
$$

The function in the integral mirrors the form of a normal distribution, such that the integral can in fact be calculated analytically and yields the inverse of the normalisation constant of a normal distribution. That is, one obtains the result

$$
\int f(y; \vartheta) f_\theta(\vartheta) d\vartheta \approx f(y; \hat{\theta}_P) f_\theta(\hat{\theta}_P) \sqrt{2\pi} (J_{P,(n)}(\hat{\theta}_P))^{-1/2}. \tag{5.3.4}
$$

While this approximation can be quite poor for small samples, it proves to be reasonably reliable if n is large.

Example 18 Let us once again take the Beta(α, β) prior for the parameter π of a Binomial distribution $B(n, \pi)$. Let $\alpha = \beta = 1$, which gives a constant prior and $\hat{\theta}_P = \bar{y} = y/n$. The second order derivative is $J_{P,(n)}(\hat{\theta}_{Pm}) = n/\{\bar{y}(1 - \bar{y})\}$ and, consequently, the Laplace approximation of the posterior is given by

$$
f_\pi(\pi|y) = \frac{\pi^y(1 - \pi)^{n-y}}{\bar{y}^y(1 - \bar{y})^{n-y}} \frac{1}{\sqrt{2\pi} \sqrt{\bar{y}(1 - \bar{y})/n}}.
$$

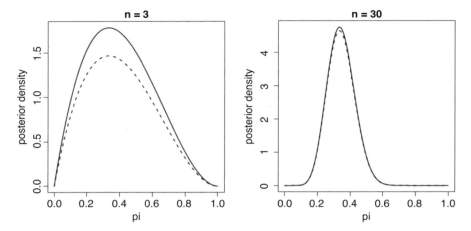

Fig. 5.4 Laplace approximation (dashed line) of posterior density (solid line) for Beta-Binomial model

In Fig. 5.4, we show the true posterior (solid line) and the Laplace approximated version (dashed line) for a sample size of $n = 3$ and $n = 30$, where $\bar{y} = \frac{1}{3}$ in both cases. The true posterior is marked by the solid line and its approximation by the dashed line. We can see that, even with a moderate sample size, the true posterior distribution is approximated nicely.   ▷

The Laplace approximation is numerically simple and also works in higher dimensions. If θ is multidimensional, the approximation in (5.3.4) becomes

$$\int f(y; \vartheta) f_\theta(\vartheta) d\vartheta \approx f(y; \hat{\theta}_P) f_\theta(\hat{\theta}_P)(2\pi)^{p/2} \left| J_{P,\theta}(\hat{\theta}_P) \right|^{-1/2},$$

where exponent p is the dimension of the parameter, the vertical bars $| \cdot |$ denote the determinant and $J_{P,(n)}(\cdot)$ is the p-dimensional version of the Fisher matrix, including the prior given in (5.3.3). Laplace approximation is today occasionally used in a more advanced form in applied Bayesian analysis, see e.g. Rue et al. (2009) or www.r-inla.org. This circumvents the more numerically demanding routines which we will discuss later in this chapter.

5.3.3 Monte Carlo Approximation

An alternative approach is inspired by comprehending the integral in the denominator of the posterior (5.3.1) as an expected value

$$E_\theta(f(y; \theta)) = \int f(y; \vartheta) f_\theta(\vartheta) d\vartheta. \tag{5.3.5}$$

Hence, we are interested in the expected value of $f(y; \theta)$ with θ randomly distributed according to the prior $f_\theta(\theta)$. The natural statistical approach to estimate this expectation would be to draw a random sample $\theta_1^*, \ldots, \theta_N^*$ from $f_\theta(\theta)$ and estimate the unknown expected value (5.3.5) with the corresponding arithmetic mean derived from the sample, i.e.

$$\hat{E}_\theta(f(y|\theta)) = \frac{1}{N} \sum_{j=1}^{N} f(y|\theta_j^*).$$

Clearly, as the number of samples $N \to \infty$ we get $\hat{E}_\theta(f(y; \theta)) \to E_\theta(f(y; \theta))$. Given that we can draw the sample computationally, the size of N is only limited by the numerical effort we are willing to make. Moreover, and this can be essential, we need to be able to actually draw a sample from the prior $f_\theta(\theta)$. If we are able to calculate the cumulative distribution function $F_\theta(\theta) = \int_{-\infty}^{\theta} f_\theta(\vartheta)d\vartheta$, that is, if we have $F_\theta(\vartheta)$ in numerical form, sampling is easy. In this case we just draw U_j^* from a uniform distribution on $[0, 1]$ and get random draws from $f_\theta(.)$ with

$$\theta_j^* = F_\theta^{-1}(u_j^*).$$

Note that θ_j^* is then distributed according to $F_\theta(.)$, because

$$P(\theta_j^* \leq \vartheta) = P(F_\theta^{-1}(U_j^*) \leq \vartheta) = P(U_j \leq F_\theta(\vartheta)) = F_\theta(\vartheta).$$

If, on the other hand, $f_\theta(\theta)$ is not a standard distribution, e.g. in the case of our hierarchical prior with hyper-hyperparameters, there is often no ad hoc routine available, with which to draw a sample from $f_\theta(\theta)$. Unfortunately, in this case, the cumulative distribution function is not available in analytic form and requires numerical integration to be calculated.

An alternative sampling routine, which does not require that we sample from $f_\theta(.)$ directly, is **rejection sampling**. First we need a distribution $f_\theta^*(.)$ from which we can easily sample, that is for which the cumulative distribution function $F_\theta^*(.)$ can be easily calculated. This distribution must satisfy the envelope (or also called umbrella) property, which means that there must exist a constant $a < \infty$ such that

$$f_\theta(\vartheta) \leq a f_\theta^*(\vartheta) \quad \text{for all} \quad \vartheta \in \Theta.$$

We can now sample from $f_\theta(.)$ as follows:

1. Draw U^* from a uniform distribution on $[0, 1]$.
2. Draw θ^* from $f_\theta^*(.)$, independent from U^*.
3. If $U^* \leq f_\theta(\theta^*)/(a f_\theta^*(\theta^*))$ then accept θ^*. Otherwise reject θ^* and jump back to Step 1.

Let us demonstrate that this strategy really gives random samples from $f_\theta(.)$, even though θ^* in Step 2 is drawn from $f_\theta^*(.)$. Let us therefore look at the distribution conditional upon acceptance in Step 3 in the sampling algorithm above, i.e.

$$P(\theta^* \le \vartheta | \theta^* \text{ is accepted}) = \frac{P(\theta^* \le \vartheta, \theta^* \text{ is accepted})}{P(\theta^* \text{ is accepted})}.$$

The numerator is given by

$$P\left(\theta^* \le \vartheta, U^* \le \frac{f_\theta(\theta^*)}{af_\theta^*(\theta^*)}\right) = \int_{-\infty}^{\vartheta} P\left(U^* \le \frac{f_\theta(\theta^*)}{af_\theta^*(\theta^*)}\right) f_\theta^*(\theta^*) d\theta^*$$

$$= \int_{-\infty}^{\vartheta} \frac{f_\theta(\theta^*)}{a} d\theta^* = \frac{F_\theta(\vartheta)}{a},$$

where $F_\theta(.)$ is the cumulative distribution function corresponding to $f_\theta(.)$. The denominator results in the same form but now with ϑ set to ∞, i.e.

$$P(\theta^* \text{ is accepted}) = P(\theta^* \le \infty, \theta^* \text{ is accepted}) = \frac{F_\theta(\infty)}{a} = \frac{1}{a}.$$

This directly proves that the rejection sampling algorithm produces a sample from $f_\theta(.)$ without actually drawing from $f_\theta(.)$. This sounds like a formidable idea, as we can sample directly from any distribution $f^*(.)$ and apply the routine above to obtain samples from our distribution of interest $f_\theta(\theta)$. However, the rejection step is a big obstacle in practice. If the proposal distribution $f_\theta^*(.)$ is chosen badly, then the proposal θ^* is only accepted with a (very) low probability, which could be very close to zero. In this case, we very rarely pass the acceptance step. This happens if a is large and $f_\theta^*(.)$ is far away from the target distribution $f_\theta(.)$. As $a \ge \max_\vartheta f_\theta(\vartheta)/f_\theta^*(\vartheta)$ the intention is to have $f_\theta^*(\vartheta)$ be as close as possible to $f_\theta(\vartheta)$ for $\vartheta \in \Theta$, i.e. we want to choose an umbrella distribution that shares the contours of our existing distribution, otherwise we will need a higher a to satisfy the umbrella property. More complex strategies, such as adaptive rejection sampling proposed by Gilks and Wild (1992), are able to elegantly address the problem of selecting an umbrella distribution. In this case, one adaptively constructs a piecewise umbrella distribution which covers $f_\theta(.)$.

An alternative strategy is **importance sampling**, see Kloek and Dijk (1978). In this case, we again draw a sample from $f_\theta^*(.)$ instead of $f_\theta(.)$ leading to independent draws $\theta_1^*, \ldots, \theta_N^* \sim f_\theta^*(.)$. We take this sample and estimate the mean value (5.3.5) with

$$\frac{1}{N} \sum_{i=1}^{N} \frac{f(y|\theta_i^*) f_\theta(\theta_i^*)}{f_\theta^*(\theta_i^*)}. \tag{5.3.6}$$

Note that this is a consistent estimate, because by taking the expectation with respect to sample $\theta_1^*, \ldots, \theta_N^*$ we get

$$E^* \left(\frac{1}{N} \sum_{i=1}^{N} \frac{f(y|\theta_i^*) f_\theta(\theta_i^*)}{f_{\theta^*}(\theta_i^*)} \right) = \int \frac{f(y|\theta^*) f_\theta(\theta^*)}{f_{\theta^*}(\theta^*)} f_{\theta^*}(\theta^*) d\theta_i^*$$

$$= \int f(y|\theta^*) f_\theta(\theta^*) d\theta^* = f(y).$$

This sampling trick again looks surprisingly simple, as we just draw samples from $f_\theta^*(.)$, whatever the distribution is and plug the samples into (5.3.6) to get an estimate for (5.3.5). One should note, however, that the terms in the sum of (5.3.6) can have a tremendous variability, which occurs if the ratio $f_\theta(\theta^*)/f_\theta^*(\theta^*)$ is far away from 1. Hence, if the proposal density $f_\theta^*(\theta)$ is far away from the target density $f_\theta(\theta)$, we need a very (!) large sample N to get a reliable estimate. Consequently, we are faced with the same problem that occurred above in rejection sampling, that is, for it to work we need $f_\theta^*(.)$ to be close to $f_\theta(.)$.

5.4 Markov Chain Monte Carlo (MCMC)

Thus far we have been trying to approximate the normalisation constant of the posterior. While this approach looks like a plausible strategy, all approximation methods turn out to be problematic for multivariate θ. Alternatively, we can pursue a different strategy and try to simulate directly from the posterior. Assume, that we were able to draw θ_j^* directly from the posterior, i.e.

$$\theta_j^* \sim f_\theta(\theta|y).$$

In this case, we could get an *i.i.d.* sample $\theta_j^*, j = 1, \ldots, N$. This sample yields consistent estimates of the empirical distribution function and other parameters of the posterior. Hence, the idea of simulating θ from the posterior makes perfect sense, although the process is generally more complicated because we are not able to directly draw independent replicates θ_j^*. Instead we will draw a Markov chain $\theta_1^*, \theta_2^*, \ldots$ where θ_j^* and θ_{j+1}^* will be correlated. The distribution of the constructed Markov chain will converge (under regularity conditions that are usually valid in this context) to a stationary distribution which is, in fact, the posterior distribution $f_\theta(.|y)$. In this case using the ergodicity of Markov chains, we have for any function $h(\theta^*)$

$$\frac{1}{N} \sum_{j=1}^{N} h(\theta_j^*) \rightarrow \int h(\vartheta) f_\theta(\vartheta|y) d\vartheta.$$

Given that we apply a **simulation based approach**, often referred to simply as Monte Carlo in statistics, we obtain a *M*arkov *C*hain *M*onte *C*arlo approach, which is abbreviated as **MCMC**.

We begin by introducing the **Metropolis-Hasting algorithm**, dating back to Metropolis et al. (1953) and Hastings (1970). The breakthrough in statistics took place when Gelfand and Smith (1990) applied the algorithm to Bayesian statistics. In the following, the reader should bear in mind that we still do *not* know the posterior probability

$$f_\theta(\theta|y) = \frac{f(y;\theta)f_\theta(\theta)}{f(y)},$$

as we do not know its denominator. However, we actually have quite a lot of information at our disposal. We know the shape of the distribution, because the posterior is proportional to the product of the likelihood and the prior, i.e.

$$f_\theta(\theta|y) \propto f(y;\theta)f_\theta(\theta).$$

In fact, if we compare the density of two values θ and $\tilde\theta$, say, we get the ratio

$$\frac{f_\theta(\theta|y)}{f_\theta(\tilde\theta|y)} = \frac{f(y;\theta)f_\theta(\theta)}{f(y)} \bigg/ \frac{f(y)}{f(y;\tilde\theta)f_\theta(\tilde\theta)} = \frac{f(y;\theta)f_\theta(\theta)}{f(y;\tilde\theta)f_\theta(\tilde\theta)}.$$

Hence, the unknown quantity $f(y)$ cancels out. It is exactly this property that is exploited by the Metropolis-Hastings algorithm. Let $\theta_{(1)}^*, \theta_{(2)}^*, \ldots, \theta_{(t)}^*$ be a Markov chain constructed as follows:

1. Given a current value $\theta_{(t)}^*$ and a proposal distribution $q(\theta|\theta_{(t)}^*)$ draw θ^*, i.e.

$$\theta^* \sim q(.|\theta_{(t)}^*).$$

2. Accept θ^* as new step in the Markov Chain with probability

$$\alpha(\theta_{(t)}^*, \theta^*) = \min\left\{1, \frac{f_\theta(\theta^*|y)q(\theta_{(t)}^*|\theta^*)}{f_\theta(\theta_{(t)}^*|y)q(\theta^*|\theta_{(t)}^*)}\right\}.$$

If θ^* is not accepted set $\theta_{(t+1)}^* = \theta_{(t)}^*$, otherwise set $\theta_{(t+1)}^* = \theta^*$.

The proposal density $q(\theta^*|\theta_{(t)}^*)$ needs to be adequately chosen as discussed below. If the proposal is symmetric, we have $q(\theta^*|\theta_{(t)}^*) = q(\theta_{(t)}^*|\theta^*)$ and the acceptance probability simplifies to

$$\alpha(\theta_{(t)}^*, \theta^*) = \min\left\{1, \frac{f_\theta(\theta^*|y)}{f_\theta(\theta_{(t)}^*|y)}\right\}.$$

In this case, the algorithm is also simply called the Metropolis algorithm. We see that if the posterior density for θ^* is higher than for $\theta^*_{(t)}$, we always accept the proposal θ^* as new step in the Markov Chain. Otherwise we accept the proposal θ^* with a probability less than one, dependent on the ratio of the posterior densities. Note that we only know the posterior $f_\theta(.|y)$ up to its normalisation constant, which however cancels out in the calculation of $\alpha(\theta^*_{(t)}, \theta^*)$, such that we can actually quite easily calculate the acceptance probability.

Property 5.1 The sequence of random numbers $\theta^*_{(j)}, j = 1, 2, \ldots$ drawn from the Markov chain with Metropolis (Hastings) as above has $f_\theta(.|y)$ as stationary distribution for $N \to \infty$.

Proof A formal proof on convergence requires deeper knowledge about Markov chains. Clearly we need some requirements to get the stationary distribution from the Markov chain. For the details we refer to classical literature like Grimmett and Stirzaker (2001). A technically rigorous discussion on requirements and properties of the Metropolis-Hastings algorithm can also be found in Robert and Casella (2004). However, for the purposes of explanation we remain on a heuristic level, instead motivating the central components of the proof. First of all, we need the proposal density to cover the full parameter space Θ, that is, all values in Θ are possible. For instance, if $\Theta = \mathbb{R}$, a normal distribution as proposal guarantees this condition. Let $K(\theta^*_{(t)}, \theta^*)$ be the transition probability (usually called the Kernel) that the Markov Chain proceeds from $\theta^*_{(t)}$ to θ^*. For the proposed Metropolis-Hastings algorithm this is given by

$$K(\theta^*_{(t)}, \theta^*) = q(\theta^*|\theta^*_{(t)})\alpha(\theta^*_{(t)}, \theta^*) + \delta_{\theta^*_{(t)}}(\theta^*)(1 - \alpha(\theta^*_{(t)})),$$

where $\delta_{\theta^*_{(t)}}(.)$ here stands for a Dirac measure taking value 1 if $\theta^* = \theta^*_{(t)}$ and 0 otherwise, and $\alpha(\theta^*_{(t)}) = \int \alpha(\theta^*_{(t)}, \theta)q(\theta|\theta^*_{(t)})d\theta$ is the overall acceptance probability. Note that we have a mixed update step, in that if θ^* is accepted it can take an arbitrary value given by a density, while if θ^* is not accepted it remains at $\theta^*_{(t)}$, giving a point mass at this parameter value. In other words, looking at the cumulative distribution function we get

$$P(\theta^* \le \vartheta|\theta^*_{(t)}) = \int_{-\infty}^{\vartheta} q(\theta|\theta^*_{(t)})\alpha(\theta^*_{(t)}, \theta)d\theta + 1_{\{\vartheta \ge \theta^*_{(t)}\}}(1 - \alpha(\theta^*_{(t)})),$$

such that a jump of size $(1 - \alpha(\theta^*_{(t)}))$ occurs at the current state $\theta^*_{(t)}$. We show first that the Markov Chain is reversible, i.e.

$$K(\theta^*, \theta^*_{(t)})f_\theta(\theta^*_{(t)}|y) = K(\theta^*_{(t)}, \theta^*)f_\theta(\theta^*|y),$$

where $f_\theta(\theta|y)$ is the posterior distribution. Note that

$$
q(\theta^*|\theta^*_{(t)})\alpha(\theta^*_{(t)}, \theta^*) f_\theta(\theta^*_{(t)}|y) = \min\left\{ f_\theta(\theta^*|y)q(\theta^*_{(t)}|\theta^*), f_\theta(\theta^*_{(t)}|y)q(\theta^*|\theta^*_{(t)}) \right\}
$$

$$
= \min\left\{ 1, \frac{f_\theta(\theta^*_{(t)}|y)q(\theta^*|\theta^*_{(t)})}{f_\theta(\theta^*|y)q(\theta^*_{(t)}|\theta^*)} \right\} q(\theta^*_{(t)}|\theta^*) f(\theta^*|y)
$$

$$
= q(\theta^*_{(t)}|\theta^*)\alpha(\theta^*, \theta^*_{(t)}) f_\theta(\theta^*|y).
$$

This implies for the point mass probability at $\theta^* = \theta^*_{(t)}$ that

$$
\delta_{\theta^*_{(t)}}(\theta^*)(1 - \alpha(\theta^*_{(t)})) f_\theta(\theta^*_{(t)}|y) = \delta_{\theta^*}(\theta^*_{(t)})(1 - \alpha(\theta^*)) f_\theta(\theta^*|y).
$$

Consequently, the Metropolis-Hastings Markov Chain is reversible. This in fact also proves that $f_\theta(\theta^*|y)$ is the stationary distribution of the Markov chain, which can be shown as follows. Assume that $\theta^*_{(t)}$ is drawn from the posterior distribution $f_\theta(\theta|y)$. Then $\theta^*_{(t+1)}$ is also drawn from $f_\theta(\theta|y)$, because

$$
P(\theta^*_{(t+1)} \le \vartheta) = \int_{-\infty}^{\vartheta} \left[\int_{-\infty}^{\infty} K(\theta^*_{(t)}, \theta^*_{(t+1)}) f_\theta(\theta^*_{(t)}|y) d\theta^*_{(t)} \right] d\theta^*_{(t+1)}
$$

$$
= \int_{-\infty}^{\vartheta} \left[\int_{-\infty}^{\infty} K(\theta^*_{(t+1)}, \theta^*_{(t)}) f_\theta(\theta^*_{(t+1)}|y) d\theta^*_{(t+1)} \right] d\theta^*_{(t)} = P(\theta^*_{(t)} \le \vartheta).
$$

$\square$

The Metropolis-Hastings algorithm allows us to sample from the posterior, even though we do not know its normalisation constant. To do this in practice, we start a Markov Chain and propose new values θ^* given the proposal distribution $q(\theta^*|\theta^*_{(t)})$. The proposal distribution should thereby fulfil two properties. First of all, it should allow us to sample the entire parameter space Θ. Secondly, it should provide a reasonable acceptance rate $\alpha(\theta^*_{(t)}, \theta^*)$, implying that we should remain somewhat close to $\theta^*_{(t)}$ to keep the density of $f_\theta(\theta^*|y)$ and $f_\theta(\theta^*_{(t)}|y)$ of similar scale. These two goals clearly contradict each other and need to be balanced. Let us demonstrate the use of the Metropolis-Hastings algorithm in a small example. In Fig. 5.5 we plot a tri-modal posterior distribution $f_\theta(\theta|y)$ for a two-dimensional parameter and attempt to sample from it.

For the proposal distribution, we use the normal distribution

$$
q(\theta^*|\theta^*_{(t)}) = N(\theta^*_{(t)}, \sigma^2 I_2)
$$

with the standard deviation set to $\sigma = 0.1$ for the first run and $\sigma = 1$ for the second run of the algorithm. We start the algorithm at $\theta^*_{(0)} = (1, 1)$ and plot the first 1000

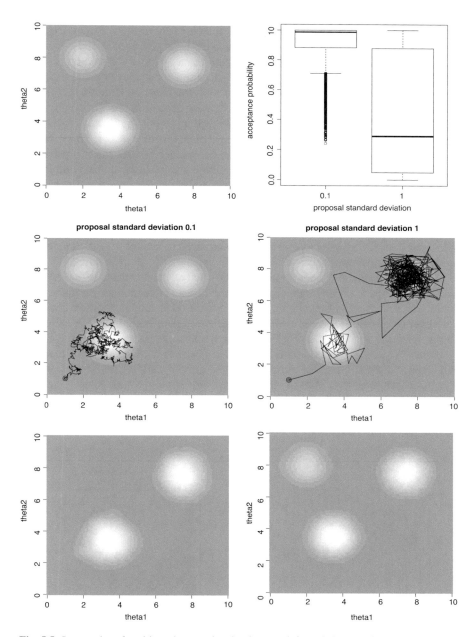

Fig. 5.5 Image plot of multi-mode posterior density (top left) and the resulting Markov Chain-Metropolis-Hastings estimates with proposal density $\sigma = 0.1$ (bottom left) and $\sigma = 1$ (bottom right). First 1000 steps of the algorithm (middle row) and acceptance probability (top right)

steps for both standard deviations. This is shown in the second row of Fig. 5.5, where the left plot corresponds to $\sigma = 0.1$ and the right plot to $\sigma = 1$.

We can see that smaller steps in the parameter space occur for a smaller proposal standard deviation, which resulted in the entire parameter space not being explored. In contrast, for the large standard deviation $\sigma = 1$, the steps are larger, which allows the algorithm to jump from the first high density area at mode $(3.5, 3.5)$ to the second mode at $(7.5, 7.5)$. The price for this exploration of the parameter space is that the acceptance rate for $\sigma = 1$ is much lower than that of $\sigma = 0.1$. This is shown in the top right of Fig. 5.5, where we plot the acceptance rate for the two proposals given 50,000 iterations. While for $\sigma = 0.1$ we have a median acceptance probability of close to 1, it is about 0.3 for $\sigma = 1$. While a high acceptance probability appears beneficial, it may lead to an incomplete exploration of the parameter space and posterior distribution. This is demonstrated in the bottom row of Fig. 5.5, where we plot the posterior density of 50,000 MCMC samples for the two proposal densities. While for $\sigma = 1$ the three modes are found and reproduced, for $\sigma = 0.1$ the top left mode is omitted, simply because the proposal density had too little range.

In practice, one also needs to pay attention to the performance of the Markov Chain with respect to convergence, autocorrelation and dependence on the starting value. The first aspect of managing these properties is to run the Markov Chain for a number of iterations until it reaches its stationary distribution. This is called the "burn-in" phase and visualised in Fig. 5.6 for a univariate parameter.

Figure 5.6 was calculated with a normally distributed posterior, with mean 10 and standard deviation 1. The proposal is also normal with $\theta^* \sim N(\theta^*_{(t)}; \sigma)$ with $\sigma = 0.1$. We start two Markov chains, one with starting value (black line), the second with 4 (grey line). Both series reach the mean value after about 500 steps. We take the first 1000 steps as burn-in, meaning that all simulated values in this burn-in phase will be ignored.

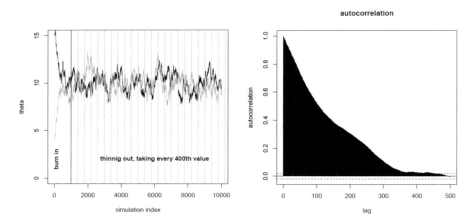

Fig. 5.6 Trajectory of Markov chain with two starting points and indicated burn-in phase (left plot). Autocorrelation of Markov chain series (right plot)

The next factor to be aware of is autocorrelation. Looking at the Markov chain after the burn-in phase, we see that successive steps are correlated over time, i.e. there is autocorrelation present. This is due to the construction of the series and a natural consequence of the Metropolis-Hastings approach and the Markov Chain model used. The autocorrelation is visualised on the right of Fig. 5.6, where we plot the empirical correlation between θ_t^* and θ_{t+lag}^*, which is strong but diminishes with larger lags. In fact, for lags of size 400 there is no empirical correlation observed. Given that $\theta_{(t)}^*$ are not uncorrelated, we may not take all simulated values $\theta_{(t)}^*$ but perhaps every 400th. The procedure is called "thinning out" and is visualised in Fig. 5.6 in the left-hand plot. Thinning out guarantees uncorrelated samples and thus an $i.i.d.$ sample from the posterior distribution.

The Metropolis-Hastings approach works well as long as the acceptance rate is reasonably high. However, if the parameter θ is high dimensional, it may occur that the acceptance probability $\alpha(\theta_{(t)}^*, \theta^*)$ is close to zero. This takes place because the density of each point in a multivariate distribution becomes smaller as dimensionality increases. Assume, for instance, that θ is p dimensional, i.e. $\theta = (\theta_1, \ldots, \theta_p)$ and let, for simplicity, the posterior distribution be given by $f_\theta(\theta|y) = f_{\theta_1}(\theta_1|y) \cdot \ldots \cdot f_{\theta_p}(\theta_p|y)$. With a symmetric proposal density

$$\alpha(\theta_{(t)}^*, \theta^*) = min\left\{1, \frac{f_{\theta_1}(\theta_1^*|y)}{f_{\theta_1}(\theta_{1(t)}^{(t)}|y)} \cdot \ldots \cdot \frac{f_{\theta_p}(\theta_p^*|y)}{f_{\theta_p}(\theta_{p(t)}^*|y)}\right\}.$$

If for every p $f_{\theta_p}(\theta_p^*|y)/f_{\theta_p}(\theta_{p(t)}^*|y) = 0.8$, that is, for every parameter component separately we have 80% acceptance rate, the overall acceptance rate is 0.8^p, which for $p = 10$ is already only 10%. Clearly, the higher the dimension of θ, the lower the acceptance rate. This in turn makes the Metropolis-Hastings algorithm as proposed infeasible if p is large.

This can be circumvented by a process called **Gibbs sampling**, which again makes use of the conditional distribution. Note that for every $k \in \{1, \ldots, p\}$ we have

$$f_{\theta_k|y,\theta_{-k}}(\theta_k|y, \theta_{-k}) \propto f(y|\theta) f_\theta(\theta), \tag{5.4.1}$$

where θ_{-k} denotes the parameter vector with (5.4.1) component k excluded, that is $\theta_{-k} = (\theta_1, \ldots, \theta_{k-1}, \theta_{k+1}, \ldots, \theta_p)$. This proportionality can now be used for sampling. The idea is that we sample only a single component of the entire parameter vector in each step, namely θ_k, and not the entire parameter vector.

To explain this more formally, we let $\theta = (\theta_1, \ldots, \theta_p)$ be a p dimensional parameter. Assume further that $f_{\theta_k|y,\theta_{-k}}$ is known and can be sampled from. Let $\theta_{(t)}^*$ be the current value and, for $t = 0$, the starting value.

1. Draw $\theta_1^* \sim f_{\theta_1|y,\theta_{-1}}(\theta_1^*|y, \theta_{2(t)}, \ldots, \theta_{p(t)})$ and set $\theta_{1(t+1)}^* = \theta_1^*$
2. Draw $\theta_2^* \sim f_{\theta_2|y,\theta_{-2}}(\theta_2^*|y, \theta_{1(t+1)}^*, \theta_{3(t)}^*, \ldots, \theta_{p(t)}^*)$ and set $\theta_{2(t+1)}^* = \theta_2^*$

$$\vdots$$

p. Draw $\theta_p^* \sim f_{\theta_p|y,\theta_{-p}}(\theta_p^*|y, \theta_{1(t+1)}^*, \ldots, \theta_{p-1(t+1)}^*)$ and set $\theta_{p(t+1)}^* = \theta_p^*$
p+1. Jump back to 1

Property 5.2 The sequence of random variables $\theta_{(j)}^*$, $j = 1, 2, \ldots$ drawn from the above Gibbs sampling scheme converges to the posterior distribution $f_\theta(.|y)$ as a stationary distribution.

Proof Let us briefly sketch why the Gibbs algorithm produces random samples from the posterior distribution. For simplicity of notation we drop the observation y in the conditioning argument and assume that we want to prove that $\theta_{(t)}^*$ drawn using Gibbs sampling is in fact drawn from $f_{\theta|y}(\theta^*)$. Note that this also applies for individual components of the parameter vector, e.g. for $\theta_{1(t)}^*$. We want to show that $\theta_{1(t)}^*$ is drawn from $f_{\theta_1|y}(\theta_1)$. Also, for simplicity, let $p = 2$. Then the density after the first step in the Markov chain is given by

$$f_{\theta_{1(1)}|\theta_{1(0)}}(\theta_{1(1)}^*|\theta_{1(0)}^*) = \int f_{\theta_1|\theta_2}(\theta_{1(1)}^*|\theta_2^*) f_{\theta_2|\theta_1}(\theta_2^*|\theta_{1(0)})d\theta_2^*,$$

where the two densities mirror the two proposal distributions in the first Gibbs sampling loop. Similarly we get

$$f_{\theta_{1(t)}|\theta_{1(0)}}(\theta_{1(t)}^*|\theta_{1(0)}) = \int f_{\theta_{1(t)}|\theta_{1(t-1)}}(\theta_{1(t)}^*|\theta_{1(t-1)}) f_{\theta_{1(t-1)}|\theta_{1(0)}}(\theta_{1(t-1)}|\theta_{1(0)})d\theta_{1(t-1)}.$$

The density $f_{\theta_{1(t)}|\theta_{1(t-1)}}(.)$ gives the one-step transition. If $t \to \infty$ it follows under appropriate regularity conditions that $f_{\theta_{1(t)}|\theta_{1(0)}}(\theta_{1(t)}^*|\theta_{1(0)})$ converges to the stationary distribution $f_{\theta_1}(\theta_1^*)$, which is in fact the distribution we aim to sample from. □

Clearly, Gibbs sampling and the Metropolis-Hastings algorithm can be combined, which is necessary if the conditional density is only known up to a constant. Such hybrid approaches are rather common and we exemplify the procedure using the same example from above, but now every update is made univariately using a Gibbs sampling scheme in combination with Metropolis-Hastings. As a proposal density we take a univariate normal distribution, i.e. $\theta_k^* \sim N(\theta_{k(t)}^*, \sigma)$. The top row of Fig. 5.7 shows the first 1000 updates, when the proposal standard deviation is either set to $\sigma = 1$ (left-hand side) or $\sigma = 2$ (right-hand side).

It can be observed that the Markov Chain moves horizontally and vertically, mirroring the fact that we simulate each parameter separately. Diagonal steps occur only if both Gibbs steps are accepted at the same time. The proposals are, however, only horizontal and vertical. The resulting density estimates for the entire sample of 50,000 simulations are shown in the bottom row of Fig. 5.7.

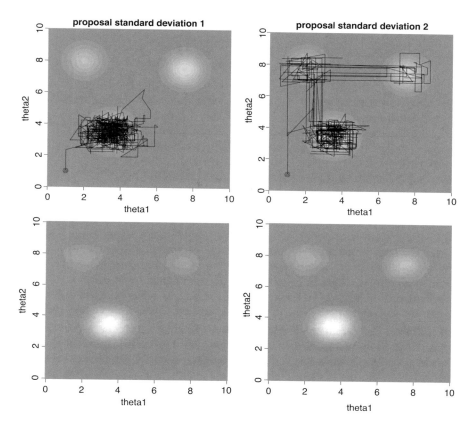

Fig. 5.7 Gibbs sampling combined with Metropolis-Hastings. The proposal density is normally distributed with $\sigma = 1$ (left side column) and $\sigma = 2$ (right side column) and centred on the previous value. The top row shows the first 1000 steps, the bottom row shows the kernel density estimate of 50,000 draws

The field of MCMC algorithms is very wide and numerous alternatives and modifications have been proposed. We refer the interested reader to Congdon (2003) or Held and Sabanés Bové (2014) for further material and references.

5.5 Variational Bayes

We now turn our attention to a recent method for the approximation of the posterior distribution, that makes use of **variational Bayes** principles. The approach intends to replace the true posterior distribution $f_\theta(.|y)$ with an approximation $q_\theta(.)$. The approximate distribution $q_\theta(.)$ belongs to a given class of distribution $\mathcal{Q}$ and is chosen such that the Kullback–Leibler divergence between the true (unknown) posterior and the approximation $\mathcal{Q}$ is minimised. We therefore look at

$$KL(f_\theta(.|y), q_\theta(.)) = \int \log \frac{q_\theta(\vartheta)}{f_\theta(\vartheta|y)} q_\theta(\vartheta) d\vartheta.$$

The idea is now to choose $q_\theta \in \mathcal{Q}$ such that it minimises the Kullback–Leibler divergence, i.e. we aim to find

$$\hat{q}_\theta = \arg\min_{q_\theta \in \mathcal{Q}} KL(f_\theta(\cdot|y), q_\theta(\cdot)).$$

Note that the Kullback–Leibler divergence decomposes to

$$KL(f_\theta(.|y), q_\theta(.)) = \int \log q_\theta(\vartheta) q_\theta(\vartheta) d\vartheta - \int \log f_\theta(\vartheta|y) q_\theta(\vartheta) d\vartheta,$$

where the first component is also called the entropy. Clearly taking $q_\theta(.) = f_\theta(.|y)$ minimises the Kullback–Leibler divergence. But instead of taking the exact posterior, the idea is to take $q_\theta(.)$ as some simple and easy to calculate distribution. For instance, we may choose $q_\theta(.)$ to be a normal distribution, i.e. $q_\theta(\theta) = \frac{1}{\sqrt{2\pi}\sigma_\theta} \exp(-\frac{1}{2}(\theta - \mu_\theta)^2/\sigma_\theta^2)$ where we now choose μ_θ and σ_θ^2 such that the Kullback–Leibler divergence is minimised. This is comparable with a Laplace approximation discussed in Sect. 5.3.2. However, the variational method has a broader scope and is suitable even if the parameter vector θ is multi- or even high dimensional. Assume therefore that $\theta = (\theta_1, \ldots, \theta_p)^T$. A possible simplification occurs if we restrict the distribution $q_\theta(.)$ to independence, such that

$$q_\theta(\theta) = \prod_{k=1}^{p} q_k(\theta_k).$$

The component-wise independence could also be replaced by block-wise independence, that is, we could group the components of θ to blocks and assume independence between the blocks. For simplicity, we present only the idea for univariate independent components here. Note that with independence we get the following Kullback–Leibler divergence

$$KL(f_\theta(.|y), q_\theta(.)) = \int \prod_{i=1}^{p} \log \frac{\prod_{j=1}^{p} q_j(\theta_j)}{f_\theta(\theta|y)} q_i(\theta_i) d\theta_i$$

$$= \int q_k(\theta_k) \Big\{ \log q_k(\theta_k) - \underbrace{\int \prod_{i=1, i\neq k}^{p} \log(f_\theta(\theta|y)) q_i(\theta_i) d\theta_i}_{=: E_k(\log f_\theta(\theta|y)) =: \log \tilde{f}_k(\theta_k|y)} \Big\} d\theta_k$$

$$+ \int q_k(\theta_k) \left\{ \int \prod_{i=1,i\neq k}^{p} \sum_{j=1,j\neq k}^{p} \log q_j(\theta) q_i(\theta) d\theta_i \right\} d\theta_k.$$

Note that only the first component depends on the true posterior. Moreover, the component $E_k(\log f_\theta(\theta|y))$ depends only on θ_k, as all of the other parameters have been integrated out. This allows us to denote this component as $\log \tilde{f}_k(\theta_k|y)$. It should also be clear that, $\tilde{f}_k(\theta_k|y)$ also depends on all marginal densities $q_j(\theta_j)$ for $j \neq k$, which is suppressed in the notation. Considering the first term of the sum above, we can comprehend this again as Kullback–Leibler divergence $KL(\tilde{f}_k(\cdot|y), q_k(\cdot))$, which is minimised if we choose

$$q_k(\theta_k) \propto \tilde{f}_k(\theta_k|y).$$

Note that $\tilde{f}_k(\theta_k|y)$ is not necessarily a density, because the integral over all values of θ_k is not necessarily equal to one. One may not even know the posterior $f_k(\theta_k|y)$ exactly, but only up to a normalisation constant. We therefore set

$$q_k(\theta) = \frac{\tilde{f}_k(\theta_k|y)}{\int \tilde{f}_k(\theta_k|y) d\theta_k},$$

where the integral is univariate only and may be calculated with numerical integration or even analytically, if possible. The idea is now to update iteratively one component after the other. Let $q_{\theta(t)} = \prod_{k=1}^{p} q_{k(t)}(\theta_k)$ be the current approximate. We then update the k-th component by

$$q_{k(t+1)}(\theta_k) = \frac{\tilde{f}_{k(t)}(\theta_k|y)}{\int \tilde{f}_{k(t)}(\vartheta_k|y) d\vartheta_k},$$

with obvious notation for $\tilde{f}_{k(t)}$. Variational Bayes has advantages in high dimensional parametric models, where the separate components are standard in structure. In this case, it can speed up the computational when compared with MCMC approaches. This happens in particular if one works with conjugated priors. We refer to Fox and Roberts (2012) for more details.

5.6 Exercises

Exercise 1

Let $Y_1, \ldots, Y_n$ be an *i.i.d.* sample from an exponential distribution $Exp(\lambda)$. As prior distribution for λ, we assume a Gamma distribution, $Ga(\alpha, \beta)$.

1. Derive the posterior distribution for λ and its mean and variance.
2. Calculate the posterior mode. For what choice of α and β are the posterior mode estimate and Maximum Likelihood estimate of λ numerically identical?

Exercise 2

Having derived the posterior distribution for a parameter θ given a sample $y = (y_1 \ldots, y_n)$, the posterior predictive distribution for a new observation y_{n+1} is defined as

$$f(y_{n+1}|y) = \int_{-\infty}^{+\infty} f(\theta|y) f(y_{n+1}|\theta) d\theta ,$$

where $f(\theta|y)$ is the posterior distribution and $f(y_{n+1}|\theta)$ is the likelihood contribution of the new observation.

Derive the posterior predictive distribution for a new observation for the case of an $i.i.d.$ sample y of a normal distribution $N(\mu, \sigma^2)$ with known variance σ^2 and flat constant prior 1 for μ.

Exercise 3 (Use R Statistical Software)

The file ch6exerc3.csv contains annual numbers Y_i, $i = 1, \ldots, 112$ of accidents due to disasters in British coal mines from 1850 to 1962 (the data are contained in the R package GeDS).

A change point model is applied to the data, which means that the parameters of the model change at a given point in time. To be specific, the model takes the following form:

$$X_i = \begin{cases} Poisson(\lambda_1) \ i = 1, \ldots, \theta, \\ Poisson(\lambda_2) \ i = \theta + 1, \ldots, 112, \end{cases}$$

where we assume as priors: $\lambda_i|\alpha \sim \Gamma(3, \alpha)$ $i.i.d., i = 1, 2$ and $\alpha \sim \Gamma(10, 10)$. We assume that θ is known.

1. Derive the univariate full conditionals and describe a Gibbs sampler to get draws from the posterior distribution $p(\lambda_1, \lambda_2, \alpha|x, \theta)$. *Hint:* all three full conditionals are Gamma distributions.
2. Implement the Gibbs sampler in R and let it run for different values of θ. Use a heuristic criterion to decide for which θ a breakpoint exists.

Chapter 6
Statistical Decisions

Data are often collected and analysed to answer a particular question. For instance, when modifying an existing product, one wants to know whether said modification has improved the product's quality. In the social or medical sciences, one wants to know whether an intervention has a positive (or negative) impact. One might also want to know whether the collected data come from a particular distribution. All of these questions can be tackled with the principles of statistical testing. It follows that testing implies deciding whether a particular conjecture holds or not, which we will introduce in the next section.

6.1 The Idea of Testing

To introduce the principles behind hypothesis testing, let us consider a binary decision problem with two disjoint possibilities, which we label as H_0 and H_1. We call H_0 the null hypothesis and H_1 the alternative hypothesis, or sometimes just the alternative. Generally, we do not know whether the hypothesis H_0 or the alternative H_1 holds, but given our data we want to choose one or the other, which may or may not be erroneous. If we decide for H_1 when H_0 is valid, and similarly for H_0 when H_1 is valid, we have made a mistake.

To exemplify this, let us assume a company questions whether they should introduce a new marketing strategy for a given product. This decision requires a comparison between the new and old strategies. Let us label with H_0 the hypothesis that the new strategy does not increase sales, while letting the alternative H_1 represent that in fact it does increase sales. The consequences of this decision depend on the true effect of the new marketing strategy. If the new strategy is better and the company introduces it, or it is not better and the company does not introduce it, then they made the correct decision. However, there are two possible types of incorrect decision. When the company decides to introduce the new strategy, despite

G. Kauermann et al., *Statistical Foundations, Reasoning and Inference*, Springer Series in Statistics, https://doi.org/10.1007/978-3-030-69827-0_6

the old strategy being better, it means they spend money implementing the new system (and also lose the benefits of the previous system). On the other hand, when the company refrains from introducing a new, better strategy, they have missed an opportunity. Both bad decisions have a completely different meaning and need to be treated as such. It should also be clear that if we try to avoid making one type of mistake, we run the risk of seeing more of the other. For instance, if we wanted to minimise the risk of falsely changing the existing marketing strategy, we could simply refrain from any change at all, no matter the results. By doing so, we clearly increase the risk that we will not take advantage of a newer, better strategy.

Let us formalise the above framework. We already introduced the two possible states as the null H_0 and alternative H_1 hypothesis. The decisions that we actually make are denoted with quotes, i.e. "H_0" and "H_1". In the example, this means that H_0 stands for the new strategy not being better, while "H_0" expresses our decision to refrain from making a change. Similarly, H_1 represents that the new strategy is actually better, while "H_1" denotes the corresponding decision to adopt the new strategy. Both of the states and decisions are binary and we may write these in the following decision matrix:

	Decision	
States	"H_0"	"H_1"
H_0	Correct decision	Type I error
H_1	Type II error	Correct decision

We can commit two types of errors that we label as type I and type II. It is clearly impossible to simultaneously avoid both. Statistical tests are built upon the idea that the type I error cannot be avoided but should only occur with a small probability. This asymmetrical approach requires a careful choice of H_0 and H_1.

In many applications, researchers aim to prove a hypothesis like "this drug has an effect" or "there are differences between these two groups". To use a statistical test for such questions, the null hypothesis is framed as the opposite of the research hypothesis, e.g. "this drug has no effect" or "there are no differences between these groups". That is to say, the hypothesis H_1 is usually the research hypothesis one wants to prove. This setting allows us to directly control the type I error, which is generally considered more harmful in research applications.

Definition 6.1 A statistical **significance level** α-**test** assumes two states H_0 and H_1 and two possible decisions "H_0" and "H_1". The decision rule is thereby constructed such that

$$P\big(\text{``}H_1\text{''}\,\big|\,H_0\big) \leq \alpha$$

for a fixed small value of α.

Example 19 We illustrate the above idea with a classical example. Assume we have observations $Y_1, \ldots, Y_n$, which are sampled *i.i.d.* and originate from a $N(\mu, \sigma^2)$ distribution. For simplicity, we assume that σ^2 is known, but μ is not. For instance, let Y_i be a quality control measurement of a process, which on average should not exceed a particular value μ_0. We want to run a test on this value and question whether $H_0 : \mu \leq \mu_0$ or $H_1 : \mu > \mu_0$ applies for a given μ_0. Note that H_0 has the inequality, while the strict inequality is in H_1. This is necessary because we need to calculate probability statements under H_0 and hence need boundedness under H_0. We take the average values of $Y_1, \ldots, Y_n$ and calculate $\overline{Y} = \sum\limits_{i=1}^{n} Y_i/n$. Clearly, if $\overline{Y}$ takes (very) large values, this is in favour of H_1. This suggests the decision rule

$$\text{``}H_1\text{''} \Leftrightarrow \overline{Y} > c,$$

where the threshold c is often called the **critical value**. To determine c, we need to relate it to our probability threshold α. Under $H_0 : \mu \leq \mu_0$, it holds that

$$P\left(\text{``}H_1\text{''}\big|H_0\right) \leq \alpha \Leftrightarrow P\left(\overline{Y} > c\big|\mu \leq \mu_0\right) \leq \alpha. \tag{6.1.1}$$

The second probability statement can be solved explicitly with

$$P\left(\overline{Y} > c\big|\mu \leq \mu_0\right) \leq P\left(\overline{Y} > c\big|\mu = \mu_0\right) \leq \alpha$$

$$\Leftrightarrow P\left(\overline{Y} \leq c\big|\mu = \mu_0\right) \geq 1 - \alpha$$

$$\Leftrightarrow P\left(\frac{\overline{Y} - \mu_0}{\sigma/\sqrt{n}} \leq \frac{c - \mu_0}{\sigma/\sqrt{n}}\bigg|\mu = \mu_0\right) \geq 1 - \alpha$$

$$\Leftrightarrow \Phi\left(\frac{c - \mu_0}{\sigma/\sqrt{n}}\right) \geq 1 - \alpha,$$

where Φ is the distribution function of the $N(0, 1)$ distribution. This gives

$$z_{1-\alpha} = \frac{c - \mu_0}{\sigma/\sqrt{n}} \Leftrightarrow c = \mu_0 + z_{1-\alpha}\,\sigma/\sqrt{n},$$

where $z_{1-\alpha}$ is the $1-\alpha$ quantile of the $N(0, 1)$ distribution. This approach can easily be extended to the **two-sided testing problem**, where we are interested in

$$H_0 : \mu = \mu_0 \text{ and } H_1 : \mu \neq \mu_0.$$

In this case, we look at the difference between $\overline{Y}$ and μ_0. If this difference is large, it speaks in favour of H_1, which can be formulated as the following decision rule:

$$\text{``}H_1\text{''} \Leftrightarrow \left|\overline{Y} - \mu_0\right| > c.$$

Following the same calculation as above gives

$$P\left(\left|\overline{Y} - \mu_0\right| > c \middle| H_0\right) \leq \alpha \Leftrightarrow P\left(\left|\overline{Y} - \mu_0\right| \leq c \middle| H_0\right) \geq 1 - \alpha,$$

and the critical value c is given by

$$P\left(-c \leq \overline{Y} - \mu_0 \leq c \middle| \mu = \mu_0\right) \qquad = 1 - \alpha$$

$$\Leftrightarrow P\left(\frac{-c}{\sigma/\sqrt{n}} \leq \frac{\overline{Y} - \mu_0}{\sigma/\sqrt{n}} \leq \frac{c}{\sigma/\sqrt{n}}\right) \qquad = 1 - \alpha$$

$$\Leftrightarrow \Phi\left(\frac{c}{\sigma/\sqrt{n}}\right) - \Phi\left(-\frac{c}{\sigma/\sqrt{n}}\right) \qquad = 1 - \alpha,$$

such that

$$\frac{c}{\sigma/\sqrt{n}} = z_{1-\alpha/2} \Leftrightarrow c = z_{1-\alpha/2}\,\sigma/\sqrt{n}.$$

We see that the critical value c can be derived similarly for both one-sided $H_0 : \mu \leq \mu_0$ and two-sided hypotheses $H_0 : \mu = \mu_0$.

$\triangleright$

6.2   Classical Tests

In this section, we will introduce a number of statistical tests that are so prevalent that they require individual attention. Statistical tests can focus not only on a single parameter but also on multiple components of a multidimensional parameter. However, in order to maintain notational simplicity, we will present our classical statistical tests for one-dimensional parameters only. Let

$$Y_i \sim f(y; \theta) \quad i.i.d.,$$

and, for simplicity, let us assume that the distribution is Fisher-regular (see Definition 3.12). We formulate our hypothesis as

$$H_0 : \theta \in \Theta_0 \text{ and } H_1 : \theta \in \Theta_1,$$

where Θ_0 and Θ_1 are a disjoint decomposition of the parameter space Θ, i.e. $\Theta = \Theta_0 \cup \Theta_1$ and $\Theta_0 \cap \Theta_1 = \varnothing$. We restrict the subsequent presentation to the cases where $\Theta_0 = \{\theta_0\}$ and $\Theta_1 = \Theta \setminus \{\theta_0\}$ or $\Theta_0 = \{\theta : \theta \leq \theta_0\}$ and $\Theta_1 = \Theta \setminus \Theta_0$.

6.2.1   t-Test

The t-test and t-distribution are used when we would like to do inference on the mean of normally distributed random variables when the variance is unknown. Let us consider normal random variables

$$Y_i \sim N(\mu, \sigma^2) \quad i.i.d., i = 1, \ldots, n,$$

and we want to test $H_0 : \mu \leq \mu_0$ versus $H_1 : \mu > \mu_0$. We obtain the decision rule

$$\text{``}H_1\text{''} \Leftrightarrow \bar{Y} = \frac{1}{n} \sum_{i=1}^{n} Y_i > c,$$

where c can be calculated with the normal distribution, i.e. $c = \mu_0 + z_{1-\alpha} \sigma / \sqrt{n}$. However, to calculate c, we need to know the variance σ^2 of the underlying distribution, which in most practical applications is unknown. In this case we can rely on the t-distribution which we will now describe. Note that under H_0, i.e. if $\mu = \mu_0$,

$$\frac{\bar{Y} - \mu_0}{\sigma / \sqrt{n}} \sim N(0, 1). \tag{6.2.1}$$

If we replace σ with its estimate

$$\hat{\sigma} = \sqrt{\frac{1}{n-1} \sum_{i=1}^{n} (Y_i - \bar{Y})^2},$$

then the standard normal distribution in (6.2.1) becomes a t-distribution, that is,

$$\frac{\bar{Y} - \mu_0}{\hat{\sigma} / \sqrt{n}} \sim t_{n-1},$$

where t_{n-1} denotes the t-distribution with $n - 1$ degrees of freedom. The t-distribution resembles the normal distribution but has heavier tails, i.e., it has a higher probability mass for more extreme observations. As n increases, that is, with increasing degrees of freedom, the t-distribution converges to the normal distribution, which can be seen in Fig. 6.1. The curious reader can also refer to Dudewicz and Mishra (1988) for more technical detail.

Example 20 One typical application of the simple t-test, also called the one-sample t-test, is to assess whether the difference between two variables has mean zero. In a psychiatric study, conducted by Leuzinger-Bohleber et al. (2019), the values of a depression score (BDI) for 149 patients were recorded at the start (baseline) and

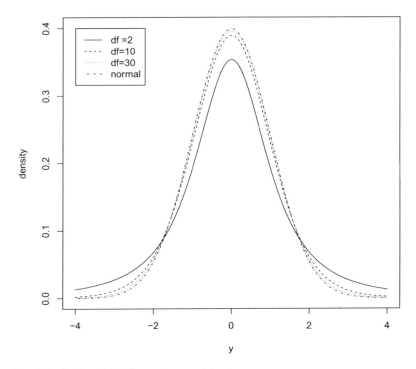

Fig. 6.1 t-Distribution with different degrees of freedom

after 3 years of a particular treatment. The boxplot of the difference between the two values for each subject can be seen in Fig. 6.2. The mean difference was -17.2, indicating a reduced average depression rating after treatment, i.e. an improvement. The standard deviation was 13. The t-value was $t = \frac{Y-0}{\sigma/\sqrt{149}} = -18.5$. Let μ_D be the mean of the difference of the depression score for each individual at baseline and after treatment. We pursue a two-sided test, i.e. $H_0 : \mu_D = 0$ against $H_1 : \mu_D \neq 0$. Then the t-value has to be compared with the critical value $t_{0.0975,149} = 1.97$. Therefore, we can reject the null hypothesis and say that we have a statistically significant change in the depression score.

$\triangleright$

In many applications, we want to compare two means, e.g. the performance of male and female students or the blood pressure of two treatment groups. In such scenarios, we consider two *i.i.d.* samples

$$Y_{1i} \sim N(\mu_1, \sigma_1^2), i = 1, \ldots, n_1 \text{ and } Y_{2j} \sim N(\mu_2, \sigma_2^2), j = 1, \ldots, n_2,$$

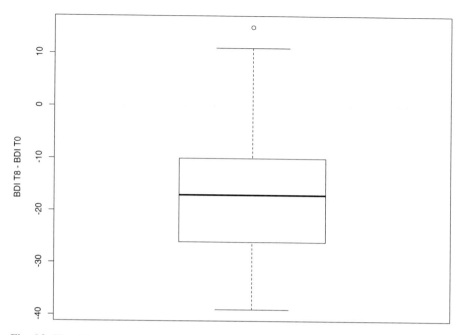

Fig. 6.2 The difference between pre-treatment and post-treatment depression score

where we allow the two groups to have different variances (variance heterogeneity). The hypothesis $H_0 : \mu_1 = \mu_2$ is tested against $H_1 : \mu_1 \neq \mu_2$. From two independent $i.i.d.$ samples, the decision rule is given by

$$\text{"}H_1\text{"} : |\bar{y}_1 - \bar{y}_2| > c.$$

As $\bar{Y}_1 - \bar{Y}_2$ has a normal distribution with mean $\mu_1 - \mu_2$ and variance $\sigma^2(1/n_1 + 1/n_2)$, the critical value c can be obtained, as in the above example, with

$$c = z_{1-\frac{\alpha}{2}} \cdot \sqrt{\frac{\sigma_1^2}{n_1} + \frac{\sigma_2^2}{n_2}}.$$

If we replace σ_k^2 with its pooled estimate

$$\hat{\sigma}_k^2 = \frac{1}{n_k - 1} \sum_{k=1,2} (y_{ki} - \bar{y}_k)^2,$$

the $z_{1-\frac{\alpha}{2}}$-quantile of the normal distribution is replaced by the quantile of a t-distribution with df degrees of freedom, where

$$df = \left(\frac{\hat{\sigma}_1^2}{n_1} + \frac{\hat{\sigma}_2^2}{n_2}\right) \Big/ \left(\frac{1}{n_1 - 1}(\frac{\hat{\sigma}_1^2}{n_1})^2 + \frac{1}{n_2 - 1}(\frac{\hat{\sigma}_2^2}{n_2})^2\right).$$

Due to the central limit theorem and the convergence of the t-distribution to a standard normal distribution, the standard normal distribution can be applied for large sample sizes. In fact, the central limit theorem also justifies the use of a normal approximation, even if the original data are not normally distributed. This is demonstrated with the following two examples.

Example 21 Let us now see how to approach a binomially distributed random variable. Assume we want to check a promotion offered by a chocolate company, which claims that every 7th chocolate contains a prize. We buy a random sample of 350 chocolates and we expect to get 50 prizes. Let p be the proportion of chocolates containing a prize. We perform a significance test for the hypothesis $H_0 : p \geq \frac{1}{7}$ versus $H_1 : p < \frac{1}{7}$. Assuming that we have an independent sample $Y_1, \ldots, Y_{350}$ where $Y_i = 1$ if we win a prize with the i-th chocolate and $Y_i = 0$ if we do not, then under the null hypothesis

$$T = \sum_{i=1}^{350} Y_i \sim B(350, \frac{1}{7}).$$

For constructing a test with significance level α, we need to find a value for c such that $P(T \leq c | H_0) \leq \alpha$. We may now make use of the central limit theorem and get

$$\bar{Y} = \frac{T}{350} \overset{a}{\sim} N(p, \frac{p(1-p)}{350}).$$

The critical value is now directly given by

$$P(\text{"}H_1\text{"}|H_0) \leq \alpha$$

$$\Leftrightarrow P(\bar{Y} \leq c | p_0 = 1/7) \leq \alpha \Leftrightarrow P\left(\frac{\bar{Y} - p_0}{\sqrt{p_0(1-p_0)/350}} \leq \underbrace{\frac{c - p_0}{\sqrt{p_0(1-p_0)/350}}}_{\approx -z_{1-\alpha} = z_\alpha}\right) \leq \alpha.$$

▷

Example 22 Let us continue with Example 20, from Leuzinger-Bohleber et al. (2019). One aim of the study was to compare different types of treatments. The first group was given cognitive behavioural therapy (CBT) and the second psychoanalytic therapy (PAT). We are interested in comparing the improvements

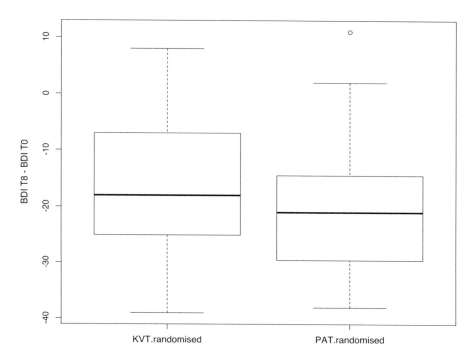

Fig. 6.3 The reduction in BDI scores after treatment with either cognitive behavioural therapy (CBT) or psychoanalytic therapy (PAT)

in depression scores of the two different groups. Denoting μ_{D1} as the difference in scores of Group 1 and μ_{D2} as the difference in scores of Group 2, we want to check the null hypothesis $H_0 : \mu_{D1}=\mu_{D2}$. The result for $n_1 = 30$ patients receiving cognitive behavioural therapy and $n_2 = 27$ patients receiving a long-term psycho analytic therapy is displayed in Fig. 6.3. The means of each group were $\bar{x}_1 = -17.5$ and $\bar{x}_2 = -20.1$. Using the sample variances gives the t-statistic

$$t = \frac{\bar{x}_1 - \bar{x}_2}{\sqrt{\frac{\hat{\sigma}_1^2}{n_1} + \frac{\hat{\sigma}_2^2}{n_2}}} = 0.81.$$

This value is lower than the corresponding critical value of 1.96, and therefore the null hypothesis cannot be rejected and a difference between the two therapies cannot be claimed.

▷

6.2.2   Wald Test

We already discovered in Sect. 4.2 that the ML estimate $\hat{\theta}$ is asymptotically normal with the true parameter θ as mean, i.e.,

$$\hat{\theta} \overset{a}{\sim} N(\theta, I^{-1}(\theta)).$$

Hence, if n is large, any hypothesis on θ can be treated as a hypothesis on the mean of a normal distribution. With the hypotheses $H_0 : \theta = \theta_0$ and $H_1 : \theta \neq \theta_0$, we obtain the following decision rule:

$$\text{``}H_1\text{''} \Leftrightarrow |\hat{\theta} - \theta_0| > c,$$

where c is given by $c = z_{1-\alpha/2}\sqrt{I^{-1}(\theta_0)}$. Tests of this form are known as **Wald tests** and were proposed in Wald (1943). The Wald test is by far the most frequently used testing principle in statistics. However, it requires that we have an estimate of a parameter and its variance. Often, the variance is calculated with the Fisher information not at its hypothetical value θ_0 but at the estimated value $\hat{\theta}$. In this case, the critical value is derived from $c = z_{1-\alpha/2}\sqrt{I^{-1}(\hat{\theta})}$. Because under H_0 we know that $\hat{\theta}$ converges to θ_0, it is plausible to use $I(\hat{\theta})$ instead of $I(\theta_0)$. Given that standard software packages also give $I(\hat{\theta})$ as an output, it is also more convenient to work with the Fisher information calculated at $\hat{\theta}$ and not at θ_0. Generally speaking, however, both versions are equivalent in large samples. Note that the Wald test runs completely comparable to test in a normal distribution, as discussed in Example 19.

6.2.3   Score Test

Another very common test in statistics is the **score test**. From Eq. (4.1.3), we know that the score $s(\theta; Y_1, \ldots, Y_n)$ has mean value zero when evaluated at the true parameter θ. Assuming $\theta = \theta_0$ allows us to calculate $s(\theta_0; y_1, \ldots, y_n)$ from the data. Clearly, if the hypothesis holds, then the random score $s(\theta_0; Y_1, \ldots, Y_n)$ is asymptotically normal with mean value zero. If the value of $s(\theta_0; y_1, \ldots, y_n)$ lies far away from zero, it speaks against the hypothesis that θ_0 is the true value. Hence, we can again apply asymptotic normality to derive a test. To be specific, we test whether $s(\theta_0; y_1, \ldots, y_n)$ is a random variable drawn from a normal distribution with zero mean. Making use of Eq. (4.2.2), we derive the decision rule

$$\text{``}H_1\text{''} \Leftrightarrow |s(\theta_0; y_1, \ldots, y_n)| > c,$$

where $c = z_{1-\alpha/2}\sqrt{I(\theta_0)}$. The score test is comparable to the Wald test, but there is a subtle difference. For the Wald test, we first need to calculate the ML estimate $\hat{\theta}$ to test whether H_0 holds. This is not the case for the score test. Here we simply have to calculate the score function, but not the ML estimate. The score test is therefore particularly useful if the calculation of the ML estimate is cumbersome or numerically demanding.

6.2.4 Likelihood-Ratio Test

Also in common use is the **likelihood-ratio test**. We saw in the previous chapter that if θ is the true parameter, the likelihood-ratio $lr(\hat{\theta}; \theta) = 2\{l(\hat{\theta}) - l(\theta)\}$ is asymptotically χ^2 distributed with p degrees of freedom, with p being the dimension of the parameter. Hence, large values of the likelihood-ratio $lr(\hat{\theta}; \theta)$ calculated at θ_0 speak against the hypothesis $H_0 : \theta = \theta_0$. This allows us to derive the decision rule

$$\text{``}H_1\text{''} \Leftrightarrow 2\{l(\hat{\theta}) - l(\theta_0)\} > c.$$

The critical value c can thereby be calculated from the $1 - \alpha$ quantiles of the χ^2 distribution with p degrees of freedom. For the normal distribution, these three tests, i.e. the Wald test, the score test and the likelihood-ratio test, are equivalent, which we will demonstrate in the following example. Generally, the three tests give different results, although for a large sample size n, all three tests are asymptotically equivalent.

Example 23 In this example, we show that the Wald, score and likelihood-ratio tests are all equivalent for normally distributed random variables. Let $Y_i \sim N(\mu, \sigma^2)$ *i.i.d.* for $i = 1, \ldots n$ and σ^2 be known. We want to test the null hypothesis $H_0 : \mu = \mu_0$. Note that the ML estimate is $\hat{\mu} = \sum_{i=1}^{n} y_i/n = \bar{y}$ with a corresponding Fisher matrix $I(\mu) = n/\sigma^2$. The Wald test gives the decision rule

$$\text{``}H_1\text{''} \Leftrightarrow |\bar{y} - \mu_0| > z_{1-\frac{\alpha}{2}}\sqrt{\frac{\sigma^2}{n}}. \tag{6.2.2}$$

The score function is given by

$$s(\mu; y_1, \ldots, y_n) = \sum_{i=1}^{n} \frac{y_i - \mu}{\sigma^2} = \frac{\bar{y} - \mu}{\sigma^2/n},$$

which gives the score test the decision rule

$$\text{``}H_1\text{''} \Leftrightarrow \left| \frac{\bar{y} - \mu_0}{\sigma^2/n} \right| > z_{1-\frac{\alpha}{2}} \sqrt{\frac{n}{\sigma^2}}$$

$$\Leftrightarrow |\bar{y} - \mu_0| > z_{1-\frac{\alpha}{2}} \sqrt{\frac{\sigma^2}{n}}.$$

Clearly, this is the same rule as that of the Wald test. Finally, the likelihood-ratio for $\mu = \mu_0$ is defined by

$$lr(\bar{y}; \mu_0) = 2 \left\{ -\frac{1}{2} \sum_{i=1}^{n} \frac{(y_i - \bar{y})^2}{\sigma^2} + \frac{1}{2} \sum_{i=1}^{n} \frac{(y_i - \mu_0)^2}{\sigma^2} \right\}$$

$$= \frac{1}{\sigma^2} \left\{ -\sum_{i=1}^{n} y_i^2 + 2n\bar{y}^2 - n\bar{y}^2 + \sum_{i=1}^{n} y_i^2 - 2n\bar{y}\mu_0 + n\mu_0^2 \right\}$$

$$= \frac{1}{\sigma^2} \{ n\bar{y}^2 - 2n\bar{y}\mu_0 + n\mu_0^2 \}$$

$$= \frac{1}{\sigma^2/n} (\bar{y} - \mu_0)^2.$$

The decision rule for the likelihood-ratio test is now

$$\text{``}H_1\text{''} \Leftrightarrow \frac{(\bar{y} - \mu_0)^2}{\sigma^2/n} > \mathcal{X}_{1,1-\alpha}^2 \tag{6.2.3}$$

$$\Leftrightarrow |(\bar{y} - \mu_0)| > \sqrt{\mathcal{X}_{1,1-\alpha}^2} \sqrt{\sigma^2/n}. \tag{6.2.4}$$

Note that for a standard normal random variable $Z \sim N(0, 1)$,

$$P(|Z| \leq z_{1-\frac{\alpha}{2}}) = 1 - \alpha$$

$$\Leftrightarrow P(\underbrace{Z^2}_{\sim \mathcal{X}_1^2} \leq z_{1-\frac{\alpha}{2}}^2) = 1 - \alpha.$$

Hence, the $(1 - \alpha)$ quantile of the chi-squared distribution is given by $\mathcal{X}_{1,1-\alpha}^2 = z_{1-\frac{\alpha}{2}}^2$, and therefore (6.2.4) is equal to the decision rule of the Wald test, and the three tests are equivalent for normally distributed random variables.

▷

6.3   Power of a Test and Neyman–Pearson Test

By this point, it should be clear that statistical tests are built with the intent of limiting our type I error to α. However, this condition can be trivially met by always choosing H_0, giving $P(\text{``}H_1\text{''}|H_0) = 0$ and $P(\text{``}H_0\text{''}|H_1) = 1$. By forcing the type I error to occur with zero probability, we have made the type II error occur with probability one. That is to say, although we have effectively bounded the type I error with α, we have not defined our chance of correctly rejecting the null hypothesis in the presence of true effects. In fact, it seems that these two objectives oppose each other in some fashion.

For fixed α, the type II error changes with the magnitude of the effect. This can be demonstrated using the normal distribution with $H_0 : \mu \leq \mu_0$ versus $H_1 : \mu > \mu_0$. If H_1 is true, but the true mean is very close to μ_0, i.e. $\mu_0 + \delta$ for some infinitely small $\delta > 0$, then the probability $P(\bar{Y} > c|H_0)$ is almost exactly 0.05 and the probability of the type II error is close to 0.95. On the other hand, the probability of the type II error is low if the true value is much larger than μ_0. To quantify how our significance level α and the true value of μ influence our probability of a type II error, we define here the power of a test.

Definition 6.2 The power of a statistical significance test is defined as $P(\text{``}H_1\text{''}|H_1)$.

That is, the power tells us how likely we are to correctly reject the null hypothesis when it is void. We will now investigate the power and derive a test that has maximal power at a given significance level. Let us demonstrate the power function with a normally distributed example, and assume that $Y_i \sim N(\mu, \sigma^2)$, where $\sigma^2 = 1$ is known. As a hypothesis, we consider $H_0 : \mu \leq \mu_0$ versus $H_1 : \mu > \mu_0$. From that, we get the test decision

$$\text{``}H_1\text{''} \Leftrightarrow \frac{\bar{Y} - \mu_0}{\sigma/\sqrt{n}} \geq z_{1-\alpha}.$$

We can now calculate the power in relation to μ, the true mean with

$$P(\text{``}H_1\text{''}|\mu) = P\left(\frac{\bar{Y} - \mu_0}{\sigma/\sqrt{n}} \geq z_{1-\alpha}|\mu\right)$$

$$= P\left(\frac{\bar{Y} - \mu}{\sigma/\sqrt{n}} \geq z_{1-\alpha} + \frac{\mu_0 - \mu}{\sigma/\sqrt{n}}|\mu\right)$$

$$= 1 - \Phi\left(z_{1-\alpha} + \frac{\mu_0 - \mu}{\sigma/\sqrt{n}}\right).$$

We can plot $P(\text{``}H_1\text{''}|\mu)$ against μ, which can be seen for $\mu \in \mathbb{R}$ in the left-hand plot of Fig. 6.4. We plot the function for two values of the sample size and see that the power increases with sample size. Despite the sample size, however, the minimum

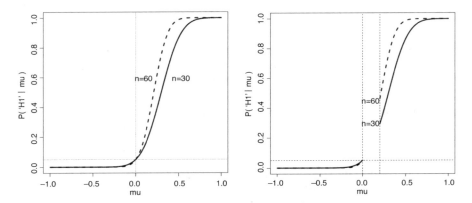

Fig. 6.4 Power function for $H_0 : \mu \leq 0$ for different sample sizes and alternatives $H_1 : \mu > 0$ (left-hand side) and $H_1 : \mu > \delta$ (right-hand side)

power is about 0.05, or to be specific

$$\sup P(\text{``}H_0\text{''}|H_1) = 1 - \inf P(\text{``}H_1\text{''}|\mu > \mu_0) = 1 - \alpha.$$

That is to say that the type II error still can occur with a high probability. Note that H_1 holds if $\mu = \mu_0 + \delta$ as long as $\delta > 0$. Hence even for $\delta = 10^{-1000}$, the hypothesis is void and H_1 holds. This brings us to the question of whether a value $\mu = \mu_0 + \delta$ is even of interest. It should be clear that the distance δ is dependent upon the problem at hand. Nanometre differences are certainly relevant in quantum physics, while in astronomy differences of thousands of kilometres might be of little interest. Thus, depending upon the problem at hand, we can set an appropriate δ and reformulate our testing problem as

$$H_0 : \mu \leq \mu_0 \text{ versus } H_1 : \mu > \mu_0 + \delta.$$

The resulting power function is shown on the right in Fig. 6.4. This time, we see that increasing the sample size has an effect on the minimum power. We also see that we can focus on two particular points in the parameter space, namely μ_0 and $\mu_1 = \mu_0 + \delta$. This simplification of the power function to just two points will be used in the following construction of an optimal test.

Property 6.1 (Neyman–Pearson Lemma) Let $H_0 : \theta = \theta_0$ be tested against $H_1 : \theta = \theta_1$ with a statistical significance test using level α. The most powerful test has the decision rule

$$\text{``}H_1\text{''} \Leftrightarrow l(\theta_0) - l(\theta_1) \leq c,$$

where c is determined such that $P(\text{``}H_1\text{''}|H_0) \leq \alpha$. This test is called the **Neyman–Pearson test**.

The lemma was published by Neyman and Pearson (1933). By defining a relationship between likelihood and optimal testing, the Neyman–Pearson lemma underlines once again the importance of the likelihood function in statistical reasoning. However, it does not specify the critical value c, which we will soon calculate. Using the results from the previous section, we know that twice the likelihood-ratio is asymptotically chi-squared. Hence, we can derive the critical value c based on quantiles of the chi-squared distribution and obtain the optimal test.

Proof Let y be the data, which gives $l(\theta) = \log f(y; \theta)$ as the log-likelihood. The above decision rule can then be rewritten as

$$l(\theta_0) - l(\theta_1) \leq c \Leftrightarrow \frac{f(y; \theta_0)}{f(y; \theta_1)} \leq k = \exp(c).$$

We define with $\varphi(y)$ the outcome of the Neyman–Pearson test with

$$\varphi(y) = \begin{cases} 1 & \text{if } \dfrac{f(y; \theta_0)}{f(y; \theta_1)} \leq k \\ 0 & \text{otherwise.} \end{cases}$$

Hence, if $\varphi(y) = 1$, we decide "H_1", and if $\varphi(y) = 0$, we conclude with decision "H_0". Now we take an arbitrary statistical significance test for which we similarly write the test outcome as a function $\psi(y) \in \{0, 1\}$, where $\psi(y) = 1$ means we decide for "H_1" and $\psi(y) = 0$ for "H_0". Let us now assume that $\theta = \theta_1$, i.e. H_1 holds. We need to prove that

$$P\left(\psi(Y) = 1; \theta_1\right) \leq P\left(\varphi(Y) = 1; \theta_1\right);$$

that is, if H_1 holds, then the Neyman–Pearson test decides for H_1 with a higher probability than any other arbitrarily chosen test with the same significance level. Note that $P(\psi(Y) = 1; \theta) = E_\theta(\psi(Y))$ and the same holds for the Neyman–Pearson test $\varphi(Y)$, such that

$$P\left(\varphi(Y) = 1; \theta_1\right) - P\left(\psi(Y) = 1; \theta_1\right) = \int \{\varphi(y) - \psi(y)\} f(y; \theta_1) dy.$$

We have to prove that this integral is greater than or equal to zero. The above integral can be labelled over three regions as

$$R_1 = \left\{ y : \frac{f(y; \theta_0)}{f(y; \theta_1)} < k \right\}$$

$$R_2 = \left\{ y : \frac{f(y; \theta_0)}{f(y; \theta_1)} > k \right\}$$

$$R_3 = \left\{ y : \frac{f(y; \theta_0)}{f(y; \theta_1)} = k \right\}.$$

For region R_1, we have $\varphi(y) \equiv 1$ and $f(y; \theta_1) > f(y; \theta_0)/k$, such that

$$\int_{y \in R_1} \Big[\varphi(y) - \psi(y)\Big] f(y; \theta_1) dy \geq \frac{1}{k} \int_{y \in R_1} \Big[\varphi(y) - \psi(y)\Big] f(y; \theta_0) dy.$$

For region R_2, we have $\varphi(y) \equiv 0$ and $-f(y; \theta_1) > -f(y; \theta_0)/k$, such that

$$\int_{y \in R_2} \Big[\varphi(y) - \psi(y)\Big] f(y; \theta_1) dy = - \int_{y \in R_2} \psi(y) f(y; \theta_1) dy$$

$$\geq -\frac{1}{k} \int_{y \in R_2} \psi(y) f(y; \theta_0) dy = \frac{1}{k} \int_{y \in R_2} \Big[\varphi(y) - \psi(y)\Big] f(y; \theta_0) dy.$$

And finally for $y \in R_3$, we have $f(y; \theta_1) = f(y; \theta_0)/k$, such that

$$\int_{y \in R_3} \Big[\varphi(y) - \psi(y)\Big] f(y; \theta_1) dy = \frac{1}{k} \int_{y \in R_3} \Big[\varphi(y) - \psi(y)\Big] f(y; \theta_0) dy.$$

Collecting the right-hand sides of the three regions, we can conclude

$$\int \Big[\varphi(y) - \psi(y)\Big] f(y; \theta_1) dy \geq \frac{1}{k} \int \Big[\varphi(y) - \psi(y)\Big] f(y; \theta_0) dy. \qquad (6.3.1)$$

As both tests have a significance level of α, one obtains

$$\alpha = P(\text{``}H_1\text{''}|H_0) = \int \varphi(y) f(y; \theta_0) dy = \int \psi(y) f(y; \theta_0) dy,$$

such that the right-hand side in (6.3.1) is equal to zero and the proof is completed.

$\square$

6.4   Goodness-of-Fit Tests

Goodness-of-fit tests are fundamentally different from the tests that we have discussed so far. Here we do not look at particular parameter values but instead test the entire distributional model itself. The hypothesis can therefore be formulated as

$$H_0 : Y \sim f(y; \theta),$$

where $f(\cdot; \cdot)$ is a known distribution, which might depend on the unknown, possibly multivariate, parameter θ. Assume now that we have *i.i.d.* data $Y_1, \ldots, Y_n$ and question whether Y_i is drawn from $f(y; \theta)$. There are two classical statistical tests that are used for this purpose: the chi-squared goodness-of-fit test and the Kolmogorov–Smirnov test.

6.4.1 Chi-Squared Goodness-of-Fit Test

The chi-squared test for goodness of fit goes all the way back to Fisher (1925). To motivate the idea, assume that Y takes discrete values $1, \ldots, K$. From a sample $Y_1, \ldots, Y_n$, we get the following contingency table:

	1	2		K
Observed	n_1	n_2	$\ldots$	n_K
Expected	e_1	e_2	$\ldots$	e_K

where $n_k = \sum_{i=1}^{n} 1_{\{Y_i = k\}}$ is the number of samples with the given value. We have $n = n_1 + n_2 + \ldots + n_K$ and define with

$$e_k = E(n_k) = nE\left(1_{\{y=k\}}\right) = n\{P(Y = k; \theta)\},$$

the expected number of elements in each cell. We assume that the data are generated by a candidate distribution $P(k; \theta)$, whose parameter θ is known or can be estimated from the data. The difference between n_k and e_k speaks to the appropriateness of the model. If the expected numbers e_k and the observed numbers n_k deviate substantially, this gives evidence that the distributional model might not hold. On the other hand, if the expected numbers e_k and the observed numbers n_k are close to each other, it speaks in favour of the model. These differences can be assessed with the chi-squared measure

$$X^2 = \sum_{k=1}^{K} \frac{(n_k - e_k)^2}{e_k}.$$

The distribution of X^2 can be approximated by a chi-squared distribution, where the degrees of freedom are the number of cells, corrected by the number of parameters in θ that have been estimated. This correction was, in fact, the basis for a controversial discussion between Pearson and Fisher, see Baird (1983). Pearson proposed the chi-squared test for measuring goodness of fit but gave a different calculation for the degrees of freedom of the chi-squared distribution. Fisher then correctly set the degrees of freedom equal to the number of cells minus the side constraints in the cells minus the number of estimated parameters in θ. In the example above, this means the degrees of freedom are

$$K - 1 - p.$$

Here, K is the number of cells, the fact that the probability of all cells sums to 1 is a side constraint and p is the number of parameters in θ. The decision rule is now given by

$$\text{``}H_1\text{''} \Leftrightarrow X^2 \geq \mathcal{X}^2_{K-1-p, 1-\alpha}.$$

The test can also be applied to continuous random variables Y. In this case, we discretise Y into "bins". To this end, let $-\infty = c_0 < c_1 < c_2 < \ldots < c_{K-1} < c_K = \infty$ be threshold values fixed on the real axis. We define the discrete-valued random variable Z with

$$Z = k \Leftrightarrow c_{k-1} < Y \leq c_k.$$

With this trick, we have discretised the continuous variable. By defining $e_k = n \int_{c_{k-1}}^{c_k} f(\tilde{y}; \hat{\theta}) d\tilde{y}$ as the (estimated) expected cell frequencies, we can simply use the previous decision rule.

It is not difficult to see that increasing K, i.e. working with a finer grid of values, makes the test more accurate as the discrete values are closer to the continuous distribution. However, this comes at the price of a reduced number of observations per bin, such that the asymptotic distribution of X^2 may no longer be valid. In other words, one needs to find a compromise when choosing K between the discrete approximation and the asymptotic approximation. A generally accepted rule of thumb is that e_k should be larger than 5 and smaller than $(n - 5)$.

6.4.2 Kolmogorov–Smirnov Test

The Kolmogorov–Smirnov test directly compares the difference between the empirical distribution function and the hypothetical distribution. Let $F(y; \theta) = \int_{-\infty}^{y} f(\tilde{y}; \theta) d\tilde{y}$ be the distribution function, and let $F_n(y) = \frac{1}{n} \sum_{i=1}^{n} 1_{\{Y_i \leq y\}}$ be its empirical counterpart. If $F(y; \theta)$ is the true distribution function, then the distance between $F(y; \theta)$ and $F_n(y)$ should be small, at least when n is large. We therefore construct a test statistic by looking at the difference between $F(y; \theta)$ and $F_n(y)$. To be specific, we consider the supremum of the absolute difference

$$D_n := \sup_y |F_n(y) - F(y; \theta)|.$$

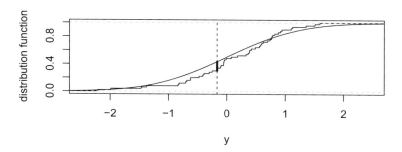

Kolmogorov–Smirnov–Distance

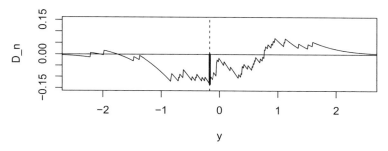

Fig. 6.5 Empirical distribution function and hypothetical distribution (top plot). Difference between these two distributions (bottom plot) with maximum (supreme) value indicated as vertical line

This quantity is also called the Kolmogorov–Smirnov distance. Figure 6.5 illustrates the construction of D_n with an example, with the empirical distribution function of a sample as $F_n(\cdot)$ and the standard normal distribution as hypothetical distribution $F(\cdot)$. This is shown in the top plot. The bottom plot gives the difference $F_n(y) - F(y)$ for all values of y, which takes its maximum (in absolute terms) at the locations indicated by the vertical line. This defines D_n. The next step is to find the distribution for D_n under H_0, i.e. if $F(y; \theta)$ is the true distribution. Let $F^{-1}(y; \theta)$ be the inverse of the distribution function, or more formally for $x \in [0, 1]$,

$$F^{-1}(x; \theta) = \min\{y : F(y, \theta) \geq x\}.$$

Then for $y = F^{-1}(x; \theta)$, we get $F(F^{-1}(x; \theta); \theta) = x$ such that

$$P(\sup_{y}|F_n(y) - F(y; \theta)| \leq t) = P(\sup_{x}|F_n(F^{-1}(x; \theta)) - x| \leq t), \qquad (6.4.1)$$

where

$$F_n(F^{-1}(x;\theta)) = \frac{1}{n}\sum_{i=1}^{n}1_{\{Y_i \leq F^{-1}(x;\theta)\}} = \frac{1}{n}\sum_{i=1}^{n}1_{\{F(Y_i;\theta)\leq x\}}.$$

Note that if $F(y;\theta)$ is the true distribution, then $F(Y_i;\theta)$ has a uniform distribution on [0,1]. This is easily seen as

$$P(F(Y_i;\theta) \leq x) = P(Y_i \leq F^{-1}(x;\theta)) = F(F^{-1}(x;\theta);\theta) = x.$$

Hence, we define $U_i = F(Y_i;\theta)$, which is uniform on [0,1], which in turn allows us to rewrite (6.4.1) as

$$P(\sup_x |\frac{1}{n}\sum_{i=1}^{n}1_{\{U_i \leq x\}} - x| \leq t).$$

Note that this probability statement does not depend on the hypothetical distribution, and thus we have formulated a pivotal distribution. It can be shown that the probability above has a limit distribution, known as the Kolmogorov–Smirnov distribution. With this result, the test decision is now given by

$$\text{``}H_1\text{''} \Leftrightarrow D_n \geq KS_{1-\alpha},$$

where $KS_{1-\alpha}$ denotes the $1 - \alpha$ quantile of the Kolmogorov–Smirnov distribution. However, the above test assumes that the parameter θ of the distribution is known, which in most practical settings is not the case. Hence, the parameter needs to be estimated, and the derivation of the exact distribution of the test statistic with estimated parameter

$$\hat{D}_n = \sup_y |F_n(y) - F(y;\hat{\theta})|$$

is thereby complicated and has only been determined for specific distributions, such as the normal or exponential distributions. For an unspecified distribution, the Kolmogorov–Smirnov distribution only holds asymptotically for D_n.

6.5   Tests on Independence

Assume multivariate data of the form

$$Y_i = (Y_{i1}, \ldots, Y_{iq}) \sim F(y_1, \ldots, y_q) \quad i.i.d. \text{ for } i = 1, \ldots, n,$$

where $F(.) : \mathbb{R}^q \to [0, 1]$ is a multivariate distribution function of any form. A central question in this setting is to assess the dependence and independence structure among the components of the random vector Y_i. We denote with $F_j(.)$ the univariate marginal distribution of the j-th component of Y_i, that is,

$$P(Y_{ij} \le y_j) = F_j(y_j) := F(\underbrace{\infty, \ldots, \infty}_{j-1 \text{ components}}, y_j, \underbrace{\infty, \ldots, \infty}_{q-j-1 \text{ components}}).$$

We can now formulate the independence of this component as a test with null hypothesis

$$H_0 : F(y_1, \ldots, y_q) = \prod_{j=1}^{q} F_j(y_j)$$

against

$$H_1 : F(y_1, \ldots, y_q) \ne \prod_{j=1}^{q} F_j(y_j).$$

This is clearly quite a general setting, but here we will limit our scope to a few classical tests. The topic is still an active research field, and we refer to Pfister et al. (2018) for an overview of modern approaches. Chapter 10 of this book also offers strategies for modelling the dependencies in multivariate data.

6.5.1 Chi-Squared Test of Independence

In Sect. 6.4.1, when testing for the goodness of fit of our distributional model, we described how to make use of a chi-squared test for discrete-valued data. This test was then extended towards discretised continuous random variables. This strategy can easily be extended towards testing for independence. For example, let us consider the bivariate case. To proceed, we discretise each random variable separately, that is, instead of $Y_i = (Y_{i1}, Y_{i2})$, we consider $Z_i = (Z_{i1}, Z_{i2})$, where

$$Z_{ij} = k \Leftrightarrow c_{j,k-1} < Y_{ij} \le c_{j,k}$$

for $j = 1, 2$ and $k = 1, \ldots, K_j$. The threshold values fulfil $-\infty = c_{j,0} < c_{j,1} < \ldots < c_{j,K_j} = \infty$, and they are chosen such that the range of the j-th variable is well covered. If Y_{i1} and Y_{i2} are independent, that is,

$$P(Y_{i1} \le y_1, Y_{i2} \le y_2) = F(y_1, y_2) = F_1(y_1)F_2(y_2)$$
$$= P(Y_{i1} \le y_1) \cdot P(Y_{i2} \le y_2),$$

then clearly Z_{i1} and Z_{i2} are also independent. We can now apply the chi-squared test to this discretised version in exactly the same fashion as in Sect. 6.4.1. This allows us to generate a table of counts:

		Z_2				
		1	2	...	K_2	
	1	n_{11}	n_{12}		n_{1K_2}	$n_{1\cdot}$
	2	n_{21}	n_{22}		n_{2K_2}	$n_{2\cdot}$
Z_1	$\vdots$					$\vdots$
	K_1	n_{K_11}	n_{K_12}		$n_{K_1K_2}$	$n_{K_1\cdot}$
		$n_{\cdot 1}$	$n_{\cdot 2}$	...	$n_{\cdot K_2}$	$n_{\cdot\cdot}$

With n_{kl}, we define the number of observations with $Z_1 = k$ and $Z_2 = l$, i.e.

$$
n_{kl} = \sum_{i=1}^{n} 1_{\{Z_{i1}=k, Z_{i2}=l\}}
$$

$$
= \sum_{i=1}^{n} 1_{\{c_{1;k-1}<Y_{i1}\leq c_{1,k}, c_{2;l-1}<Y_{i2}\leq c_{2,l}\}}.
$$

Moreover, we define $n_{\cdot l} = \sum_k n_{kl}$ as the column sum, $n_{k\cdot}$ as the row sum and $n_{\cdot\cdot} = n$ as the sample size. We now compare the observed counts to their expected versions under independence. Let $\pi_{kl} = P(Z_{i1} = k, Z_{i2} = l)$, which in case of independence decomposes to $\pi_{k\cdot}\pi_{\cdot l} = P(Z_{i1} = k) \cdot P(Z_{i2} = l)$. It is not difficult to see that the Maximum Likelihood estimates equal

$$
\hat{\pi}_{k\cdot} = n_{k\cdot}/n_{\cdot\cdot} \text{ and } \hat{\pi}_{\cdot l} = n_{\cdot l}/n_{\cdot\cdot}
$$

such that

$$
e_{kl} = n_{\cdot\cdot}\hat{\pi}_{k\cdot}\hat{\pi}_{\cdot l} = \frac{n_{k\cdot}n_{\cdot l}}{n_{\cdot\cdot}}
$$

gives the expected number of observations in the (k,l)-th cell of the table. This leads us to the chi-squared statistic

$$
X^2 = \sum_{k=1}^{K_1}\sum_{l=1}^{K_2} \frac{(n_{kl} - e_{kl})^2}{e_{kl}}.
$$

The distribution of X^2 can again be approximated with a chi-squared distribution, but the degrees of freedom need to be calculated. Following the results of Sect. 6.4.1, we get $K_1 K_2$ as the number of cells and $(K_1 - 1) + (K_2 - 1)$ as the number of fitted parameters. Together with the side constraint that all cell probabilities sum up to 1, this allows us to calculate the degrees of freedom

$$K_1 K_2 - 1 - (K_1 - 1) - (K_2 - 1) = (K_1 - 1)(K_2 - 1).$$

Note that the application and validity of the test require the same conditions as already discussed in Sect. 6.4.1. A rule of thumb to guarantee that the chi-squared approximation is valid is that e_{kl} should be larger than 5 and smaller than $(n_{..} - 5)$. If this does not hold, a coarser discretisation is advisable.

6.5.2 Fisher's Exact Test

In the case where we have two populations with a binary response, e.g. two samples under two different conditions, Fisher's exact test can be used as an alternative to the chi-squared test, if the sample sizes are small. Let us assume that the binary outcomes are coded as "success" and "failure". We consider two samples from the two populations:

$$X_1, X_2, \ldots, X_{n_1} \sim B(1; p_1) \quad i.i.d.$$
$$Y_1, Y_2, \ldots, Y_{n_2} \sim B(1; p_2) \quad i.i.d.$$

Fisher's exact test is a test of the null hypothesis

$$H_0 : p_1 = p_2 = p$$

against the alternative hypothesis

$$H_1 : p_1 \neq p_2.$$

For the sums of these random variables, we get

$$X = \sum_{i=1}^{n_1} X_i \sim B(n_1; p_1), \quad Y = \sum_{i=1}^{n_2} Y_i \sim B(n_2; p_2).$$

For the following, we define $Z = X + Y$. Fisher's exact test uses the fact that the row sums n_1 and n_2 in the following 2×2 contingency table are treated as fixed.

	Success	Failure	Total
Sample 1	X	$n_1 - X$	n_1
Sample 2	$Z - X = Y$	$n_2 - (Z - X)$	n_2
Total	Z	$(n_1 + n_2 - Z)$	$n = n_1 + n_2$

Conditional on the total number of successes, $Z = z$, and thus also conditional on the column sums, the only remaining random variable is X. With $n = n_1 + n_2$ and under H_0, X is distributed hypergeometrically,

$$X \sim H(z, n_1, n),$$

i.e.

$$P(X = x | Z = z) = \frac{\binom{n_1}{x} \binom{n - n_1}{z - x}}{\binom{n}{z}}.$$

This allows us to derive critical values. Note that also one-sided hypotheses can be tested.

Proof The proof uses $Z = z$ and the assumption of independence between the two samples:

$$P(X = x | Z = z) = \frac{P(X = x, Z = z)}{P(Z = z)} = \frac{P(X = x, Y = z - x)}{P(Z = z)}$$

$$= \frac{P(X = x) P(Y = z - x)}{P(Z = z)} = \frac{\binom{n_1}{x} p^x (1 - p)^{n_1 - x} \binom{n_2}{z - x} p^{z - x} (1 - p)^{n_2 - (z - x)}}{\binom{n}{z} p^z (1 - p)^{n - z}}$$

$$= \frac{\binom{n_1}{x} \binom{n_2}{z - x}}{\binom{n}{z}} = \frac{\binom{n_1}{x} \binom{n - n_1}{z - x}}{\binom{n}{z}}.$$

Here we also made use of the fact that under H_0, $Z = X + Y$ is $B(n, p)$, which comes simply from the additivity of the binomial distribution. □

Example 24 In a controlled experiment, the taste of mineral water was assessed, see Clausnitzer et al. (2004). The research question was whether the addition of oxygen changes the subjective taste of the water. There were two groups: the control group, who tasted the same water twice (type W1), and the treatment group, who tasted water of type W1 and then water with extra oxygen (type W2). The participants were asked whether there was a difference between the two samples. The experiment was double blind, i.e. both the participants and the observers did not know whether the second sample was of type W1 or type W2. The result was the following:

	"differ"	"equal"
Control group	76	24
Treatment group	89	11
	100	100

We test the hypothesis $H_0 : p_C = p_T$ vs. $H_1 : p_C \neq p_T$, where p_C and p_T denote the probability of a member of the control group and of the treatment group saying that the samples are different. The estimated probabilities are $\hat{p}_C = 0.76$ and

$\hat{p}_T = 0.89$. Fisher's exact test shows a p-value of 0.025. This indicates a significant effect on the 5% level.

▷

Let us demonstrate the idea of one-sided tests by considering 2×2 tables, which are more "extreme" than the one calculated from the observed data. In the case of a two-sided test, one calculates the corresponding probabilities for all possible entries of x, $x \in \{0, \ldots, \min(n_1, z)\}$. Now, all probabilities that are equal to or lower than the probability of the observed table are summed up to calculate the p-value.

Example 25 We want to assess whether a treatment works by testing if the response and treatment are independent (in which case the treatment has no effect). We look at 2 treatments (success or failure) and 10 patients per treatment:

	Success	Failure	Total
Treatment 1	0	10	10
Treatment 2	4	6	10
Total	4	16	20

Under H_0, the observed table ($x = 0$) can be evaluated using the hypergeometric distribution $H(4, 10, 20)$, which gives the probability 0.04334. Another possible table with $x = 4$, e.g.

	Success	Failure	Total
Treatment 1	4	6	10
Treatment 2	0	10	10
Total	4	16	20

has exactly the same probability (and thus belongs to the more "extreme" tables). All other fictitious and possible tables (with $x = 1, 2, 3$) fixing the row and column sums have higher probabilities. The two-sided p-value is therefore approximately 0.08669. As a consequence, H_0 would not be rejected in this example at $\alpha = 0.05$.

▷

Fisher's exact test can also be seen as a test of independence of two binary outcomes in a 2×2 table, i.e. the null hypothesis $H_0 : p_1 = p_2 = p$ is equivalent to the hypothesis that success and failure are independent from the conditions under which the samples have been drawn.

6.5.3   Correlation-Based Tests

A simple test on independence is carried out by directly looking at the correlation between two variables. A word of warning is required here, as independence implies zero correlation but the reverse most certainly is not true. Hence, having two uncorrelated variables does not imply that they are independent, which only holds if we additionally assume normality. In general, we can calculate the empirical correlation between two variables with

$$\hat{\rho} = \frac{\sum_{i=1}^{n}(Y_{i1} - \bar{Y}_1)(Y_{i2} - \bar{Y}_2)}{\sqrt{\left(\sum_{i=1}^{n}(Y_{i1} - \bar{Y}_1)^2\right)\left(\sum_{i=1}^{n} Y_{i2} - \bar{Y}_2\right)^2}},$$

where $\bar{Y}_j = \sum_{i=1}^{n} Y_{ij}/n$ for $j = 1, 2$. As $\hat{\rho}$ is a statistic of the data, we can calculate its variance to derive confidence intervals. In the case of uncorrelated normally distributed random variables Y_{i1} and Y_{i2}, it can be shown (see Dudewicz and Mishra 1988) or Kendall and Stuart 1973) that

$$t = \hat{\rho}\sqrt{\frac{n-2}{1-\hat{\rho}}}$$

follows a t-distribution with $n - 2$ degrees of freedom. Fisher (1915) more than 100 years ago proposed the transformation of t with

$$z = \frac{1}{2}\log\left(\frac{1+\hat{\rho}}{1-\hat{\rho}}\right),$$

to obtain a more stable inference; see also Zimmerman et al. (2003) for a deeper discussion. A common alternative is to make use of resampling techniques to test the hypothesis $H_0 : \rho = 0$ directly. This has the advantage of being less dependent on the assumption of normality. We propose such techniques later in Chap. 8.

6.6 p-Value, Confidence Intervals and Test

6.6.1 The p-Value

When presenting the results of a statistical test, one usually does not report the outcome of the decision, that is, "H_0" or "H_1", but instead reports the **p-value**. This is also the case with results reported by statistical software. The p-value is closely connected to statistical tests. In fact, the idea is even older than statistical testing and was first proposed by Fisher (1925). Its calculation begins with the null hypothesis H_0. With this assumption in place, one then calculates the probability of seeing data that contradict the null hypothesis even more than the observed data. For this purpose, one also needs a discrepancy measure that quantifies how far the observed data lie from the hypothesis.

Let us make the concept more clear with a simple example. Assume we have $Y_i \sim N(\mu, \sigma^2)$ *i.i.d.* and consider the hypothesis $H_0 : \mu \leq \mu_0$. Taking $\bar{Y} = \frac{1}{n} \sum Y_i$, we can see that large values of $\bar{Y}$ speak against the hypothesis. Assume now that we have observed $\bar{y}_{obs} = \frac{1}{n} \sum y_i$. Then, we can calculate the probability that $\bar{Y}$ is larger than $\bar{y}_{obs}$, which is exactly the desired property, that is, the probability of observing data that contradict the hypothesis even more than the current data. We can now define the *p*-value as

$$p\text{-value} = P(\bar{Y} \geq \bar{y}_{obs} | \mu \leq \mu_0)$$

$$\leq P(\bar{Y} \geq \bar{y}_{obs} | \mu = \mu_0)$$

$$= P\left(\frac{\bar{Y} - \mu_0}{\sigma/\sqrt{n}} \geq \frac{\bar{y}_{obs} - \mu_0}{\sigma/\sqrt{n}} \Big| \mu = \mu_0\right)$$

$$= 1 - \Phi\left(\frac{\bar{y}_{obs} - \mu_0}{\sigma/\sqrt{n}}\right).$$

Fisher argued that the smaller the *p*-value, the stronger the evidence against the null hypothesis H_0 and even proposed the following thresholds:

$$p\text{-value} \leq 0.1 \Leftrightarrow \text{weak evidence against } H_0$$

$$p\text{-value} \leq 0.05 \Leftrightarrow \text{increased evidence against } H_0$$

$$p\text{-value} \leq 0.01 \Leftrightarrow \text{strong evidence against } H_0.$$

Note that the *p*-value exclusively expresses the evidence *against* the null hypothesis H_0, which is the core of statistical significance. In other words, a large *p*-value does not give evidence in favour of H_0; it simply means that for a large *p*-value, there is little evidence that H_0 might not hold. In contrast, the smaller the *p*-value, the more evidence there is against the validity of the hypothesis. Therefore, the *p*-value contains quantifiable information. In fact, *p*-values and statistical significance tests are closely connected. Neyman and Pearson (1933) developed the idea of testing further, leading to the significance test given in Definition 5.1. This is related to the *p*-value as follows.

Property 6.2 Assume a statistical significance test with significance level α, and then it holds

$$p\text{-value} \leq \alpha \Leftrightarrow \text{``}H_1\text{''}.$$

Proof The proof of this statement is rather simple. Assume we have a test statistic $t()$ such that we decide in favour of H_1 if $t(y_{obs}) > c$, where y_{obs} are the observed data, and the critical value c is determined such that

$$P(t(Y) > c | H_0) \leq \alpha.$$

Because

$$p\text{-value} = P(t(Y) > t(y_{obs})|H_0),$$

it is directly clear that if $t(y_{obs}) > c$, then the p-value $\leq \alpha$ and vice versa.   □

Hence, p-values and significance tests are deeply related. However, the p-value, in explicitly quantifying how much evidence there is against the hypothesis, gives even more information. In fact, Efron (2010) writes "Fisher's famous $\alpha = 0.05$ direction for 'significance' has been overused, but has served a crucial purpose nevertheless in bringing order to scientific reporting". In Fig. 6.6, we visualise the connection and consequences between p-values and significance tests using the example of a traffic light. While the p-value is a continuous measure from green (large) to red (small), a significance test discretises this range with a fixed threshold (α), such that one either rejects or does not reject a hypothesis.

Arguments against $\alpha = 0.05$ and their misuse in scientific reporting have been collected by Goodman (2008). The p-value is unfortunately also often misinterpreted. For a recent discussion of p-values, see the American Statistical Association's statement on the issue (Wasserstein and Lazar 2016). One major misconception is that the p-value is sometimes interpreted as the size of an effect. As the p-value is also a function of the sample size, a small effect can be shown with

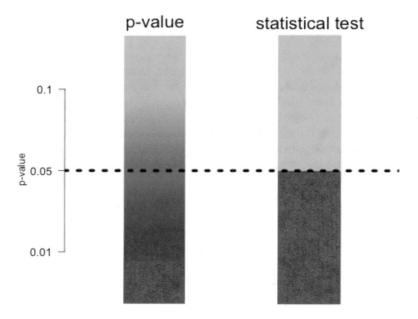

Fig. 6.6 p-value (left) and test decision (right) with threshold $\alpha = 0.05$

strong evidence ($p < 0.01$) when the sample size n is large. On the other hand, a
p-value $p < 0.05$ can indicate a strong effect when the sample size is small.

Note that we can also comprehend the p-value as random variable, if we consider
the observed data as random. Let us make this clear with the normal distribution
example from above. In this case we denote with $\bar{Y}_{obs}$ the resulting random variable
for the (later) observed data. We can calculate the random p-value for the hypothesis
$H_0 : \mu \leq \mu_0$ with

$$p\text{-value} = P(\bar{Y} \geq \bar{Y}_{obs}|\mu = \mu_0) = 1 - \Phi\left(\frac{\bar{Y}_{obs} - \mu_0}{\sigma/\sqrt{n}}\right)$$

with Φ as distribution function of the $N(0, 1)$ distribution. The distribution of the
p-value is then given by

$$P(p\text{-value} \leq p) = P\left(1 - \Phi\left(\frac{\bar{Y}_{obs} - \mu_0}{\sigma/\sqrt{n}}\right) \leq p\right)$$

$$= P\left(\Phi\left(\frac{\bar{Y}_{obs} - \mu_0}{\sigma/\sqrt{n}}\right) \geq 1 - p\right)$$

$$= P\left(\frac{\bar{Y}_{obs} - \mu_0}{\sigma/\sqrt{n}} \geq \underbrace{\Phi^{-1}(1 - p)}_{z_{1-p}}\right) = p,$$

where z_{1-p} is the $1 - \alpha$ quantile of the standard normal distribution. In other words,
if the hypothesis holds, then the p-value has a uniform distribution on [0,1].

6.6.2   Confidence Intervals and Tests

Confidence intervals and statistical tests are deeply related. In fact, given a
confidence interval, one can directly construct a corresponding test and vice versa.
To demonstrate this, we require Definition 3.14, where a confidence interval was
defined as $[t_l(y), t_r(y)]$ such that

$$P_\theta(t_l(Y) \leq \theta \leq t_r(Y)) \geq 1 - \alpha,$$

where $Y = (Y_1, \ldots, Y_n)$. We define the corresponding test on the hypothesis $H_0 :
\theta = \theta_0$ through the decision rule

$$\text{``}H_1\text{''} \Leftrightarrow \theta \notin [t_l(y), t_r(y)].$$

Hence, for all parameter values in the confidence interval, we accept the hypothesis
and vice versa. This gives us a statistical test with significance level α. To see this,

we can also define the function

$$\varphi_\theta(Y) = \begin{cases} 0 & \text{if } \theta \in [t_l(y), t_r(Y)] \\ 1 & \text{otherwise.} \end{cases}$$

Then,

$$\begin{aligned} P(\text{``}H_1\text{''}|H_0) &= 1 - P(\text{``}H_0\text{''}|H_0) \\ &= 1 - P(\varphi(Y) = 0|\theta = \theta_0) \\ &= 1 - \underbrace{P(t_l(Y) \le \theta_0 \le t_r(Y))}_{\ge 1-\alpha} \le \alpha. \end{aligned}$$

The opposite works in the same fashion. Assume we have a statistical test, which we can define as

$$\varphi_\theta(Y) = \begin{cases} 0 & \text{if ``}H_0\text{''} \\ 1 & \text{if ``}H_1\text{''}. \end{cases}$$

The corresponding confidence interval can then be calculated with

$$CI(y) = \{\theta : \varphi_\theta(y) = 0\}.$$

As $P(\text{``}H_1\text{''}|H_0) = P(\varphi_\theta(Y) = 1|\theta) \le \alpha$, we have

$$\begin{aligned} P(\theta \in CI(Y)|\theta) &= 1 - P(\theta \notin CI(Y)|\theta) \\ &= 1 - P(\varphi_\theta(Y) = 1|\theta) \\ &= 1 - P(\text{``}H_1\text{''}|H_0) \ge 1 - \alpha, \end{aligned}$$

which demonstrates the connection between tests and confidence intervals.

6.7 Bayes Factor

Although significance testing and the Bayesian paradigm have little in common, there is a plausible link based on what is called the Bayes factor. In some way, the Bayes factor can be seen as the Bayesian version of a p-value. Without putting too much emphasis on this link, we want to demonstrate how the Bayes factor can be used for decision-making. The Bayesian idea allows for a number of extensions by taking the Bayesian paradigm quite rigorously and making use of it in decision analytic questions. In fact, using the Bayesian view, we can also calculate the posterior probabilities of whole models to assess their validity. To demonstrate, let

M_0 and M_1 represent two different models. These models may differ because of different parameterisations or may be completely different. The former refers to the Bayesian parameter selection, while the latter relates to goodness-of-fit questions.

Generally, model selection is discussed in more detail in Chap. 9. Therefore, we will focus here on exploring a Bayesian view of parameter selection. To begin, assume that

$$Y \sim f(y|\theta),$$

where we have two different models for parameter θ. We either assume that $\theta \in \Theta_0$, which we call model M_0, or alternatively $\theta \in \Theta_1$, which we denote as model M_1. With

$$P(M_0|y) \quad \text{and} \quad P(M_1|y),$$

we can express the posterior probability that model M_0 or model M_1 holds. If we restrict ourselves to two models, we are able to calculate posterior model probabilities with

$$P(M_1|y) = 1 - P(M_0|y).$$

Using the Bayes theorem, we have

$$P(M_0|y) = \frac{f(y|M_0)P(M_0)}{f(y)} \tag{6.7.1}$$

and accordingly for $P(M_1|y)$, where $P(M_0)$ denotes the prior belief in model M_0. Clearly, each model is specified with parameters, which affect the above calculation through

$$f(y|M_0) = \int f(y|\vartheta) f_\theta(\vartheta|M_0) d\vartheta,$$

where $f(y|\theta)$ is the likelihood, as before, and $f_\theta(\theta|M_0)$ is the prior distribution of the parameter if model M_0 holds. The same holds for model M_1, i.e.

$$f(y|M_1) = \int f(y|\vartheta) f_\theta(\vartheta|M_1) d\vartheta, \tag{6.7.2}$$

such that different parameter spaces enter the comparison through different prior distributions $f_\theta(\theta|M_0)$ and $f_\theta(\theta|M_1)$. In the calculation of (6.7.1), one also has the prior distribution of the models, which expresses our prior belief in model M_0 relative to model M_1 before seeing any data. Dependent on the situation, it may be useful to set $P(M_0) = P(M_1) = \frac{1}{2}$, but this is not a requirement. Finally, we have the marginal distribution $f(y)$ in (6.7.1), which, as we have seen, is difficult to evaluate. Fortunately, we do not need to calculate $f(y)$, because instead of looking at (6) we look at the ratio

$$\frac{P(M_1|y)}{P(M_0|y)} = \underbrace{\frac{f(y|M_1)}{f(y|M_0)}}_{\text{Bayes-Factor}} \frac{P(M_1)}{P(M_0)}. \tag{6.7.3}$$

The first component in (6.7.3) is the **Bayes factor**, which gives the evidence for model M_1 relative to M_0. The larger the Bayes factor, the more evidence there is for model M_1. A rough guideline on how to interpret the values of the Bayes factor is given by Kass and Raftery (1995), who declare the following evidence in favour of M_1 based on the Bayes factor:

Bayes factor	Interpretation
1 to 3	Not worth mentioning
3 to 20	Evidence
20 to 150	Strong evidence
> 150	Very strong evidence

The calculation of the Bayes factor does require the numerical tools introduced in Chap. 6, because we need to integrate out the parameters. Looking at (6.7.2), we can solve the integral by sampling from the prior. The problem with this approach is that $f(y|\theta)$ might be (very) small for a range of values of θ, such that only a few values of θ determine the integral. It is therefore more useful to apply a sampling scheme from the posterior to calculate the integral. There is a close connection between the Bayes factor and the p-value discussed in Sect. 6.6.1. We refer to Held and Ott (2015) for a deeper discussion.

6.8 Multiple Testing

In many data analytic situations, we are not testing a single hypothesis but multiple hypotheses concurrently. A common case is when the parameter θ is multivariate of the form $\theta = (\theta_1, \ldots, \theta_p)$, in which case we test p individual hypotheses

$$H_{0j} : \theta_j = \theta_{0j}$$

for $\theta_0 = (\theta_{01}, \ldots, \theta_{0p})$ and $j = 1, \ldots, m$. For example, in a genetic experiment, we may be testing the influence of thousands of genes separately, with a single test for each individual gene. Equally, in a multicolumn dataset, we may be testing the mean of each individual column. This confronts us with the problem of multiple testing. Let us assume that we have a single test and its resulting p-value. If the hypothesis H_0 holds, then we know that the p-value is uniform on $[0,1]$ and

$$P(p\text{-value} \leq \alpha | H_0) = \alpha.$$

Hence, if we use the decision rule "H_1" $\Leftrightarrow$ p-value $\leq \alpha$, we make a type I error with the small probability α. Now, let us move on to two tests on the hypotheses H_{01} and H_{02}, with p-values p_1 and p_2. Assume that both H_{01} and H_{02} hold. If we define the decision rules

$$\text{"}H_{1j}\text{"} \Leftrightarrow p_j \leq \alpha \qquad j = 1, 2,$$

we can derive the probability of a type I error for either of our two hypotheses, that is, rejecting one or both of them when they both are, in fact, true. This probability is given by

$$P\Big((p_1 \leq \alpha) \vee (p_2 \leq \alpha)|H_{01} \wedge H_{02}\Big) \geq P\Big((p_1 \leq \alpha) \wedge (p_2 \leq \alpha)|H_{01} \wedge H_{02}\Big),$$

where "$\vee$" denotes the logical "or" and "$\wedge$" is the logical "and". If the two tests are independent, we can calculate the final probability with

$$P((p_1 > \alpha) \wedge (p_2 > \alpha)|H_{01} \text{ and } H_{02}) = P(p_1 \geq \alpha|H_{01})P(p_2 \geq \alpha|H_{02})$$
$$= (1 - \alpha)(1 - \alpha)$$
$$= 1 - 2\alpha + \alpha^2. \qquad (6.8.1)$$

Consequently, we get that

$$P(\text{"}H_{11}\text{" and/or "}H_{12}\text{"}|H_{01} \wedge H_{02}) = 2\alpha - \alpha^2 = \alpha(2 - \alpha).$$

As $0 < \alpha < 1$, we get $\alpha(2 - \alpha) > \alpha$. That is to say, applying two (independent) tests at significance level α leads to a significance level larger than α. In fact, one can show that the result of (6.8.1) is a lower bound, such that

$$\alpha \leq P\Big((p_1 \leq \alpha) \text{ and/or } (p_2 \leq \alpha)|H_{01} \wedge H_{02}\Big) \leq 2\alpha - \alpha^2 \leq 2\alpha. \qquad (6.8.2)$$

By definition, α is small and therefore α^2 is negligible, and we can safely work with the limit 2α. In contrast, the leftmost limit occurs if the p-values are exactly the same, e.g. when we apply the same test twice to the same data. In reality, the true probability lies between these two limits.

Assume now that we have $m \geq 2$ tests for hypotheses H_{0j}, $j = 1, \ldots, m$. Then for p tests with resulting p-values p_j, we have

$$\alpha \leq P\left(\bigcup_{j=1}^{m}(p_j \leq \alpha)|H_{0j}, j = 1, \ldots, m\right) \leq m\alpha.$$

This stacking of individual tests can have powerful consequences. It means that if we make many individual tests (m large), we will falsely reject at least one hypothesis with high probability. To correct for this drawback, Bonferroni (1936) suggested the adjustment of the α value in case of multiple testing. To motivate his idea, we must first define the **family-wise error rate** as the probability of rejecting at least one of the hypotheses

$$FWER := P(\text{``}H_{11}\text{''} \text{ and/or } \text{``}H_{12}\text{''} \text{ and/or } \ldots \text{ and/or } \text{``}H_{1m}\text{''}|H_{01} \wedge H_{02} \wedge \ldots \wedge H_{0m}).$$

The intention is now to control the FWER instead of the significance level of each individual test.

Property 6.3 (Bonferroni Adjustment) The FWER is limited to α, if we set the significance level of each individual test to α/m, where m is the number of tests. The adjusted significance level is defined by $\alpha_{adjust} = \alpha/m$.

This correction is easily motivated, as we simply need to use the right-hand boundary in (6.8.2). Clearly, this correction is an approximation, and by again considering (6.8.1), we can find a less restrictive adjusted $\tilde{\alpha}_{\text{adjust}}$, by realising that under independence

$$P\left(\bigcap_{j=1}^{m}(p_j \geq \tilde{\alpha}_{adjust})|H_{01} \wedge H_{02} \wedge \ldots \wedge H_{1m}\right) = (1 - \tilde{\alpha}_{adjust})^m.$$

Hence, the FWER is given by

$$FWER = 1 - (1 - \tilde{\alpha}_{adjust})^m.$$

If we set $FWER = \alpha$, this gives

$$\alpha = 1 - (1 - \tilde{\alpha}_{adjust})^m \Leftrightarrow \tilde{\alpha}_{adjust} = 1 - (1 - \alpha)^{1/m}.$$

This approach is also known as the **Šidák procedure**, which improves the Bonferroni correction in the case of independent tests. Note that Bonferroni's adjustment works even for dependent tests, where the Šidák approach can fail.

However, both corrections do not take the explicit p-values of the various tests into account. On the other hand, **Holm's procedure**, originally presented in Holm (1979), makes use of the information contained in the p-values. For m tests, we derive m corresponding p-values, which we order such that

$$p_{(1)} \leq p_{(2)} \leq \cdots \leq p_{(m)},$$

where $p_{(1)}$ is the smallest p-value and $p_{(m)}$ is the largest p-value obtained from the m tests. The ordering corresponds to the decision rule that if we reject the test on hypothesis $H_{0(i)}$ with p-value $p_{(i)}$, we also reject the hypotheses of all tests with p-values $p_{(j)}$ for $j = 1, \ldots, i - 1$. Holm's testing procedure now proceeds as follows:

1. If $p_{(1)} > \alpha/m$, accept all hypotheses $H_{0(1)}, H_{0(2)}, \ldots, H_{0(m)}$ and stop. If $p_{(1)} \leq \alpha/m$, reject hypothesis $H_{0(1)}$ and proceed to step 2.
2. If $p_{(2)} > \alpha/(m-1)$, accept all hypotheses $H_{0(2)}, \ldots, H_{0(m)}$ and stop. If $p_{(2)} \leq \alpha/(m-1)$, reject hypothesis $H_{0(2)}$ and proceed to step 3.
3. If $p_{(3)} > \alpha/(m-2)$, accept all hypotheses $H_{0(3)}, \ldots, H_{0(m)}$ and stop. If $p_{(3)} \leq \alpha/(m-2)$, reject hypotheses $H_{0(3)}$ and proceed to step 4.

$\vdots$

m. stop

With this procedure, we tend to first reject the tests with the smallest p-value while still keeping the FWER limited to significance level α. As $P(p_{(1)} \leq \alpha/m) = P(\min p_j \leq \alpha/m)$, the Bonferroni bound tells us that the FWER is bounded by α. In this respect, Holm's procedure does not appear to be an improvement. To fully understand the advantages of the procedure, we also need to look at the power of the test. So far we have assumed that all hypotheses are valid. In practice, however, some of the m hypotheses may not hold and, in fact, we *want* the tests on these hypotheses to lead to rejections. With that in mind, let us look at our matrix of errors again, but this time with counts of the number of times we made each decision. The following contingency table summarises the relevant values:

	Non rejected hypotheses i.e. "H_{0j}"	Rejected hypotheses i.e. "H_{1j}"	
True hypotheses H_{0j}	U	V	m_0
False hypotheses, i.e. H_{1j}	T	S	m_1
	$m - R$	R	m

Note that such tables are also used in classification problems. We stress, however, that the common arrangement of the cells in classification is different and the reader should not get confused. The table above is arranged to mimic the statistical decision matrix from above, but now for m tests. On the other hand, in classification, one labels hypothesis H_0 as "negative", while H_1 is labelled as "positive". This leads us to the quantities:

- true positive (TP) = S,
- true negative (TN) = U,
- false positive (FP) = V and
- false negative (FN) = T.

Although these are commonly used terms in classification, we stick to the notation used in the table above to exhibit the relation to statistical testing. We will return to this point in Sect. 6.9.2. Hence we assume that for m_0 of the m tests, the hypothesis is correct, while for m_1 tests, the hypothesis should be rejected. We reject R hypotheses, out of which V hypotheses are rejected incorrectly. Note that the capital letters in the inner part of the matrix are random variables, which we cannot observe, i.e. we can observe R, but neither V nor S is observable. We have defined the FWER with

$$FWER = P(V \geq 1| \text{ true hypotheses } H_{0j}).$$

Assume now that $\Lambda_0 \subseteq \{1, \ldots, m\}$ is the index set of the true hypotheses, such that $|\Lambda_0| = m_0$. If $m_0 < m$, we have some hypotheses that are not true, and clearly we want tests on these hypotheses to reject them with high probability, that is, S should be positive and preferably as large as possible.

Coming back to Holm's procedure, we can see that the ordering of the p-values has the welcome effect that hypotheses with small p-value are more likely to be rejected, i.e. S may increase, while V is still controlled. This can be demonstrated as follows. Let j^* be the index of the smallest p-value for the valid hypotheses. That is, $p_{(j^*)} = \min_{j \in \Lambda_0} \{p_j\}$. An incorrect rejection of the true hypotheses occurs if the p-value leads to a rejection in the j^*-th step of Holm's procedure, i.e. if

$$p_{(j^*)} \leq \frac{\alpha}{m - j^* + 1}.$$

As $j^* \leq m_1 + 1 = m - m_0 + 1$, we have $m_0 \leq m - j^* + 1$, such that

$$\frac{\alpha}{m - j^* + 1} \leq \frac{\alpha}{m_0}.$$

As a consequence, the Bonferroni bound applies, but with the ordering of the p-values, it is more likely to reject false hypotheses compared to the simple Bonferroni adjustment and is therefore always recommended. The procedure has been extended in various ways, and we refer to Efron (2010) for further details.

Statistical testing overall is a conservative strategy, which means that the priority is to limit the probability of a type I error. Consequently, the probability of a type II error is uncontrolled in multiple testing. In particular, in a setting where one has a limited number of observations, but hundreds or thousands of hypotheses to test, multiple testing procedures do not suffice. A typical example is genetic data, where the sample size is usually limited, but the number of genes to be analysed is large. Statisticians sometimes call this the $n << p$ problem, meaning that the sample size n is much smaller than the number of variables p.

Benjamini and Hochberg (1995) introduced the **False Discovery Rate (FDR)** as a way to balance the type I and type II errors. The idea behind the FDR is that it is reasonable to accept a small number of false detections among the rejected hypotheses. This is expressed with the false discovery rate, which is defined as the proportion of falsely rejected hypotheses from the total. With the notation from the contingency table above, this means we are looking at the ratio

$$Q = \frac{V}{\max(R, 1)}.$$

This is an empirical number and hence random. Taking the expectation defines the FDR as

$$FDR = E(Q) = E\left(\frac{V}{R}\right).$$

The idea is now, instead of controlling V, to control Q. This is especially useful if R is large, that is, we are rejecting a substantial number of the hypotheses we are testing. It is also useful to allow V to be greater than 0, as long as we are also detecting a reasonable number of true positives.

The procedure was suggested by Benjamini and Hochberg (1995) and proceeds as follows. We again order the p-values of the m tests and get

$$p_{(1)} \leq p_{(2)} \leq \cdots \leq p_{(m)}.$$

We then fix $\alpha \in (0, 1)$ as the targeted FDR, where α should be small. The largest index j is then chosen such that $p_{(j)} \leq \alpha j / m$ and therefore $p_{(j+1)} > \alpha(j + 1)/m$. Then all hypotheses $H_{0,(i)}$ are rejected, for which $i \leq j$. Clearly, this procedure is quite simple, which certainly explains why it is often used in practice. We omit the proof that the FDR resulting from the above strategy is in fact keeping the α level, i.e. $E(V/R) \leq \alpha$, but Benjamini and Hochberg (1995) show, in fact, that $E(V/R) \leq m_0 \alpha / m$ with m_0 as the number of true hypotheses.

This still leaves the choice of α for the FDR. Note that the purpose of the FDR is *not* to control the type I error. Instead, it is to control the proportion of falsely rejected hypotheses (type I error) with respect to all rejected hypotheses. In other words, α percentage of the rejected hypotheses which are in fact false discoveries. It follows that α as FDR has a completely different interpretation from α as a significance level. It is important that one keeps this fact in mind when working with the FDR.

6.9 Significance and Relevance

6.9.1 Significance in Large Samples

It is becoming clear that in the age of "Big Data", the application of tests and confidence intervals on very large datasets might need to be reexamined. To demonstrate how the sample size affects our results, let us once again look at the power function. Assume $Y_i \sim N(\mu_0 + \delta_n, \sigma^2)$ i.i.d., $i = 1, \ldots, n$, where our hypothesis is $H_0 : \delta_n = 0$ versus $H_1 : \delta_n > 0$. If we allow our effect size δ_n to be related to our sample size, we get some interesting results. To demonstrate, we set $\delta_n = \delta_1/\sqrt{n}$, where δ_1 is some constant. It is not difficult to show that this setting results in a test with constant power. Let the critical value c be defined by

$$P((\bar{Y} - \mu_0) \geq c | \delta_1 = 0) = \alpha \Leftrightarrow c = z_{1-\frac{\alpha}{2}} \sigma / \sqrt{n}$$

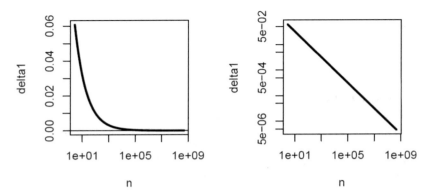

Fig. 6.7 Quantity δ_1 which leads to a power of order 0.25. Left-hand plot shows $\delta_1/\sqrt{n}$ with n on the log scale. The right-hand side shows both axes on the log scale

such that $P(\text{``}H_1\text{''}|H_0) = \alpha$. Then,

$$P(\text{``}H_1\text{''}|H_1) = P\left((\bar{Y} - \mu_0) \geq z_{1-\alpha}\frac{\sigma}{\sqrt{n}}|\delta_1\right)$$

$$= P\left(\frac{\bar{Y} - \mu_0 - \delta_1/\sqrt{n}}{\sigma/\sqrt{n}} \geq z_{1-\alpha} - \frac{\delta_1/\sqrt{n}}{\sigma/\sqrt{n}}\right)$$

$$= 1 - \Phi(z_{1-\alpha} - \frac{\delta_1}{\sigma}),$$

which does not depend on n.

The behaviour of $\delta_1/\sqrt{n}$ is visualised in Fig. 6.7, where on the left the sample size is given on a log scale and on the right both the sample size and $\delta_1/\sqrt{n}$ are shown on a log scale. We see that for increasing sample sizes, $\delta_1/\sqrt{n}$ goes to zero quite fast. This implies that, with a large enough sample, one could in principle detect any discrepancy from the hypothesis. If the database is large, nearly everything becomes significant. In this case, one must pose the question whether significant parameters are also relevant. We will tackle this problem in more depth in Chap. 9, when we talk about model validation. We will also look at this question from the perspective of data quality and quantity in Sect. 11.3. For now, we simply emphasise that principles of statistical testing have their limits when the sample size is very large.

6.9.2   Receiver Operating Characteristics

Let us return to the multiple testing problem and consider the decision matrix again. Note that the decision depends upon a decision rule, which is determined by the test-specific significance level $\tilde{\alpha}$. We test the j-th hypothesis H_{0j} with the j-th decision rule

$$\text{"}H_1\text{"} \Leftrightarrow p_j \leq \tilde{\alpha} \text{ for } j = 1, \ldots, m.$$

So far, we have chosen the test-specific level $\tilde{\alpha}$ either with the intention of controlling the FWER, i.e. postulating for V in the above contingency table

$$P(V \geq 1 \mid \text{true hypotheses } H_{0j}) \leq \alpha,$$

or by limiting the false discovery rate, i.e.

$$FDR \leq \alpha.$$

But what happens if we increase α? Clearly, two things will happen in opposite directions. If we allow for a higher rate of the false positive cases V, we will also increase the number of true positives S. In the same fashion, decreasing the number of true negatives U will decrease the number of false negatives T. In other words, increasing the significance level α, which is the probability of a type I error, will decrease the probability of a type II error, that is, $P(\text{"}H_0\text{"} \mid H_1)$. To tackle this problem, let us now look at the two quantities at the same time and define the following terms.

Definition 6.3 The **specificity** of a test, or **true negative rate**, is given by

$$TNR = E\left(\frac{U}{m_0}\right) = 1 - E\left(\frac{V}{m_0}\right) = 1 - FPR,$$

which is the proportion of correct null hypotheses that are not rejected. The term $E(V/m_0)$ is also called the false positive rate (FPR). The **sensitivity** of a test, or **true positive rate**, is given by

$$TPR = E\left(\frac{S}{m_1}\right).$$

This is the proportion of incorrect null hypotheses that are correctly rejected.

One can now relate the false positive rate $FPR = 1 - TNR = E(V/m_0)$ to the true positive rate by changing the value of α. This leads us to a **Receiver Operating Curve (ROC)**, which is commonly used in classification. A typical ROC is shown in Fig. 6.8. Different values of α lead to different decisions, visualised in the behaviour of the curve. **The Area Under the Curve (AUC)** describes the quality of the test situation. Clearly, if the area is exactly $\frac{1}{2}$, the test is not recommendable, as it does not have any power. The larger the AUC, the better the test situation, which also means the higher the overall power for different values of the significance level α.

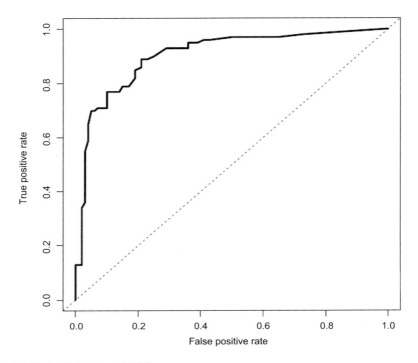

Fig. 6.8 Typical behaviour of ROC

One should note that all quantities involved are random numbers, and, of course, in reality, one cannot derive the ROC in an analytic form. We will see in Chap. 8, however, that resampling methods allow us to estimate the ROC.

6.10 Exercises

Exercise 1
We consider an *i.i.d.* sample $Y_1, \ldots, Y_n$ from an exponential distribution $Exp(\lambda), \lambda > 0$, with density function $f(y|\lambda) = \lambda \exp(-\lambda y), y \geq 0$, and want to construct a statistical test for the hypotheses

$$H_0 : \lambda = 1 \quad \text{versus} \quad H_1 : \lambda \neq 1.$$

Construct the Wald, score and likelihood-ratio tests with the appropriate critical values and decision rules, i.e. when one decides for H_1.

Exercise 2 (Use R Statistical Software)
Simulate N vectors ("the samples") of length $n = 50$ of independent $N(0, 1)$
random numbers. Then modify the vectors by adding constants $c_1, \ldots, c_N$ to each
of the vectors.

1. Use appropriate tests ($\alpha = 0.05$) for each of the N test problems

$$H_{0j} : \mu_j = 0 \quad \text{versus} \quad H_{1j} : \mu_j = c_j , j = 1, \ldots, N .$$

   Simulate the distribution of the p-values for the case that $c_j = 0, \forall j = 1, \ldots, N$,
   and the case that some $c_j \neq 0, \forall j = 1, \ldots, N$, by increasing the number N
   ($N \to \infty$).
2. Now set $N = 10$ and $\alpha = 0.05$. Repeat the generation of samples 100 times, i.e.
   1000 vectors of length 50 are generated. Estimate the false discovery rate (FDR).
   What is the FDR after correcting the p-values with the Bonferroni procedure?
   Repeat the process for different values of the constants c_j.

Chapter 7
Regression

Thus far, we have looked at different approaches to modelling random variables with parametric distributions. Let us now address the very common situation where additional variables are associated with the distribution of our random variable. In other words, we have input variables x that influence the distribution of our output variable Y. To mediate this influence, we allow the parameters θ of the distribution of Y to depend upon x. More formally, we attempt to model the conditional distribution of Y given x

$$Y|x \sim f(y; \theta(x)),$$

where the parameter θ is influenced by x, possibly in a complex manner. This setting is generally called **regression** in statistics. Often, the mean value of Y is the parameter of interest and is modelled in response to the input variables x, i.e. $E(Y|x)$. However, other quantities, e.g. quantiles, can be considered as well. In the following chapter, we will present commonly used statistical regression models and end with an extended example to demonstrate their flexibility and usability.

From here on, we call the output variable Y the dependent variable and the input variable x the independent variable. In econometrics, Y is often called the endogenous variable and x the exogenous variable and in other strands of literature Y the response variable and x the covariate. We use these terms interchangeably in the subsequent chapters and have included Table 7.1 as a reference.

7.1 Linear Model

The problem of modelling a quantity x that influences the outcome Y is rather broad and regression models as described above are just one of the many possible approaches. Regression analyses have the advantage that they allow for model

G. Kauermann et al., *Statistical Foundations, Reasoning and Inference*,
Springer Series in Statistics, https://doi.org/10.1007/978-3-030-69827-0_7

Table 7.1 Alternative terms for Y and x in the regression setting	Y	x
	Endogenous variable	Exogenous variable
	Output variable	Input variable
	Response variable	Covariate
	Dependent variable	Independent variable
	Label	Feature
	Target	Covariable

interpretation. The regression coefficient β quantifies the influence of x on Y and allows its prediction when only x is known. We stress, however, that regression models are only one tool for modelling the influence of x on Y and a swathe of alternative techniques are available. This includes approaches like classification and regression trees, neural networks, support vector machines and so on. We will not cover these tools here but refer to the literature, e.g. Hastie et al. (2009).

7.1.1 Simple Linear Model

The classical linear model involves quantifying how the mean of the response variable Y depends on the covariate x. This is modelled as follows:

$$Y = \beta_0 + x\beta_x + \varepsilon. \qquad (7.1.1)$$

The quantity ε is called the **error term** and the parameters are the intercept β_0 and the slope β_x. Model (7.1.1) is known as the linear regression model and even traces back to Gauß (1777–1855), who was the first to propose estimates of β_0 and β_x. The relationship between Y and x can be causal in nature, that is, one can assume that there is a direct influence of x on y. We will explicitly discuss this in Chap. 12. For now, we simply consider Model (7.1.1) as relating the distribution of Y to the independent variable x. To make this more explicit, we assume that covariate x is known and that the distribution of Y depends conditionally upon x. Secondly, we assume that the error term ε has zero mean, i.e. $E(\varepsilon|x) = 0$, which in turn implies that

$$E(Y|x) = \beta_0 + x\beta_x.$$

Hence, the conditional mean of Y is given by the linear predictor $\beta_0 + x\beta_x$. In other words, we have a linear relation between Y and x, which is disturbed by an error ε. These are the two essential assumptions in a regression model. The latter is often replaced by postulating normality of the error terms, that is, one assumes

$$\varepsilon|x \sim N(0, \sigma^2)$$

or equivalently

$$Y|x \sim N(\beta_0 + x\beta_x, \sigma^2).$$

Additionally, it is necessary that the distribution of ε does not depend on x, which is clearly stronger than simply assuming a vanishing mean $E(\varepsilon|x) = 0$. This also implies that the variance of ε (and hence of Y) does not depend on x. This lack of influence is called **variance homogeneity** or synonymously **(variance) homoskedasticity**. In contrast, if the variance depends on x, we call this **variance heterogeneity** or equivalently **(variance) heteroskedasticity**. We will demonstrate methods for addressing homoskedasticity later in this chapter.

An applied example of a regression model is given in Fig. 7.1, where x is the floor size of an apartment and Y the corresponding rent, with data taken from the Munich rental guide 2015. We see that the larger the apartment, the higher the rent. We also include the fitted linear line in the plot, where the intercept β_0 is 135.50 and

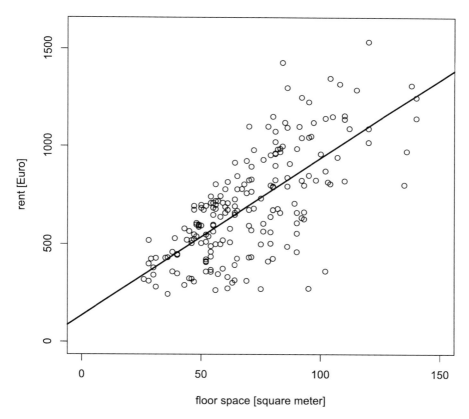

Fig. 7.1 Rent (in euros) for apartments with a given floor size. Taken from the data used to produce the Munich rental table 2015

the slope β_x is 8.04. This implies that the average rent increases by 8.04 euros per square metre.

Let us now turn our attention to estimating the parameters of our model that best fit the data. Suppose that data pairs $(y_i, x_i), i = 1, \ldots, n$, are available, as in Fig. 7.1. Assuming normality and independence for the error terms ε_i gives the log-likelihood (ignoring constant terms)

$$l(\beta_0, \beta_x, \sigma^2) = \sum_{i=1}^{n} \left\{ -\frac{1}{2} \log \sigma^2 - \frac{1}{2} \frac{(y_i - \beta_0 - x_i \beta_x)^2}{\sigma^2} \right\}. \tag{7.1.2}$$

Differentiating the log-likelihood with respect to β_0, β_x and σ^2 gives

$$\frac{\partial l(\beta_0, \beta_x, \sigma^2)}{\partial \beta_0} = \sum_{i=1}^{n} \frac{(y_i - \beta_0 - x_i \beta_x)}{\sigma^2}$$

$$\frac{\partial l(\beta_0, \beta_x, \sigma^2)}{\partial \beta_x} = \sum_{i=1}^{n} x_i \frac{(y_i - \beta_0 - x_i \beta_x)}{\sigma^2}$$

$$\frac{l(\beta_y, \beta_x, \sigma^2)}{\partial \sigma^2} = -\frac{n}{2\sigma^2} + \frac{1}{2} \sum_{i=1}^{n} \frac{(y_i - \beta_0 - x_i \beta_x)^2}{\sigma^4}.$$

Setting these to zero gives the Maximum Likelihood estimates

$$\hat{\beta}_0 = \frac{1}{n} \sum_{i=1}^{n} y_i - \frac{1}{n} \sum_{i=1}^{n} x_i \hat{\beta}_x$$

$$\hat{\beta}_x = \sum_{i=1}^{n} x_i (y_i - \hat{\beta}_0) \Big/ \sum_{i=1}^{n} x_i^2$$

$$\hat{\sigma}^2 = \frac{1}{n} \sum_{i=1}^{n} (y_i - \hat{\beta}_0 - x_i \hat{\beta}_x)^2. \tag{7.1.3}$$

The estimates $\hat{\beta}_0$ and $\hat{\beta}_x$ are also known as least square estimates $\hat{\beta}_0$ and $\hat{\beta}_x$, which minimise the squared distance between prediction and true values $\sum_{i=1}^{n} (y_i - \beta_0 - x_i \beta_x)^2$. This is clear in Eq. (7.1.2), where exactly this expression must be minimised. The calculation of the estimates becomes much simpler if we rewrite the whole model in matrix notation. To do so, let

$$y = (y_1, \ldots, y_n)^T \in \mathbb{R}^{n \times 1}$$

$$X = \begin{pmatrix} 1 & x_1 \\ \vdots & \vdots \\ 1 & x_n \end{pmatrix} \in \mathbb{R}^{n \times 2}$$

$$\varepsilon = (\varepsilon_1, \ldots, \varepsilon_n)^T \in \mathbb{R}^{n \times 1}.$$

We can then write Model (7.1.1) in the form

$$Y = X\beta + \varepsilon,$$

where $\beta = (\beta_0, \beta_x)^T$ and $Y = (Y_1, \ldots, Y_n)^T$. Matrix X is also called the **design matrix**. Note that the column with entries "1" corresponds to including the intercept in the model and is the default setting. Given the data y, the likelihood is written as

$$l(\beta, \sigma^2) = -\frac{n}{2} \log \sigma^2 - \frac{1}{2\sigma^2} (y - X\beta)^T (y - X\beta). \tag{7.1.4}$$

Using matrix notation to get the derivative with respect to β gives us

$$\frac{\partial l(\beta, \sigma^2)}{\partial \beta} = X^T (y - X\beta).$$

This gives the estimate

$$\hat{\beta} = (X^T X)^{-1} X^T y, \tag{7.1.5}$$

which is identical to the estimates in (7.1.1) but clearly has a more convenient form than the individual derivatives given above. The estimate (7.1.5) has a number of welcome properties, which can be easily derived. Firstly, the estimate is unbiased because

$$E(\hat{\beta}) = (X^T X)^{-1} X^T E(Y|X) = (X^T X)^{-1} X^T X \beta = \beta.$$

Secondly, as $Var(\varepsilon) = \sigma^2 I_n$, where I_n is the diagonal matrix of dimension n, it can be seen that

$$Var(\hat{\beta}) = \sigma^2 (X^T X)^{-1}. \tag{7.1.6}$$

It is also not difficult to show that $\sigma^2 (X^T X)^{-1}$ is equal to the inverse Fisher information matrix. Hence, the variance of the estimate is minimal. The estimate is also the best linear unbiased estimate (BLUE), i.e. the best unbiased linear estimator of the type $\hat{\beta} = AY$ with $A \in \mathbb{R}^{2 \times n}$. This property is known as the **Gauß–Markov theorem**. The correlation of the estimates $\hat{\beta}_0$ and $\hat{\beta}_x$ is expressed in (7.1.6), where $X^T X$ is the 2×2 matrix

$$X^T X = \begin{pmatrix} n & \sum_{i=1}^{n} x_i \\ \sum_{i=1}^{n} x_i & \sum_{i=1}^{n} x_i^2 \end{pmatrix}.$$

This demonstrates that if $\sum x_i = 0$, then $\hat{\beta}_0$ and $\hat{\beta}_x$ are uncorrelated. Note that for a single covariate, we can always achieve this by redefining

$$\tilde{x}_i = x_i - \bar{x} \Leftrightarrow x_i = \tilde{x}_i + \bar{x},$$

such that the model becomes

$$Y_i = \beta_0 + x_i \beta_x + \varepsilon_i = \beta_0 + \tilde{x}_i \beta_x + \bar{x} \beta_x + \varepsilon_i = \tilde{\beta}_0 + \tilde{x}_i \beta_x + \varepsilon_i.$$

The intercept changes from β_0 to $\tilde{\beta}_0$, but the slope remains unchanged. For $\tilde{x}_i$, we find that $\sum_{i=1}^{n} \tilde{x}_i = \sum_{i=1}^{n} x_i - n\bar{x} = 0$, such that the Fisher information matrix for $\tilde{\beta}_0$ and β_x is given by

$$Var\begin{pmatrix} \hat{\tilde{\beta}}_0 \\ \hat{\beta}_x \end{pmatrix} = \sigma^2 \begin{pmatrix} n & 0 \\ 0 & \sum \tilde{x}_i^2 \end{pmatrix}^{-1}.$$

Furthermore, the matrix formulation enables us to make various extensions of the simple model (7.1.1). For instance, let x be a binary covariate of the form $x \in \{0, 1\}$. Because we did not make any assumptions about the values of the x variable, the different model formulation (7.1.1) remains valid. This also applies for (7.1.5).

7.1.2 Multiple Linear Model

Let us now assume that x is a discrete covariate which takes k different values labelled $\{1, \ldots, k\}$. Model (7.1.1) can then be rewritten as

$$Y = \beta_0 + 1_{\{x=1\}} \beta_1 + 1_{\{x=2\}} \beta_2 + \ldots + 1_{\{x=k-1\}} \beta_{k-1} + \varepsilon,$$

where $1_{\{.\}}$ is the indicator function, which takes value 1 if the statement in the brackets is true and zero if it is false. Note that we only need $k-1$ indicator variables, because the intercept covers the case where x takes value k. We can construct the matrix X as

$$X = \begin{pmatrix} 1 & 1_{\{x_1=1\}} & \cdots & 1_{\{x_1=k-1\}} \\ \vdots & \vdots & & \vdots \\ 1 & 1_{\{x_n=1\}} & \cdots & 1_{\{x_n=k-1\}} \end{pmatrix}$$

with y and ε defined as above. This gives

$$Y = X\beta + \varepsilon,$$

where $\beta = (\beta_0, \beta_1, \ldots, \beta_{k-1})^T$. Clearly, the estimate $\hat{\beta}$ is given by (7.1.5), but now has dimension k, i.e. $k-1$ coefficients β_1 to β_{k-1} plus the intercept β_0. The same model formulation applies if we have more than one covariate. Let, for instance, z be a second covariate, such that Model (7.1.1) extends to

$$Y = \beta_0 + x\beta_x + z\beta_z + \varepsilon.$$

We then define X as

$$X = \begin{pmatrix} 1 & x_1 & z_1 \\ \vdots & \vdots & \vdots \\ 1 & x_n & z_n \end{pmatrix}.$$

Using data (y_i, x_i, z_i) again gives (7.1.5) as an estimate. This demonstrates the generality of the matrix formulation of the model, which does not need to be altered when including more covariates or parameters.

Given the structure of $\hat{\beta}$, we see that the Maximum Likelihood estimate $\hat{\beta}$ fulfils the distributional property

$$\hat{\beta} \sim N(\beta, \sigma^2(X^T X)^{-1}). \tag{7.1.7}$$

For the existence of the estimator $\hat{\beta}$, we need $(X^T X)$ to be invertable, which holds if X has full rank.

Note that σ^2 also needs to be estimated and $\hat{\sigma}^2$ in (7.1.3) can easily be written in matrix notation. It appears, however, that the estimate should be bias corrected. Note that the Maximum Likelihood estimate fulfils

$$\hat{\sigma}^2 = (y - X\hat{\beta})^T(y - X\hat{\beta})/n.$$

The observable values $\hat{\varepsilon} = y - X\hat{\beta}$ are called fitted residuals, such that $\hat{\sigma}^2 = \hat{\varepsilon}^T\hat{\varepsilon}/n$.

We define with

$$H = X(X^T X)^{-1} X^T$$

the **hat matrix**, which is idempotent, meaning that $HH = H$ and $H^T = H$. Given the structure of the estimate $\hat{\beta}$, we can then write $y - X\hat{\beta} = (I_n - H)y$ with I_n as n dimensional identity matrix. If we now replace the observed values y with random variables Y, we can calculate the mean value of $\hat{\sigma}^2$. In fact, using linear algebra rules for the trace $tr(.)$ of a matrix, we get

$$
\begin{aligned}
E((Y - X\hat{\beta})^T (Y - X\hat{\beta})) &= E(Y^T (I_n - H)^T (I_n - H)Y) \\
&= E(\text{tr}(Y^T (I_n - H)Y)) \\
&= E(\text{tr}((I_n - H)YY^T)) \\
&= \text{tr}((I_n - H)E(YY^T)) \\
&= \text{tr}((I_n - H)(X\beta\beta^T X^T + \sigma^2 I_n)) \\
&= \sigma^2 \text{tr}(I_n - H),
\end{aligned}
$$

where the final simplification follows as $(I_n - H)X\beta\beta^T X^T = 0$. Because

$$
\begin{aligned}
\text{tr}(I_n - H) &= n - \text{tr}(X(X^T X)^{-1} X^T) \\
&= n - \text{tr}((X^T X)^{-1} X^T X) = n - p,
\end{aligned}
$$

we find the variance estimate to be biased, where p is the number of columns in **X**. It is therefore common to replace the variance estimate with its unbiased counterpart

$$s^2 = \frac{(y - X\hat{\beta})(y - X\hat{\beta})}{n - p}.$$

If we now also replace σ^2 in (7.1.7) with its unbiased estimate s^2, we obtain a multivariate t-distribution for $\hat{\beta}$ with $n - p$ degrees of freedom

$$\hat{\beta} \sim t_{n-p}(\beta, s^2 (X^T X)^{-1}). \tag{7.1.8}$$

However, if n is large, it is reasonable to use the normal distribution described in (7.1.7), even if σ^2 is estimated with s^2.

The multiple linear regression model described by the linear equation

$$E(Y|x) = \beta_0 + \beta_1 x_1 + \beta_2 x_2 + \ldots + \beta_p x_p$$

is a very useful and flexible tool for modelling associations between Y and regressor variables $x_1, \ldots, x_p$. The regression coefficients β_k can be interpreted as follows. If x_k is increased by 1, then the expectation of Y is increased by β_k, if all other covariates remain fixed. This is a typical "ceteris paribus" interpretation. The coefficient β_k describes the pure association when adjusting for all other covariates.

Furthermore, we make no assumptions about the nature of the variables $x_1, \ldots, x_p$ and therefore have many alternative model structures at our disposal. These include

- indicator variables $x_k \in \{0, 1\}$ to model the association between nominal variables and Y;
- transformed variables, i.e. defining for example $x_k = \log(z)$, which indicates a logarithmic relationship between z and Y;
- using polynomials or splines, e.g. $x_1 = z, x_2 = z^2, x_3 = z^3$ for polynomial regression;
- using products of variables, e.g. use x_1, x_2 and $x_3 = x_1 x_2$. In this case, the term $x_1 x_2$ is called an **interaction term**.

Regression is an essential tool in statistics and there are many extensions, even beyond the material discussed in this chapter. We refer to Fahrmeir et al. (2015) for an extensive discussion of the field.

Example 26 In a clinical trial, patients suffering from high blood pressure were treated psychologically. Patients were randomly assigned to three groups: the control group (S1) did not receive psychological treatment. The other two groups of patients received either 1 or 2 therapy sessions (S2) or more than 2 therapy sessions (S3). Blood pressure was measured at the beginning and at the end of the study. We use the following linear regression model

$$Y_i = \beta_0 + BPS_i \beta_1 + 1_{\{i \in S2\}} \beta_2 + 1_{\{i \in S3\}} \beta_3 + \varepsilon_i \qquad (7.1.9)$$

to model the relationship between the final blood pressure Y and the type of therapy. The variables $1_{\{i \in S2\}}$ and $1_{\{i \in S3\}}$ indicate whether person i belongs to group S2 or group S3, respectively. The control group is the reference and β_2 and β_3 can be interpreted as the difference in blood pressure between the control group and their respective groups, for a given starting blood pressure. The covariate BPS_i gives the blood pressure at onset, which should have an effect on the final blood pressure Y. The fitted regression coefficients are given in Table 7.2. The intercept is difficult to interpret, but we can see that the blood pressure at onset has a significant effect. This

Table 7.2 Regression estimates and standard errors

	Estimate	Std. error	p-Value
β_0	76.4	9.8	<0.001
β_1	0.31	0.08	<0.001
β_2	−4.7	1.9	0.114
β_3	−6.07	1.8	0.0014

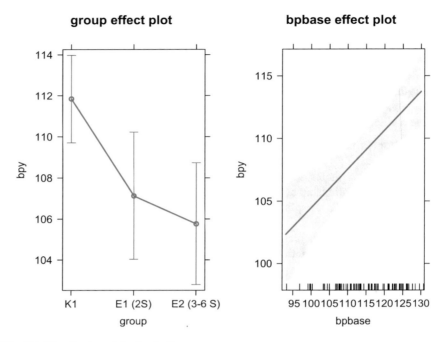

Fig. 7.2 Visualisation of the fitted effects

can be seen from the given p-value, which tests $H_0 : \beta_1 = 0$ with the test statistic in (7.1.8). As a simple rule of thumb, the standardised estimate

$$\frac{\hat{\beta}_1}{\sqrt{\widehat{Var}(\hat{\beta}_1)}} \tag{7.1.10}$$

can also be used, which asymptotically follows a standard normal distribution. Hence, if the absolute value of the ratio (7.1.10) is larger than the $1 - \alpha/2$ quantile of the $N(0, 1)$ distribution, this indicates a rejection of the hypothesis $H_0 : \beta_1 = 0$, with significance level α. The result shows a significant reduction in blood pressure in both treatment groups, meaning that the estimated effect in the treatment group S3 is significant. The results are visualised in Fig. 7.2.

▷

7.1.3   Bayesian Inference in the Linear Model

Thus far, we have treated β as an unknown model parameter, which justifies the use of Maximum Likelihood estimation. We could also take a Bayesian view and impose a prior distribution on the parameters β and σ^2. Taking this perspective will also give us an insight into the similarities between Bayesian estimation and Maximum Likelihood estimation.

To begin with, we need to choose a prior. We will assume, for simplicity, a flat prior for β, that is, $f_\beta(\beta) = \text{const}$. This prior is degenerate, meaning that $\int f_\beta(\beta)d\beta = \infty$ and is, therefore, not a proper density. However, as discussed previously, prior distributions may be degenerate but still lead to proper posterior distributions, which happens to be the case here. For the prior of the variance, we assume that $f_\sigma(\sigma^2) \propto \sigma^{-2}$, which is a flat prior on $\log \sigma^2$ and also degenerate. The posterior is given by

$$f_{\beta,\sigma^2}(\beta, \sigma^2|y) \propto (\sigma^2)^{-\frac{n}{2}+1} \exp\left(-\frac{1}{2\sigma^2}(y - X\beta)^T(y - X\beta)\right).$$

Let us now determine the conditional posterior of β given σ^2, i.e. $\beta|\sigma^2, y$. We rewrite $y - X\beta$ as $(y - X\hat{\beta}) - (X\beta - X\hat{\beta})$ with $\hat{\beta}$ as Maximum Likelihood estimate as defined above. This gives

$$(y - X\beta)^T(y - X\beta) = (y - X\hat{\beta})^T(y - X\hat{\beta}) + (\hat{\beta} - \beta)^T X^T X(\hat{\beta} - \beta)$$
$$\underbrace{- 2(y - X\hat{\beta})^T X(\hat{\beta} - \beta)}_{0}.$$

The final component vanishes, with the definition of $\hat{\beta}$ in (7.1.5). This can be seen because

$$(y - X\hat{\beta})^T X(\hat{\beta} - \beta) = (y - X(X^T X)^{-1}X^T y)^T (X(X^T X)^{-1}X^T y - X\beta)$$
$$= y^T X(X^T X)X^T y - y^T X(X^T X)^{-1}X^T X(X^T X)^{-1}X^T y -$$
$$y^T X\beta + y^T X(X^T X)^{-1}X^T X\beta$$
$$= 0.$$

Defining $s^2 = (y - X\hat{\beta})(y - X\hat{\beta})/(n - p)$ as above, with p as the dimension of β, the posterior can then be rewritten as

$$f_{\beta,\sigma^2}(\beta, \sigma^2|y) \propto (\sigma^2)^{-(\frac{n-p}{2}+1)} \exp\left\{-\frac{n-p}{2\sigma^2}s^2\right\} (\sigma^2)^{-\frac{p}{2}} \exp\left\{-\frac{1}{2\sigma^2}(\beta - \hat{\beta})^T X^T X(\beta - \hat{\beta})\right\}.$$
$$(7.1.11)$$

This shows that, conditional on σ^2, the parameter vector β has a normal posterior, i.e.

$$\beta|\sigma^2, y \sim N(\hat{\beta}, (X^T X)^{-1}\sigma^2). \qquad (7.1.12)$$

Note that (7.1.12) is the same as the distribution of the Maximum Likelihood parameter estimate (7.1.7), but with estimate and parameter reversed. We will take a closer look at this interesting phenomenon in a moment.

The marginal posterior for σ^2 can be obtained by integrating out β in the latter component of (7.1.11), which is proportional to $|\sigma^2 \mathbf{X}^T \mathbf{X}|^{\frac{1}{2}} \propto (\sigma^2)^{-\frac{p}{2}}$. Consequently, the last two multiplicative components in (7.1.11) cancel out and no longer depend upon σ^2. The first two components are proportional to an inverse gamma distribution (see Chap. 5), such that

$$\sigma^2|\mathbf{y} \sim \text{Inv Gamma} \left(\frac{n-p}{2}, \frac{s^2(n-p)}{2} \right).$$

These statements hold for the given prior distributions. We refer to Box and Tiao (1973) for more details on the above calculations. Note that the posterior factorises to

$$f_{\beta,\sigma^2}(\beta, \sigma^2|\mathbf{y}) = f(\beta|\sigma^2, \mathbf{y}) f(\sigma^2|\mathbf{y})$$

and integrating out σ^2 gives the marginal posterior distribution for β. It can be shown that this is, in fact, a p-dimensional t-distribution such that

$$\beta|\mathbf{y} \sim t_{n-p}(\hat{\beta}, s^2(X^T X)^{-1}).$$

We have reached a rather interesting relation between the parameter posterior and the ML estimate. The posterior of β is equal to the distribution of the estimate $\hat{\beta}$, just with β and $\hat{\beta}$ reversed. We observed the same similarity in the previous section, when we looked at the mean of the normal distribution.

We may now draw inference about β from either the Frequentist or the Bayesian perspective. The former takes β as a fixed, but unknown, parameter, while the latter looks at its posterior distribution. In fact, taking the above flat prior with $f_\beta(\beta)$ and $f_\sigma(\sigma^2) = \sigma^{-2}$ (which is again flat for $\log(\sigma^2)$), we have the same distribution for $\beta - \hat{\beta}$, regardless of whether we consider β as random and $\hat{\beta}$ as fixed (the Bayesian view) or, conversely, $\hat{\beta}$ as random and β as fixed (the Frequentist view).

We can use the example in Fig. 7.1, which related the rent of an apartment to its floor size, to help explain how the model can be interpreted. The parameter estimates $\hat{\beta}$ are listed in Table 7.3. The intercept is estimated with $\hat{\beta}_0 = 135.47$ and the fitted

Table 7.3 Parameter estimates and variance estimates for rental data

	Estimate	Std. error	t value	p value p-Value
(Intercept)	135.47	43.29	3.13	0.002
fspc	8.04	0.58	13.76	$< 2e-16$

slope with $\hat{\beta}_x = 8.04$. Hence, one could say that the average rent increase per square metre is in the order of $8.04 \, €/qm^2$. The Fisher matrix is given by

$$\hat{\sigma}^2(X^T X)^{-1} = \begin{pmatrix} 1873.97 & -23.95 \\ -23.65 & 0.34 \end{pmatrix}.$$

Assuming β_0 and β_x are parameters with estimates $\hat{\beta}_0$ and $\hat{\beta}_x$ gives the standard errors listed in the second column above, e.g. $\sqrt{1873.97} = 43.29$. The third column gives the standardised value of the ratio $\hat{\beta}_0/\sqrt{Var(\hat{\beta}_0)}$ and $\hat{\beta}_x/\sqrt{Var(\hat{\beta}_x)}$. The fourth column lists the p-values that result from testing hypotheses $H_0 : \beta_0 = 0$ and $H_0 : \beta_x = 0$. Clearly, we can conclude that floor space significantly influences the rent of an apartment (which is no surprise).

Let us use the above results to derive confidence intervals or credibility regions in the Bayesian case. Note that

$$\hat{\delta}_0 = \hat{\beta}_0 - \beta \quad \text{and} \quad \hat{\delta}_x = \hat{\beta}_x - \beta$$

follow a multivariate t-distribution with $n-2$ degrees of freedom. The corresponding isolines of this density are visualised in the left-hand plot of Fig. 7.3. We may now derive a credibility region R such that

$$P(\hat{\delta} \in R) = 0.95$$

with $\hat{\delta} = (\hat{\delta}_0, \hat{\delta}_x)$. This region is indicated by the solid ellipse in Fig. 7.3, which is in fact both a confidence region and a posterior credibility region. Hence, $\hat{\beta}_0 - \beta$ can take values between approximately -100 and 100, while $\hat{\beta}_x - \beta_x$ can take values

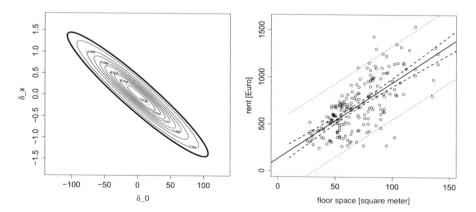

Fig. 7.3 Left plot: isolines of multivariate t-distribution resulting from rental guide data. Right plot: confidence region or credibility region for linear influence of floor space on rent (α=0.05, dashed lines) and prediction interval ($1 - \alpha$=0.95, dotted lines)

between approximately -1.5 and 1.5. Taking values (δ_0, δ_x) within the confidence region R gives different intercepts and slopes for the trend line. This is mirrored in the right-hand plot of Fig. 7.3, where the dashed lines represent the bounds of R, the confidence region of δ_0 and δ_x. They also represent the confidence bounds on β_0 and β_x.

Note that this confidence interval represents the uncertainty of both the Frequentist and the Bayesian approaches on the *linear relation* between floor space and rent. It does not mirror the range that the actual rent might take. This quantity is called the **prediction interval** and, in this case, would depict the range of possible rents for a given floor space. With the prediction interval, we aim to predict the rent of an apartment which is not in the database. Assume for a given floor space x_0, we want to predict the rent. We know that the rent is given by

$$Y = \beta_0 + x_0\beta_x + \varepsilon. \tag{7.1.13}$$

If we replace the parameters in the model above with their estimates, this gives

$$\hat{Y} = \hat{\beta}_0 + x_0\hat{\beta}_x + \varepsilon, \tag{7.1.14}$$

where clearly the error term ε is unobserved. Taking the expectation gives

$$E(\hat{Y}|x_0) = \hat{\beta}_0 + x_0\hat{\beta}_x.$$

We can also derive the variance of the prediction, which is given by

$$Var(\hat{Y}|x) = Var(\hat{\beta}_0 + x\hat{\beta}_x) + \sigma^2.$$

The first component is the estimation variability, which is expressed in the confidence interval, as previously discussed. Because the (new) residual ε is independent of the observed values, the variance decomposes additively to the estimation variance and the residual variance. In the Bayesian formulation, this looks very similar, but now we do not replace β_0 and β_x in (7.1.13) with estimates but work with the posterior distribution directly. Consequently, the variance in the rent of a new apartment is $Var(\hat{Y}|x) = Var(\beta_0 + x\beta_x|\mathbf{y}) + Var(\varepsilon|\mathbf{y})$, where we condition on the data $\mathbf{y}$. In both cases, the variance decomposes additively, as the inference on β_0 and β_x is based on the data, i.e. the data points seen in Fig. 7.1. We include the prediction interval as a dotted line in the right-hand plot of Fig. 7.3. Both the confidence and prediction intervals are shown for a $(1 - \alpha)$ level with $\alpha = 0.05$.

7.2 Weighted Regression

This section will cover how the idea of weighting can be used in linear regression. Let us first assume that the variance of Y_i and hence the variance of the error terms ε_i are different for each observation. This is usually called variance heterogeneity, which we denote as

$$\varepsilon_i \sim N(0, \sigma_i^2).$$

σ_i^2 can now depend on some of the covariates, i.e. $\sigma_i^2 = \sigma^2(x_i)$. This could, for instance, be modelled as $\sigma_i = \exp(x_i \gamma)$. In this case, we get a regression model where both the mean *and* the variance depend upon the covariates. Estimation can then be carried out with the standard Maximum Likelihood approximation, see Kneib (2013). We will not fully exploit the flexibility of this model here, but just consider σ_i as known to an individual multiplicative constant, i.e. $\sigma_i^2 = a_i \sigma^2$, where a_i is known. Note that the model is not identifiable, which means we need to put an extra constraint on a_i to have a unique representation. The common setting is that $\sum_{i=1}^{n} a_i = n$, such that the variance homogeneous model results from the special case $a_i = 1$ for $i = 1, \ldots, n$. The log-likelihood is then given by

$$l(\beta, \sigma^2) = -\frac{n}{2} \log \sigma^2 - \frac{1}{2} \sum_{i=1}^{n} \frac{(y_i - x_i \beta)^2}{a_i \sigma^2}$$

$$= -\frac{n}{2} \log \sigma^2 - \frac{1}{2\sigma^2} (y - X\beta)^T W (y - X\beta),$$

where $W = diag(\frac{1}{a_1}, \ldots, \frac{1}{a_n})$. The ML estimate for β is given by the weighted least squares estimate

$$\hat{\beta} = (X^T W X)^{-1} (X^T W y). \tag{7.2.1}$$

It is easy to see that $\hat{\beta}$ is unbiased, because $E(Y|X) = X\beta$, which inserted in (7.2.1) shows $E(\hat{\beta}) = \beta$. Moreover, the variance is given by

$$Var(\hat{\beta}) = (X^T W X)^{-1} (X^T W Var(Y) W X)(X^T W X)^{-1} \tag{7.2.2}$$

$$= \sigma^2 (X^T W X)^{-1}, \tag{7.2.3}$$

as the variance of the observations is $Var(Y) = \sigma^2 W$. In this case, we can rely on the inverse Fisher matrix as variance estimate for $\hat{\beta}$.

A different form of weighting occurs if we focus on the mean of Y, which often occurs in **survey weighting**. This becomes an issue when the data themselves are biased, for instance, by the over-representation of particular groups of individuals or when observations are not identically distributed. To make this clear, let us look at the following simplified example. Let, as above, Y be the rent of an apartment of floor size x. The data come from two different groups: 50% from city-run housing societies, which are known to provide comparably cheap accommodation, and 50% from private landlords, whose apartments are relatively expensive.

However, in reality, only 25% of the apartments are rented out by housing societies, while 75% are offered by private landlords. Clearly, the data are biased as the 50% of the data represent only 25% of the rental market, while the other 50% represent 75%. Under these circumstances, weighting can again play a useful role. The weights in this case are commonly called survey weights and would be $w_i = 75/50 = 3/2$ for observations from private and commercial landlords and would be set to $w_i = 25/50 = 1/2$ for city-run apartments. A correct use of survey weights in regression is deceptively complex and one must first assume a model where rents of the one group of landlords behave differently than the other. Hence, we need to in fact assume two separate regression models. This idea extends to stratified sampling. In our example, this refers to the fact that the apartments in the city fall into two different groups, or "strata". See DuMouchel and Duncan (1983) and Gelman (2007) for a deeper discussion and also Hu and Zidek (2002), who generalise the idea towards weighted likelihoods. In this example, we will keep things simple and just include weights in the least squares. This leads to (7.2.1), which is however a biased estimate.

To understand this bias, let us look again at the rental example. We assume that apartments from city-run housing organisations have a different (average) monthly rent than those of private and commercial landlords. Let z_i be an indicator variable for the i-th apartment expressing whether the apartment is city run ($z_i = 1$) or commercially run ($z_i = 2$). Then, assuming the same variance in the two groups, we have

$$Y_i | x_i, z_i \sim N(x_i \beta_{(z_i)}, \sigma^2),$$

where $\beta_{(1)}$ and $\beta_{(2)}$ are the slope parameters in the two groups. The population parameter, that is the slope when the groups are omitted, is

$$\beta = 0.25\beta_{(1)} + 0.75\beta_{(2)} = P(Z_i = 1)\beta_{(1)} + P(Z_i = 2)\beta_{(2)}.$$

With these prerequisites, we can now calculate the expectation of $\hat{\beta}$, which leads us to

$$E(\hat{\beta}) = (X^T W X) \sum_{i=1}^{n} w_i x_i^T x_i \beta_{(z_i)},$$

which clearly is not equal to β. Hence, the bias is difficult to calculate. If, however, the distribution of the covariate x_i is independent of z_i, then the estimate is asymptotically unbiased. Instead of looking at the bias of the estimate, let us focus on its variance, which, assuming the same variance in both groups, is given by

$$Var(\hat{\beta}) = (X^T W X)^{-1} (X^T W Var(\varepsilon) W X)(X^T W X)^{-1}$$

$$= (X^T W X)^{-1} (\sum_{i=1}^{n} w_i^2 x_i^T Var(Y_i - x_i \beta) x_i)(X^T W X) \tag{7.2.4}$$

$$= \sigma^2 (X^T W X)^{-1} (X^T W^2 X)(X^T W X)^{-1}. \tag{7.2.5}$$

The variance has a sandwich-type structure and clearly differs from (7.2.3) unless all weights are equal. Note also that the weights do not need to sum up to 1, as any normalisation of the weights cancels out in (7.2.5). A direct estimate of (7.2.4) was proposed by Huber (1967), which is better known as the Eicker–White estimator (see White, 1980, or Mage, 1998). The idea is to replace $Var(Y_i - x_i \beta)$ with its empirical counterpart (assuming unbiasedness), i.e.

$$\widehat{Var}(\hat{\beta}) = (X^T W X)^{-1} (\sum_{i=1}^{n} w_i^2 x_i x_i^T (y - x_i \hat{\beta})^2)(X^T W X)^{-1}.$$

By default, standard software packages that allow for weighted regression do not provide the variance estimate (7.2.5) but instead (7.2.3). Hence, weighting is considered to account for variance heterogeneity but not for survey weights, which should be kept in mind when applying survey weighting in regression. This approach using weights can also be extended to the case of dependent error terms. Then the weighting matrix is not a diagonal matrix and reflects the dependence structure within the error terms. For details, see Fahrmeir et al. (2015), Chap. 4.

7.3   Quantile Regression

The above regression model plays a central role in statistics and a number of extensions and generalisations have been proposed over the last few decades. One of these is quantile regression, which has proven itself to be both practical and powerful in many situations. The most essential piece of literature in the field is "Quantile Regression" by Koenker (1996). To motivate the idea behind quantile regression, let us first repeat the definition of a quantile. A quantile can be comprehended as the inverse of the distribution function. The τ-quantile of a distribution function $F(y)$ is defined as

$$Q(\tau) = \inf\{y : F(y) \geq \tau\} \tag{7.3.1}$$

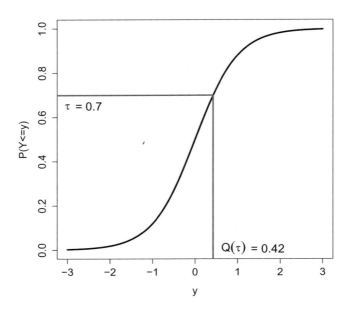

Fig. 7.4 Sketch of the definition of a quantile

for $\tau \in (0, 1)$. This can be seen in Fig. 7.4. Note that if $F(.)$ is invertable, even if only on a subset of $\mathbb{R}$, then $Q(\tau)$ is the inverse of $F(.)$ such that $F(Q(\tau)) = \tau$. A commonly used quantile is the median, defined as $Q(0.5)$, but the quartiles $Q(0.25)$ and $Q(0.75)$ are also used in practice. The idea of quantile regression is now to model the relationship between x and the quantile of the variable Y, in contrast to linear regression which models the *mean* of Y. To do so, we condition everything in (7.3.1) on x, that is

$$Q(\tau|x) = \inf\{y : F(y|x) \geq \tau\},$$

where $F(y|x)$ denotes the conditional distribution of Y given x. We make the dependence on x explicit by modelling $Q(\tau|x)$ in a regression framework of the type

$$Q(\tau|x) = \beta_{0,\tau} + x\beta_{x,\tau}. \tag{7.3.2}$$

This gives a linear quantile regression model, which clearly can be generalised in the same way as linear regression models, i.e. the covariate x can be multivariate or discrete.

Assume now that we have observed the data (y_i, x_i), $i = 1, \ldots, n$ with x_i as the covariates. In order to demonstrate the estimation of the parameters in (7.3.2), let us first look at median regression, i.e. $\tau = 0.5$. Remember that the least squares estimates in linear models were derived by minimising the sum of squared residuals

$\sum_{i=1}^{n}(y_i - x_i\beta)^2$ for the given data $(y_i; x_i), i = 1, \ldots, n$. If we replace the squared distance with the absolute distance, we get the framework for median regression. Setting $Q(0.5|x) = \beta_{0,0.5} + x\beta_{x,0.5}$, the parameters are found with

$$(\hat{\beta}_{0,0.5}, \hat{\beta}_{x,0.5}) = \arg\min \sum_{i=1}^{n} |y_i - \beta_{0,0.5} - x_i\beta_{x,0.5}|. \tag{7.3.3}$$

We will now demonstrate how these estimates can be computed with linear programming. To do so, we will ignore the regression framework for the moment and give a general definition of quantiles. For $\tau \in (0, 1)$, we define the **check function** $\delta_\tau(y)$ with

$$\delta_\tau(y) = \begin{cases} \tau y & \text{if } y \geq 0 \\ (\tau - 1)y & \text{if } y \leq 0, \end{cases}$$

which can also be written as

$$\delta_\tau(y) = y(\tau - \mathbb{1}_{\{y<0\}}).$$

The typical shape of a check function is visualised in Fig. 7.5 for different values of τ. The check function allows the definition of the quantile function $Q(\tau)$ with

$$Q(\tau) = \arg\min_{q} E\left\{\delta_\tau(Y - q)\right\}$$

$$= \arg\min_{q} \left\{(\tau - 1)\int_{-\infty}^{q}(y - q)f(y)dy + \tau\int_{q}^{\infty}(y - q)f(y)dy\right\}. \tag{7.3.4}$$

If we now differentiate $E\left\{\delta_\tau(Y - q)\right\}$ with respect to q, we obtain

$$(1 - \tau)\int_{-\infty}^{q}f(y)dy - \tau\int_{q}^{\infty}f(y)dy = (1 - \tau)F(q) - \tau(1 - F(q)) = F(q) - \tau.$$

Clearly, this defines q as the τ-quantile.

The quantiles can be derived using the method of moments estimation. To do so, one replaces the expectation in (7.3.4) with its empirical counterpart. This gives

$$\hat{Q}(\tau) = \arg\min_{q_\tau} \left(\sum_{i=1}^{n}\delta_\tau(y_i - q_\tau)\right). \tag{7.3.5}$$

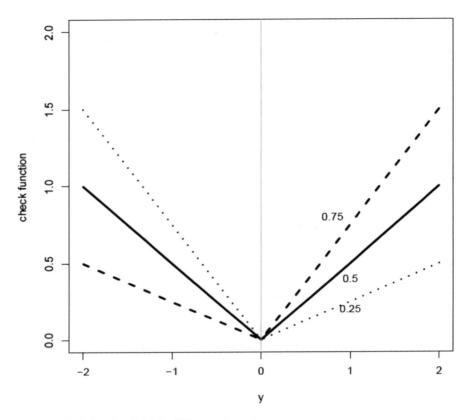

Fig. 7.5 Check function $\delta_\tau(y)$ for different values of τ

Replacing q_τ with a linear regression, for example,

$$q_\tau = \beta_{0,\tau} + x\beta_{x,\tau}$$

gives the general formula for quantile regression. To derive the estimates numerically, we can rewrite the minimisation problem as

$$\delta_\tau(y_i - q_\tau) = \tau u_i + (1 - \tau)v_i,$$

where

$$u_i = \max\{y_i - q_\tau, 0\} \text{ and } v_i = \max\{-(y_i - q_\tau), 0\}$$

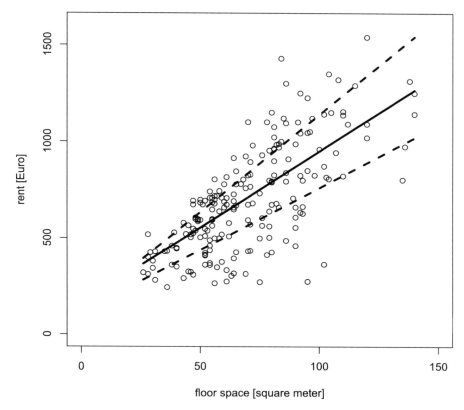

Fig. 7.6 Quantile regression functions for $\tau = 0.25$ (bottom dashed line), $\tau = 0.5$ (centre line) and $\tau = 0.75$ (upper dashed line)

such that $x_i - q_\tau = u_i - v_i$. Hence, minimising (7.3.5) is equivalent to minimising the linear function

$$\min \sum_{i=1}^{n} (\tau u_i + (1 - \tau) v_i) \qquad (7.3.6)$$

subject to the linear constraints $u_i \geq 0$ and $v_i \geq 0$ for $i = 1, \ldots, n$. Clearly, (7.3.6) is a linear programming problem. For an overview, see e.g. Danzig and Thapa (1997)

Figure 7.6 shows the resulting quantile regression functions for the rental data with $\tau_1 = 0.25$, $\tau_2 = 0.5$ and $\tau_3 = 0.75$. The estimated parameters are given in Table 7.4. The plot shows the advantages of quantile regression, in that it allows the modelling and visualisation of variance heteroskedasticity. The range of the observed rents gets larger with increasing floor space. Hence, it seems that not only the median and mean but also the variance of the rents depends upon x.

Table 7.4 Parameter estimates for quantile regression

	Value	Std. error	t value	Pr(>|t|)
$\tau = 0.25$				
(Intercept)	114.49	65.36	1.75	0.08
fspc	6.46	0.88	7.32	< 0.001
$\tau = 0.5$				
(Intercept)	159.96	43.86	3.64	< 0.0001
fspc	7.90	0.59	13.34	< 0.001
$\tau = 0.75$				
(Intercept)	132.33	38.07	3.47	< 0.01
fspc	10.07	0.51	19.59	< 0.01

To draw proper statistical inference, we need to take the estimation variability into account. Hence, we consider the distributional properties of our estimates. It is shown in Koenker (1996) that

$$\hat{\beta}_\tau - \beta_\tau \sim N\left(0, n^{-1} H^{-1}(\tau) J(\tau) H^{-1}(\tau)\right),$$

where $J(\tau)$ and $H(\tau)$ can be approximated with

$$J(\tau) = \frac{\tau(1-\tau)}{n} \sum_{i=1}^{n} (1, x_i)^T (1, x_i)$$

$$H(\tau) = \frac{1}{n} \sum_{i=1}^{n} (1, x_i)^T (1, x_i) f(\beta_{0\tau} + x_i \beta_{1\tau})$$

with $f(.)$ as density of Y. The asymptotic normality is slightly more complicated to prove, because the check function $\delta_\tau(.)$ is not differentiable. We therefore do not go into more detail here.

7.4   Nonparametric Smooth Models

So far, we have only considered parametric models, which are simple in structure and are often merely defined by a linear relationship $\beta_0 + x\beta_x$. Let us now extend these models towards **nonparametric models**, sometimes also labelled as **semiparametric models**. The term nonparametric here refers to functional, nonparametric components in the model. Returning to the original linear regression model (7.1.1), we assume a metric covariate x that can take arbitrary values on the real axis. The idea is now to replace the linear relationship between covariate and response with a more flexible model of the type

$$Y = m(x) + \varepsilon,$$

where the influence of x on y is mediated by $m(.)$, a smooth, i.e. differentiable, but otherwise unspecified function. Like above we assume that ε is a vector of normally distributed random variables with mean zero and variance σ^2. Given the data (y_i, x_i), the intention is to estimate $m(x)$ in a smooth form, meaning a sufficiently differentiable function. The literature on smooth estimation of $m(x)$ is vast and we refer to Hastie and Tibshirani (1990), Fan and Gijbels (1996) or Ruppert et al. (2003) for a comprehensive discussion or see Fahrmeir et al. (2015) for a more recent work.

Various methods have been proposed for the estimation of $m(.)$, including kernel regression and spline smoothing. Here, we demonstrate the use of penalised splines, originally proposed in Eilers and Marx (1996). The idea is closely related to regression splines, where one replaces the unknown function $m(x)$ with a linear combination of known basis functions. To begin, we start with

$$m(x) = \boldsymbol{B}(x)\boldsymbol{\theta} = \sum_{k=1}^{K} B_k(x)\theta_k.$$

There are many possibilities for choosing the basis function. A convenient and practical choice is to work with B-splines as proposed in de Boor (1972). A B-spline basis is constructed by first locating knots on the x-axis. Between the knots, the B-spline is a piecewise polynomial function. A linear B-spline is thereby linear between the knots and a quadratic B-spline is built from polynomials of order 2. In Fig. 7.7, we show a linear B-spline basis (top row) and quadratic B-splines (bottom row). Linear B-splines lead to a continuous piecewise linear function $m(x)$ and a piecewise differentiable quadratic function is given by the quadratic B-spline. In the left-hand plots, we show the B-spline for equidistant knots. Often, however, the covariate measurements x_i are not uniformly distributed and it is advisable to instead allocate the knots based on the quantiles of x_i. This is sensible, because it gives the spline basis more structure where the x_i are dense and there is more information content, while giving less structure in areas with less information. Given the rent data, the resulting bases for non-equidistant knots are shown on the right of Fig. 7.7. The B-spline basis now replaces the original linear relationship and therefore provides more structure and flexibility. Note that we can now in principle follow the arguments from Sect. 7.1 and write the design matrix X as

$$X = \begin{pmatrix} B_1(x_1) & \dots & B_K(x_1) \\ \vdots & & \vdots \\ B_1(x_n) & \dots & B_K(x_n) \end{pmatrix}.$$

The intercept column is incorporated into the B-splines, and hence an extra column of ones is not necessary here. This becomes more clear in Fig. 7.7. The overall level, which is usually included in the intercept, can be expressed by the overall level of the B-splines.

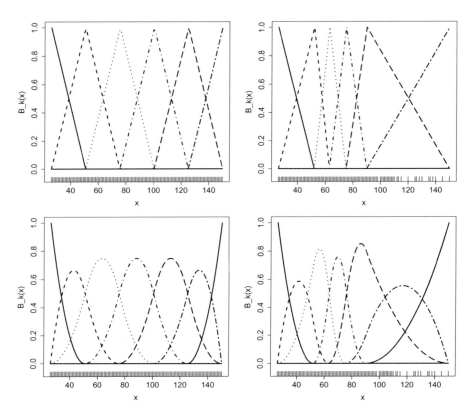

Fig. 7.7 B-spline basis for equidistant (left-hand side) and non-equidistant knots (right-hand side). The top row shows linear B-splines, and the bottom row shows quadratic B-splines

As previously, the estimate of the regression parameter θ is given by

$$\hat{\boldsymbol{\theta}} = (\boldsymbol{X}^T \boldsymbol{X})^{-1} \boldsymbol{X}^T \boldsymbol{y}.$$

We have plotted the resulting fit for different values of the spline dimension K in Fig. 7.8. As before, this is based on the Munich rent data, but this time with Y as the rent divided by the number of square metres of the apartment. The top row shows the fit with linear B-splines and the bottom with quadratic B-splines. The corresponding basis $\boldsymbol{B}(x) = (B_1(x), \ldots, B_K(x))$ is shown in the bottom of the plot, where the basis functions $B_k(x)$ are weighted by their corresponding estimated coefficient $\hat{\theta}_k$. The resulting fit is included in the plot as a line. For $K = 5$, shown in the left column, we obtain a reasonably smooth fit, but for $K = 15$, shown in the right-hand column, the fit gets too "wiggly" and appears unrealistic. This suggests that the basis for $K = 15$ is too complex and the estimation variability too large. To overcome this problem, we impose a penalty on the spline coefficients $\boldsymbol{\theta}$. Note

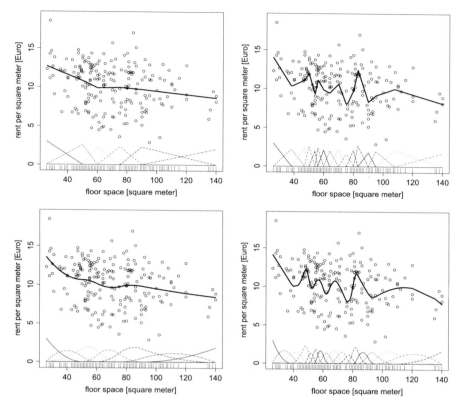

Fig. 7.8 Fitted functional shape for different spline dimensions. The top row shows linear B-splines, and the bottom row shows quadratic B-splines. $K = 5$ is given in left-hand plots and $K = 15$ in right-hand plots

that the wiggliness occurs when two neighbouring spline coefficients have highly differing values, which seems implausible given our assumption of smoothness. We therefore favour solutions where neighbouring coefficients have similar values, that is $|\theta_k - \theta_{k-1}|$ should be small. We enforce this by imposing a penalty on neighbouring spline coefficients, i.e. $\sum_{j=2}^{K}(\theta_j - \theta_{j-1})^2$. An alternative is to use second order differences, i.e.

$$\sum_{j=2}^{K}\left((\theta_j - \theta_{j-1}) - (\theta_{j-1} - \theta_{j-2})\right)^2 = \sum_{j=2}^{K}(\theta_j - 2\theta_{j-1} + \theta_{j-2})^2 \rightarrow \text{ small.}$$

To formulate this in matrix notation, we define the difference matrix

$$
L = \begin{pmatrix}
1 & -2 & 1 & 0 & \ldots & 0 \\
0 & 1 & -2 & 1 & & \\
\vdots & \ddots & \ddots & \ddots & & \vdots \\
0 & & \ldots & 1 & -2 & 1
\end{pmatrix}
$$

and replace the least squares with a penalised version. To do so, we add a penalty to the log-likelihood (7.1.4) and write

$$
l_p(\boldsymbol{\theta}, \sigma^2, \lambda) = -\frac{n}{2}\log\sigma^2 - \frac{1}{2\sigma^2}(\boldsymbol{y} - \boldsymbol{X}\boldsymbol{\theta})^T(\boldsymbol{y} - \boldsymbol{X}\boldsymbol{\theta}) - \frac{1}{2}\frac{\lambda}{\sigma^2}\boldsymbol{\theta}^T \boldsymbol{L}\boldsymbol{L}^T\boldsymbol{\theta}
$$

$$(7.4.1)$$

$$
= l(\boldsymbol{\theta}, \sigma^2) - \frac{1}{2}\frac{\lambda}{\sigma^2}P(\boldsymbol{\theta}, \lambda),
$$

where $P(\boldsymbol{\theta}, \lambda)$ defines the penalty term. The scalar term λ defines the desired level of smoothness and will be discussed later. Simple differentiation gives the estimate as

$$
\hat{\boldsymbol{\theta}} = (\boldsymbol{X}^T\boldsymbol{X} + \lambda \boldsymbol{L}\boldsymbol{L}^T)^{-1}(\boldsymbol{X}^T\boldsymbol{y}).
$$

$$(7.4.2)$$

The effect of penalisation can be seen in Fig. 7.9, which has the same format as Fig. 7.8. The left-hand column shows the unpenalised and the right the penalised fit. The penalisation makes the fit smooth, even if the basis is complex. In this respect, penalisation compensates for a basis that is too complex and forces the resulting fit to be smooth.

The **smoothing parameter** λ plays a central role in the above estimation process. If we set $\lambda = 0$, we obtain an unpenalised, wiggly estimate. On the other hand, if we set $\lambda \to \infty$, given our current difference matrix L, the coefficients must satisfy $\theta_j - 2\theta_{j-1} + \theta_{j-2} \equiv 0$ for $3 \le j \le K$. In practice, λ needs to be chosen based on the data and there are well-established routines that allow its numerical calculation. These are, for instance, cross validation or the AIC and BIC criteria, which we will discuss more generally in Chaps. 8 and 9 and therefore only briefly sketch out here. The key idea behind the **Akaike Information Criterion (AIC)** (see Akaike, 1973) is to balance the complexity of a model and the goodness of fit. The more complex a model, the better the fit, but if we allow the model to become too complex, then parsimony is violated. Thus, we want a model that is as simple as possible, but no simpler. We measure the fit using the log squared error

$$
\text{fit}(\lambda) := n \log\left\{\sum_{i=1}^{n}(y_i - \hat{m}(x_i))^2\right\},
$$

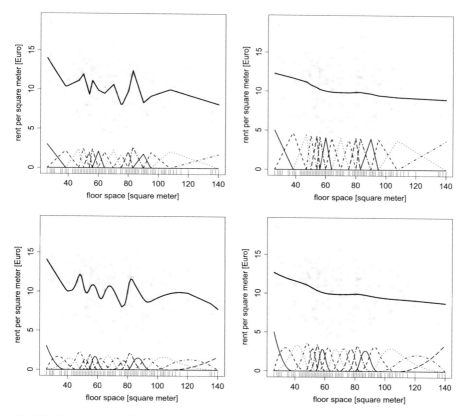

Fig. 7.9 Unpenalised (left) and penalised fit (right) for rental data. Top row is for linear B-splines and the bottom row for quadratic B-splines

where $\hat{m}(x_i) = \boldsymbol{B}(x_i)\hat{\boldsymbol{\theta}}$ with $\hat{\boldsymbol{\theta}}$ as in (7.4.2). Note that the fit $\hat{m}(x) = \boldsymbol{B}(x)\hat{\boldsymbol{\theta}}$ still depends on the smoothing parameter λ, because our estimated $\hat{\boldsymbol{\theta}}$ in (7.4.2) is calculated for a given λ. This is suppressed in the notation for simplicity but should be kept in mind. Clearly, the more complex the model, i.e. the smaller the smoothing parameter λ, the better the fit and the closer $\hat{m}(x_i)$ is to y_i. However, the resulting function is wiggly and complex. As a measure of the complexity of a model, we can define its dimension with

$$\dim(\lambda) = tr\left\{ \boldsymbol{X}(\boldsymbol{X}^T\boldsymbol{X} + \lambda\boldsymbol{L}\boldsymbol{L}^T)^{-1}\boldsymbol{X} \right\} = tr\left\{ (\boldsymbol{X}^T\boldsymbol{X} + \lambda\boldsymbol{L}\boldsymbol{L}^T)^{-1}(\boldsymbol{X}^T\boldsymbol{X}) \right\}.$$
$$(7.4.3)$$

Note that if $\lambda = 0$, then $\dim(\lambda) = K$, the number of splines in the basis, and the matrix in the first component of (7.4.3) becomes the hat matrix. If $\lambda \to \infty$, the dimension decreases, and hence for general λ, the matrix in the first component of

AIC

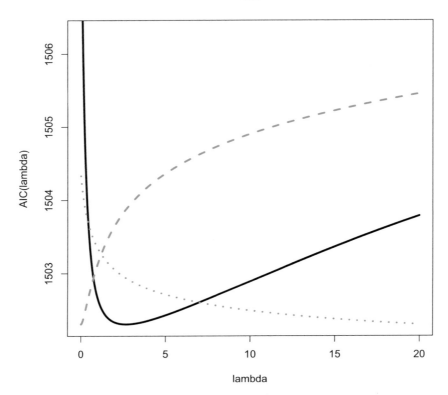

Fig. 7.10 AIC(λ) (solid line) and corresponding functions fit(λ) (dashed) and dim(λ) (dotted)

(7.4.3) can be seen as a generalisation of the hat matrix, and in fact for $\lambda = 0$ it is the hat matrix. These two components are now combined in the Akaike Information Criterion and give

$$\text{AIC}(\lambda) = \text{fit}(\lambda) + 2\,\text{dim}(\lambda).$$

A deeper motivation for this formula is given in Chap. 8. For now let us look at the criterion in the rent example from before, which is shown in Fig. 7.10. We show the curve AIC(λ) for different values of λ, which has a minimum at approximately $\lambda = 4$. We also include the curves for the fit(λ) (dashed grey line) and dim(λ) (dotted grey line).

Minimising the AIC(λ) to obtain an optimal λ requires its calculation for a dense grid of possible values. This can be a computational burden or even infeasible if multiple functions are fitted and thus several smoothing parameters need to be optimised simultaneously. In recent years, a numerically more feasible routine to select λ has been developed, which relies on a Bayesian approach. To understand this idea, we must first look at the penalty term in (7.4.1), which has a quadratic form in the log-likelihood. Taking the exponential of the penalty gives

$$\exp(-\frac{1}{2\sigma_\theta^2}\theta^T D\theta),$$

where $D = LL^T$ and $\sigma_\theta^2 = 1/\lambda$. This term mirrors the structure of a multivariate normal distribution. In fact, taking a Bayesian viewpoint, we can impose a prior on the parameter vector $\boldsymbol{\theta}$ in the form

$$\boldsymbol{\theta} \sim N(0, \sigma_\theta^2 \boldsymbol{D}^-), \tag{7.4.4}$$

where $\boldsymbol{D}^-$ is the generalised inverse of $\boldsymbol{LL}^T$, that is $(D^-)^- = \boldsymbol{LL}^T$. Because $\boldsymbol{LL}^T$ is not invertable, the prior (7.4.4) is degenerate and not proper. We have seen, however, that improper priors may still lead to proper posterior distributions. Note that the log posterior is now, up to an additive constant, equal to

$$\log f(\boldsymbol{\theta}, \sigma^2; \sigma_\theta^2|y) = const + l(\boldsymbol{\theta}, \sigma^2) - \frac{\tilde{K}}{2}\log(\sigma_\theta^2) - \frac{1}{2\sigma_\theta^2}\boldsymbol{\theta}^T \boldsymbol{D}^- \boldsymbol{\theta},$$

where $\tilde{K}$ is the rank of $\boldsymbol{D}^-$. We can see now that the penalty parameter λ is equal to the inverse of the prior variance, i.e. $\lambda = \sigma_\theta^{-2}$. This is, in fact, a (hyper)parameter in the Bayesian model, which allows the derivation of its posterior distribution. Instead of using the posterior, it is more common to simply set σ_θ^2 to the posterior mode estimate or, equivalently, estimate the hyperparameters using empirical Bayes. In fact, one can comprehend the whole model as a linear mixed model (see e.g. Fahrmeir et al. (2015)) and estimate σ_θ^2 with Maximum Likelihood estimation to derive a penalised fit for $\boldsymbol{\theta}$. The resulting fit is shown in Fig. 7.9 in the right-hand column. Clearly, the above results have only been sketched and need to be explained in much more depth to be fully understandable. We consider this to be beyond the scope of this chapter but refer to Ruppert et al. (2003) or Wood (2017) for a more comprehensive introduction to nonparametric regression.

7.5   Generalised Linear Models

Thus far, we have assumed that Y is metric with normal residuals. Let us now relax this assumption to include any exponential family distribution for Y given the covariates x, see Sect. 2.1.5. Moreover, we assume that the parameter θ in the exponential family distribution depends on some covariates x. To be specific, let

$$Y|x \sim \exp\{t(y)\theta(x) - \kappa(\theta(x))\}h(y),$$

where, for simplicity, we limit ourselves to a univariate model. Parameter θ is assumed to depend on covariates x in the following way: first, we define what is called the linear predictor η, which, in the univariate case, is given by $\eta = \beta_0 + x\beta_x$. This is the linear part of the model. Second, let

$$\mu = \frac{\partial \kappa(\theta)}{\partial \theta} = E\left(t(Y); \theta\right)$$

be the expectation of the statistic $t(Y)$ of the distribution. Note that in the regression framework, parameter θ depends on x, which is however suppressed in the notation. We now link μ with the linear predictor η through a link function $g(.)$, such that $\mu = g^{-1}(\eta)$. This results in the relationship

$$g(E(Y_i|x_i)) = \beta_0 + x_i\beta_x.$$

The link function $g(.)$ needs to be suitably chosen, which becomes more clear in concrete examples given below. In fact, given that μ and θ have a one-to-one relation, a mathematically convenient setting is to link η and μ by taking

$$\theta = \eta. \tag{7.5.1}$$

This is also called the **canonical link function** and makes for simpler derivation of the log-likelihood. Let $(y_i, x_i), i = 1, \ldots, n$, be the observed data points and assume a canonical link (7.5.1). The log-likelihood is then given by

$$l(\beta) = \sum_{i=1}^{n} \{t(y_i)\eta_i - \kappa(\eta_i)\} + \log(h(y)),$$

where $\eta_i = \beta_0 + x_i\beta_x$. Note that $\log(h(y))$ can be dropped, as it has no influence on the maximum of $l(\beta)$. Taking derivatives gives the score function

$$\frac{\partial l(\beta)}{\partial \beta} = \sum_{i=1}^{n} \binom{1}{x_i} \{t(y_i) - \mu_i\}$$

with $\mu_i = E(t(y_i)|x_i)$. We can reformulate this in matrix notation by setting

$$t(\mathbf{y}) = (t(y_y), \ldots, t(y_n))^T$$

and

$$t(\mathbf{Y}) = (t(Y_1), \ldots, t(Y_n))^T$$

and

$$X = \begin{pmatrix} 1 & x_1 \\ \vdots & \vdots \\ 1 & x_n \end{pmatrix}.$$

The score function can then be written as

$$X^T \{t(y) - E(t(Y); \eta)\},$$

where $\eta = (\eta_1, \ldots, \eta_n)$. The Maximum Likelihood estimate is given by the fix-point equation

$$X^T E(t(Y); \hat{\eta}) = X^T t(y).$$

Note that the Fisher matrix is given by

$$I(\beta) = X^T W X,$$

where W is a diagonal matrix with

$$\frac{\partial^2 \kappa(\eta_i)}{\partial \eta \partial \eta} = \text{Var}(t(Y_i), \eta_i), i = 1, \ldots, n$$

on the diagonal. The above introduction to generalised linear models is certainly not directly illuminative. It shows, however, the principles behind the construction of the model class. Looking at a few examples should help to gather some intuition.

Example 27 The linear model discussed in Sect. 7.1 is a special case of a generalised linear model. We take the results from Sect. 2.1.5 and write the normal distribution as exponential family model, where we consider the variance σ^2 as given. For simplicity of notation, we set $\sigma^2 = 1$. We can then write the normal distribution model as

$$f(y; \mu) = \exp\left\{y\mu - \frac{\mu^2}{2}\right\} h(y^2)h(y^2; \sigma^2).$$

We get $E(t(y)) = \mu = \partial \kappa(\theta)/\partial\theta$. Setting $\theta = \beta_0 + x\beta_x$ leads to the linear model as previously discussed.

▷

Example 28 We now move on to logistic regression. Assume that $Y_i \in \{0, 1\}$ and consider the model $P(Y_i = 1|x_i)$. Using the results from Sect. 2.1.5, we obtain as canonical link the logit function and get the following regression model:

$$\text{logit } P(Y_i = 1|x_i) = \log\left(\frac{P(Y_i = 1|x_i)}{1 - P(Y_i = 1|x_i)}\right) = \beta_0 + x_i \beta_x.$$

Note that $\text{Var}(Y_i|x_i) = P(Y_i = 1|x_i)(1 - P(Y_i = 1|x_i))$, which defines the weights in the weight matrix $\boldsymbol{W}$ from above.

$\triangleright$

Example 29 If Y_i are count data, it is suitable to apply a Poisson model. In this case, the log is the canonical link, such that we model

$$\log E(Y_i|x_i) = \beta_0 + x_i\beta_x.$$

Note that $\text{Var}(Y_i|x_i) = E(Y_i|x_i) = \exp(\beta_0 + x\beta_x)$, which defines the weights in the weight matrix $\boldsymbol{W}$.

$\triangleright$

Generalised linear models are a powerful tool and have seen multiple extensions since their introduction by Wedderburn and Nelder (1972). The canonical reference is Nelder and McCullagh (1989), see also Myers et al. (2010).

7.6   Case Study in Generalised Additive Models

As generalised linear models are a very central tool in statistical modelling, let us demonstrate their power and flexibility with a case study from the e-commerce sector. An internet retailer ran a marketing campaign for 30 days, with TV advertisements broadcast on various channels. The number of visits to their website was tracked for this period. The data are visualised in Fig. 7.11, where we can see striking peaks and clearly some diurnal variation, both of which need to be addressed in our statistical model. The daily pattern is also visualised in Fig. 7.12, where we plot the number of clicks (i.e. visits) as boxplots for each individual hour of the day. This plot shows that the data is highly skewed, with a large number of upper outliers. In fact, the boxes, which cover 50% of the data points, cover only a negligible portion of the data's range.

Before exploring and modelling the data, we zoom in and look arbitrarily at day 15, shown in Fig. 7.13. In this plot, we are better able to observe the nature of the peaks which will be explained shortly. First of all, however, we need to select a suitable probability model for the data. As we are recording counts, namely counts of clicks, the Poisson model appears suitable. In Sect. 2.1.4, we saw that the Poisson distribution has equal mean and variance. This feature is common for count data and thankfully our data also satisfy this assumption. To visualise this, we zoom once again into two separate hours, 5–6 am and 6–7 pm. These are indicated in grey in Fig. 7.13 and were chosen because they exhibit no extreme outliers. In Fig. 7.14, we plot the number of clicks in these two windows and report the (arithmetic) mean and the (empirical) variance of the data. Between 5 am and 6 am, we observe only low traffic with an average of 4.11 clicks per minute and a similar variance of 4.3. In the evening window, the traffic is ten times as large, with about 44.39 visits per

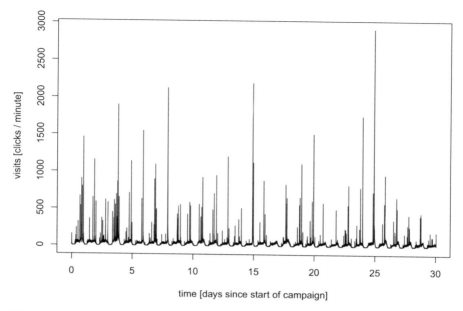

Fig. 7.11 Number of visits per minute over 30 days

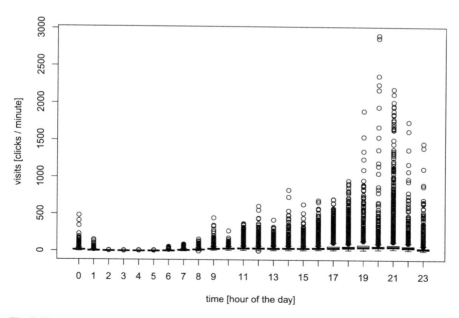

Fig. 7.12 Number of visits per minute plotted against hour of the day

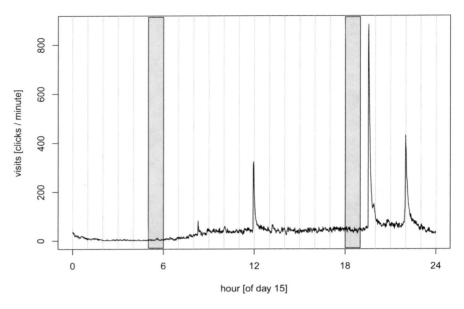

Fig. 7.13 Number of visits for day 15 with two hours being highlighted

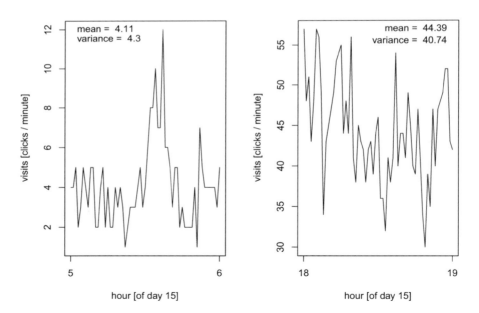

Fig. 7.14 Number of clicks for two hours of day 15 with mean and variance

minute and, again, a similar variance of 40.74. Thus, we can conclude that a Poisson distribution appears suitable for analysing the data. To start with, we define Y_i as the number of visits per minute and assume it is Poisson distributed

$$Y_i \sim Po(\lambda_i). \tag{7.6.1}$$

The intensity λ_i now needs to be modelled appropriately and, to that end, we define with t_i the timepoint at which Y_i was observed. t_i ranges from 0 to 30 in step size $1/(24*60)$, covering the time range of the campaign in minute intervals. With h_i we denote the hour of the day corresponding to t_i, where $h_i = (t_i \mod 60)$. We follow the framework of a generalised additive model (GAM) as discussed in Wood (2017), which combines smoothing models with generalised linear models, which we covered in Sects. 7.4 and 7.5, respectively. To be specific, we make use of the distribution in (7.6.1) and set

$$\lambda_i = \exp\{\beta_0 + t_i \beta_t + \beta_{\text{weekday}_i} + m(d_i)\}. \tag{7.6.2}$$

The exp guarantees that the intensity of the Poisson distribution is positive and β_t expresses the overall trend in the data, i.e. the long-term trend of the campaign. $m(d)$ carries the diurnal variation, and we postulate that it is cyclic, that is $\lim_{d \to 0}(m(d)) = \lim_{d \to 24}(m(d))$. The coefficient vector β_{weekday_i} captures differences between the different days of the week. The model can be easily fitted with the GAM procedure provided in the mgcv package in R. Again, the technically inclined reader can refer to Wood (2017) for more detail as the aim here is to demonstrate the flexibility of the model class without going into too much technical detail.

The fit of $m(D)$ is shown in Fig. 7.15, where the function is shown on a log scale and centred around zero, which can be interpreted as the average, and quantifies the deviation of the visits per minute throughout the day. We include horizontal lines, for instance, $\log(2)$ and $-\log(2)$, which delineate traffic at double or half the average rate, respectively. We can see that at around 3 am we observe only about 5% of the average traffic, while around 10 pm the traffic is 4 times greater than average.

The other parameter estimates are shown in Table 7.5. On Mondays, the reference category, we average about 31 ($= \exp(3.439)$) visits per minute. We see a slight variation over the weekdays and the negative time coefficient shows that there is a small downward trend. We do not want to interpret these effects at this stage and instead focus on the spikes shown in Figs. 7.11 and 7.13. The spikes occur, as expected, at the same time as TV advertisements. To model this effect, we make use of a second dataset that specifies the exact minute that each advertisement was broadcast. This is visualised for day 15 in Fig. 7.16. The plot shows the same data as in Fig. 7.13, but now the vertical lines indicate the exact time of each advertisement. In particular, for the timepoints after midday, we see that the peaks occur exactly at the same time as the advertisements. Consumers view the advertisement on TV and immediately visit the advertised website. This explains the spikes and the

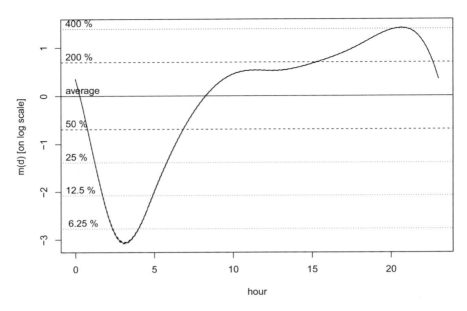

Fig. 7.15 Diurnal effect of the number of clicks

Table 7.5 Parameter estimates for quantile regression

	Value	Std. error	t value	Pr(>|t|)
(Intercept)	3.439e+00	2.650e−03	1297.68	<2e−16 ***
t	−1.057e−02	8.574e−05	−123.22	<2e−16 ***
wdTuesday	−7.028e−02	2.915e−03	−24.11	<2e−16 ***
wdWednesday	−1.489e−01	2.988e−03	−49.82	<2e−16 ***
wdThursday	6.398e−02	2.667e−03	23.99	<2e−16 ***
wdFriday	−1.154e−01	2.782e−03	−41.47	<2e−16 ***
wdSaturday	−1.012e−01	2.919e−03	−34.68	<2e−16 ***
wdSunday	3.852e−01	2.613e−03	147.45	<2e−16 ***

All coefficients are significantly different from zero, which is indicated with *** in the table

next modelling step is to include this phenomenon appropriately in the model. We propose a simple yet effective approach here. To motivate this we look at a single peak, namely the peak occurring at approximately half past 7 on the evening of the 15th day. The advertisement was sent out at 7:31 pm, leading to a sharp increase and subsequent exponential decay, which is clearly visible in Fig. 7.17. We model (or approximate) this behaviour by introducing a stimulus function.For the k-th advertisement broadcast, this is defined as

$$z_k(t) = \begin{cases} 0 & \text{for } t \le t_{(k)} \\ \exp(-(t - t_{(k)})\delta) & \text{for } t > t_{(k)}, \end{cases}$$

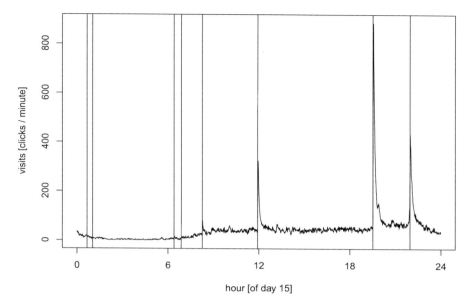

Fig. 7.16 Visits for day 15 with timepoints of advertisements indicated as vertical lines

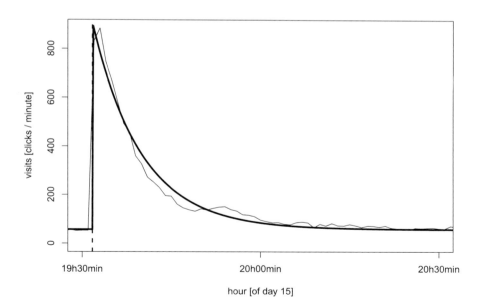

Fig. 7.17 Visualisation of function $z_k(t)$ (thick line) and the data (thin line) for one broadcasting timepoint

where $t_{(k)}$ is the timepoint of broadcast of the k-th advertisement. The decay parameter δ could in principle be estimated from the data, but, for simplicity, we use a calibrated value, which gives the function shown in Fig. 7.17. In actuality, $z_k(t)$ decays from a peak value of 1 but was scaled in the diagram to better demonstrate its fit. By defining for each of the $K = 254$ advertisements a corresponding stimulus function, we can extend Model (7.6.2) to

$$\lambda_i = \exp\{\beta_0 + t_i\beta_t + \beta_{\text{weekday}_i} + \sum_{k=1}^{K} z_k(t_i)\beta_k + m(d_i)\}. \qquad (7.6.3)$$

We fit Model (7.6.3) and obtain parameter estimates and a fitted curve $\hat{m}(d)$. The curve looks comparable to the one shown in Fig. 7.15 and is therefore not shown again. The weekday effects change slightly, but our focus is generally more on the fitted effects $\hat{\beta}_k$. Before drawing conclusions from the model, it is however necessary to look at goodness of the model fit. We take a rudimentary approach here and compare the observed values Y_i with the fitted values $\hat{\lambda}_i$. If the model is appropriate, then the difference between Y_i and $\hat{\lambda}_i$ should be random. In Fig. 7.18, we plot the original data in the upper plot (this is a repetition of Fig. 7.11) and the fitted values $\hat{\lambda}_i$ in the bottom row. Without going into depth here, we see sufficient concordance. Now that we are more confident in our model, we can look at the fitted advertisement effects $\hat{\beta}_k$. The coefficients β_k can be interpreted as the efficacy of the advertisement. Given that the maximum value of $z_k(t)$ is 1, we can interpret β_k on a log scale., i.e. the value of $\beta_k \geq \log(2) = 0.693$ means that the traffic was temporarily more than doubled. This also means that $\beta_k \leq 0$ indicates timepoints where the advertisement was not at all effective. We fit Model (7.6.3) and plot the corresponding fitted advertisement effects in Fig. 7.19. We also include the confidence intervals calculated by adding and subtracting twice the standard deviation of the estimates. We can see that most fitted values of $\hat{\beta}_k$ are positive and, in fact, the largest fitted values are in the order of 4.5, indicating a 90-fold temporary increase to the number of website visits ($90 \approx \exp(4.5)$). The next step in our analysis is to investigate and explore the coefficients $\hat{\beta}_k$, to determine the timepoints when advertisements are most effective. To do so, we first look again at day 15 and plot both the data and the corresponding advertisement effects $\hat{\beta}_k$ for this day. There were 8 advertisements broadcast, which can be seen in Fig. 7.20. In the top plot, we again show the data for day 15, and in the bottom plot we show the fitted advertisement effects $\hat{\beta}_k$. We see that advertisements shown in the early morning had hardly any effect, and in fact one fitted effect is even negative, while there is a clear benefit to advertising in the evening. This is a general pattern, which can be seen from Fig. 7.21, where we plot the fitted advertisement effect $\hat{\beta}_k$ against the hour of the day. It becomes obvious that evening advertisements, broadcast between 8 pm and 11 pm, are the most effective.

While there was plenty of room for improvement in each of the individual steps of the analysis, we hope to have demonstrated the flexibility of generalised regression models with nonparametric and parametric effects.

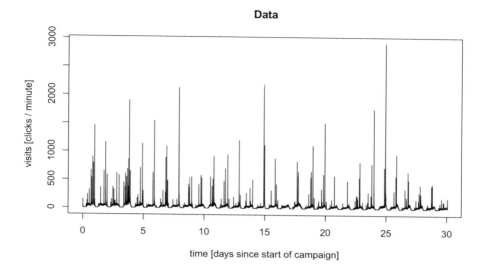

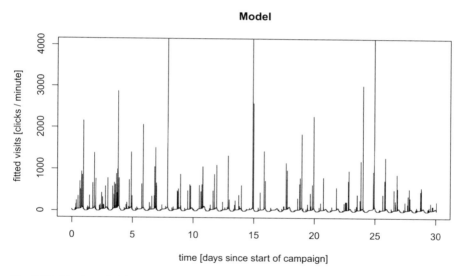

Fig. 7.18 Number of visits (top plot) and fitted model (bottom plot)

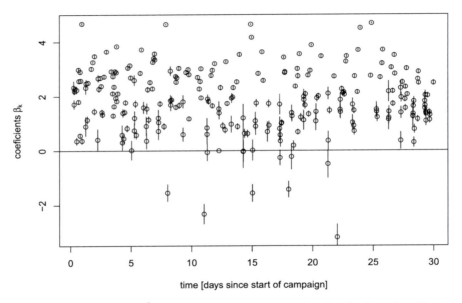

Fig. 7.19 Fitted coefficients $\hat{\beta}_k$ with confidence intervals plotted against the timepoint of broadcasting the advertisement

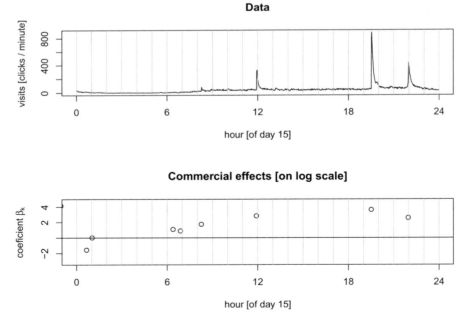

Fig. 7.20 Data for day 15 (top plot) and fitted advertisement effects $\hat{\beta}_k$ for that day

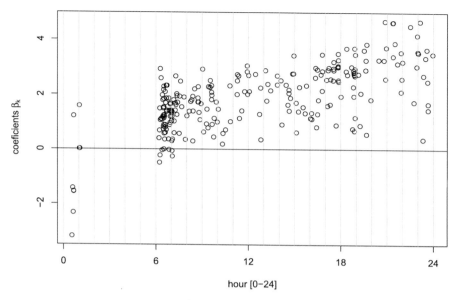

Fig. 7.21 Fitted advertisement effects $\hat{\beta}_k$ plotted against the hour of the day of the timepoint when the advertisement was broadcast

7.7 Exercises

Exercise 1 (Data Analysis and Linear Models in R)

The dataset teengamb in the **R** package faraway contains information on the gambling behaviour of teenagers in Great Britain:

Variable	Description
gamble	Money spent on gambling in pounds per year (response)
status	Socioeconomic status (based on parents' occupation)
income	Income (in pounds per week)
verbal	Test score based on the number of properly defined words (max. 12)
sex	Gender (0 = male and 1 = female)

1. Define and fit an ordinary linear regression model (including an intercept β_0) to explore the relationship between gambling and socioeconomic status and the income and verbal test score (verbal). Use the lm function in R.
2. Give the distribution of the estimated coefficient vector and compute the parameters if possible.
3. Perform an in-depth interpretation of the estimates of your coefficients.
4. Test if the model as a whole is useful, i.e. for the hypothesis $\beta_1 = \ldots = \beta_p = 0$. Interpret the result.

5. Make a linear model with all four covariates and also include the interactions between all variables and gender. Appropriately visualise the estimates of each effect. Finally, test whether we can reduce the set of covariates to `income` and `sex`.

Exercise 2 (Quantile Regression)
We revisit the dataset of the previous exercise. Instead of modelling the mean, we will now apply median regression.

1. Use the `quantreg` package in R to perform a quantile regression for the median, i.e. $\tau = 0.5$, to explore the relationship between gambling and socioeconomic status and the income and verbal test score. Try to interpret the results and compare them to the regression for the mean performed in Exercise 7.1 point 1.
2. What are possible advantages of quantile regression in comparison to ordinary linear regression?

Exercise 3 (Logistic Regression and Classification)
In credit scoring, banks want to check whether a client will pay back a credit in the specified time frame. We consider a binary outcome variable $Y_i \in \{0, 1\}$, which indicates whether client i pays back the credit ($Y_i = 0$) or not ($Y_i = 1$). In our example, further variables about the credit and the client are available: x_1 = duration of the credit, x_2 = amount of the credit, x_3 = previous payment behaviour (1 = good and 0 = bad), x_4 = intended use (1 = private and 0 = business) and x_5 = running account (0 = good running account and 1 = no or bad account running). The data can be downloaded from www.uni-goettingen.de/de/551625.html, see also the discussion in Fahrmeir et al. (2015).

1. Fit a logistic regression model $logit\, P(Y_i = 1 | x) = \beta_0 + \beta_1 x_1 + \beta_2 x_2 + \beta_3 x_3 + \beta_4 x_4 + \beta_5 x_5$ resulting in estimations $\hat{\beta}_0, \ldots, \hat{\beta}_5$.
2. Using the fitted model, the probability of a failure of a credit can be estimated by

$$P(Y_i = 1 | x) = G(\hat{\beta}_0 + \hat{\beta}_1 x_1 + \ldots + \hat{\beta}_5 x_5)$$

with $G(t) = logit^{-1}(t) = (1 + \exp(-t))^{-1}$. Use the result from 1 and give probabilities for different values of x: $x_1 = (0, 1)$, $x_2 = (0, 1)$, $x_3 = (0, 1)$, $x_4 = (12, 24, 36)$ and $x_5 = (4, 6, 8)$. The calculation can be used to decide whether the bank offers the credit or not by using a threshold for the failure probability.
3. Explain why logistic regression can be seen as a tool for binary classification based on information in the variables x.

Chapter 8
Bootstrapping

In the early days of statistics, calculations had to be done by hand or by unwieldy mainframe computers with very limited memory and computational power. Consequently, statisticians were permanently looking for ways to simplify calculations through approximation. Time consuming calculations, e.g. computation of quantiles and distribution functions, were performed once at high precision and the results were published in tables. Even today, tables for distributions, such as the standard normal or the t-distribution, are available in statistics textbooks. The generation of random numbers was also not a trivial task and thick books containing sequences of random numbers were published, e.g. for the purpose of survey sampling.

The rise in readily available computing power led to a corresponding increase in the range of techniques available to statisticians. In particular, in complex data analytic situations, where the derivation of asymptotic formulae is too cumbersome or where their validity or robustness may be questionable, these new resources could be exploited with a range of new statistical methods built upon sampling and resampling. Early papers by Quenouille (1956) and Tukey (1958) introduced the jackknife estimator, a predecessor to the nowadays more popular bootstrap and subsampling procedures. The bootstrap as we know it today was introduced in the seminal paper written by Efron (1979). The general idea is to use the actual sample to calculate the properties of statistic of interest. The idea is simple, but very powerful and extends readily to complex data analyses and will be explored throughout this chapter.

8.1 Nonparametric Bootstrap

8.1.1 Motivation

To introduce the basic idea of the nonparametric bootstrap, we begin with an independent and identically distributed sample of size n from a distribution function $F(.)$, i.e. we assume that $Y_1, \ldots, Y_n$ such that

$$Y_i \sim F(y) \quad i.i.d.. \tag{8.1.1}$$

Usually, $F(.)$ is unknown, at least up to some parameter θ. We will keep things as general as possible here and postulate a generic function $F(.)$ that is neither necessarily indexed by a parameter θ, nor defined by a particular distributional model, such as normality. Therefore, we address this approach as nonparametric bootstrap, i.e. we do not use any assumptions about $F(.)$. It proceeds as follows. First, we have our observed sample $y_i, i = 1, \ldots, n$, which we denote as $y = (y_1, \ldots, y_n)$. The bootstrap algorithm is given by

1. Calculate a statistic of interest, $t(y)$, e.g. the median.
2. Take n draws *with replacement* from the original data $y = (y_1, \ldots, y_n)$ to create a new dataset of the same size and call it $y^* = (y_1^*, \ldots, y_n^*)$. The sample y^* is called a bootstrap sample. Calculate the statistic of interest, $t(y^*)$. Store the result.
3. Repeat the previous step B times with indices $b = 1 \ldots, B$, such that y^{*b} is the b-th bootstrap sample from which $t(y^{*b})$ is calculated and stored.
4. Use the B bootstrap samples $t(y^{*1}), \ldots t(y^{*B})$ to obtain information about the statistical variation of $t(y)$.

The bootstrap process is sketched in Fig. 8.1. The collection of bootstrapped statistics $t(y^{*b})$, $b = 1, \ldots, B$ can be used to estimate properties of the statistic

Fig. 8.1 Illustration of the bootstrap procedure

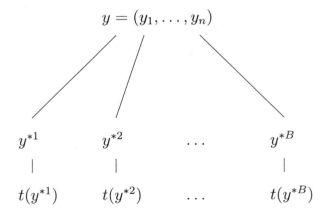

of interest, e.g. its variance. For instance, an estimate of the unknown variance $\mathrm{Var}_F(t(Y))$ is given by its bootstrap estimate

$$\mathrm{Var}_{\mathrm{Boot}}(t(Y)) \approx \frac{1}{B-1} \sum_{b=1}^{B} \left(t(y^{*b}) - \bar{t}_{\mathrm{Boot}} \right)^2 , \qquad (8.1.2)$$

where

$$\bar{t}_{\mathrm{Boot}} = \frac{1}{B} \sum_{b=1}^{B} t(y^{*b}) .$$

Let us demonstrate the bootstrap idea with a simple example, shown in Fig. 8.2. In the top left are 50 simulated observations drawn from a standard normal

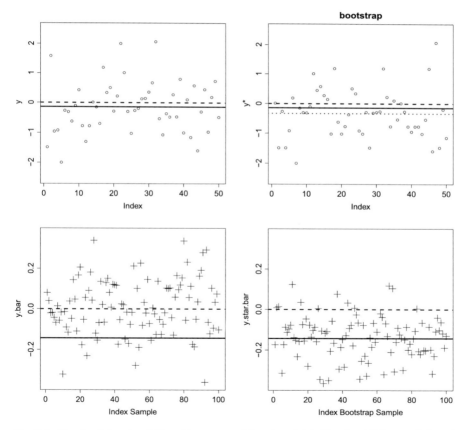

Fig. 8.2 Simulated data (top left) and bootstrapped data (top right). Simulated $\bar{y}$ (bottom left) and bootstrapped $\bar{y}^*$ (bottom right)

distribution, with the arithmetic mean of the 50 data-points represented by the solid line and the true mean by the dashed line. From the 50 observations y_i, $i = 1, \ldots, 50$, we draw with replacement a single bootstrap sample of 50 observations y_i^*, $i = 1, \ldots, 50$ shown in the top right. The arithmetic mean of this bootstrap sample $\bar{y}^*$ is represented by the dotted line. We repeat this step $B = 100$ times and calculate the arithmetic mean for each bootstrap sample, which we denote with $\bar{y}^{*b}$ for $b = 1, \ldots, B$. These values are shown in the bottom right plot, with the solid line representing the arithmetic mean of the bootstrapped means and the dashed line representing the true mean. Note that the mean of the original sample is slightly less than the true mean, and thus, the fact that the mean of the bootstrapped means is also less than the true mean is to be expected. The bootstrap aims to mimic the true variation, which we show in the bottom left plot. Here we simulated 100 times $Y_1, \ldots, Y_{50}$ from a $N(0, 1)$ distribution and calculated the arithmetic mean. Clearly, this is only possible if we know the true distribution, while the bootstrap makes use of only the available data and can therefore be applied without knowledge of the true distribution function. The distribution of the two sets of points appears similar, just with the mean shifted for the bootstrap samples.

It is important to note that we make two approximations above. Firstly, the bootstrap estimate $\mathrm{Var}_{\mathrm{Boot}}(t(Y))$ is only an approximation (we could also call it an estimate) of the true variance $\mathrm{Var}_F(t(Y))$ of the statistic of interest which clearly depends on the true but unknown distribution function $F(.)$. Secondly, we only use a limited number of bootstrap samples, i.e. B is finite, which itself only gives an approximation of the true or *ideal* bootstrap distribution. This is the distribution of the statistic of interest using all possible bootstrap samples.

8.1.2 Empirical Distribution Function and the Plug-In Principle

In this section, let us motivate the nonparametric bootstrap a little more formally, which will hopefully also help us to understand why bootstrapping works in general. Let $Y = (Y_1, \ldots, Y_n)$ with $Y_i \sim F(.)$ *i.i.d.* from an unknown distribution function $F(.)$ and let $y = (y_1, \ldots, y_n)$ be the observed data. For simplicity, we also assume that all values of y are unique, i.e. there are no two indices i and j for which $y_i = y_j$. The empirical distribution function $\hat{F}_n(y)$ is given by

$$\hat{F}_n(y) = \frac{1}{n} \sum_{i=1}^{n} 1(y_i \leq y).$$

Let us start our explanation with the **plug-in principle**. To do this, we need to explain the concept that any statistic $t(Y)$ can be written as functional. This means that the behaviour of $t(Y)$ depends on the (unknown) distribution function $F()$. For example, assume we are interested in the variance of $t(Y)$. This variance depends on the function $F()$. We denote this functional dependence in mathematical terms as a statistical functional which we motivate later. The concept of a functional is now combined with the plug-in principle as follows: Whenever the true distribution function F is involved in a statistical functional, it should be replaced by its empirical analogue $\hat{F}_n$. For the sake of simplicity, we avoid mathematical details and instead motivate the idea with an example.

Example 30 For example, let us consider the expectation of the random variable Y. The expectation of Y depends on the unknown distribution function. Altogether, this can be expressed as

$$T(F) = \mu = \int y \, dF(y).$$

The notation $\int y \, dF(y)$ refers, as previously mentioned, to an integral if $F(.)$ is both continuous and differentiable and a sum if $F(.)$ is discrete valued. Note that if $F(.)$ is continuous and differentiable we can rewrite the functional as

$$T(F) = \int y \, dF(y) = \int y \frac{dF(y)}{dy} dy = \int y f(y) dy$$

with $f(.)$ as the density function. If, in contrast, Y is discrete valued with $Y \in \{a_1 < a_2 < a_3, \ldots\}$, then $F(.)$ is a step function and the functional is given by

$$T(F) = \int y \, dF(y) = \sum_k a_k P(Y = a_k) = \sum_k a_k \{F(a_k) - F(a_{k-1})\},$$

where $F(a_0) = 0$.

Given that we do not know F, the plug-in principle states that we can replace F with its empirical approximation $\hat{F}_n$ to obtain $T(\hat{F}_n)$ as the statistic that estimates $T(F)$. For example, for the mean this gives

$$T(\hat{F}_n) = \int y \, d\hat{F}_n(y) = \sum_{i=1}^n y_{(i)} \{\hat{F}_n(y_{(i)}) - \hat{F}_n(y_{(i-1)})\} = \frac{1}{n} \sum_{i=1}^n y_i = \bar{y},$$

where $y_{(1)}, \ldots, y_{(n)}$ is the ordered sample. The plug-in principle is rather flexible and can be broadly applied to many other statistical quantities. It also allows us to calculate further properties of our estimates. If, for example, we want to calculate the variance of $\bar{Y} = \frac{1}{n} \sum_{i=1}^n Y_i$, then

$$T(F) = \text{Var}(\bar{Y}) = \int (\bar{y} - \mu)^2 dF(y) = \frac{1}{n} \sigma^2,$$

where $\sigma^2 = Var(Y_i), i = 1, \ldots, n$. Taking the sample $y_1, \ldots, y_n$ we can again apply the plug-in principle which gives

$$T(\hat{F}_n) = \int (\bar{y}^* - \bar{y})^2 d\hat{F}_n(y^*),$$

where $\bar{y}^* = (\bar{y}_1^*, \ldots, \bar{y}_n^*)$ is a sample drawn from $\hat{F}_n(.)$. A little bit of calculation shows that

$$T(\hat{F}_n) = \frac{1}{n}(\frac{1}{n}\sum_{i=1}^{n}(y_i - \bar{y})^2),$$

where the term in the inner brackets is the empirical variance of the sample.   ▷

The plug-in principle is used when no further information about F is available, excluding the information contained in the sample y. As the explicit calculation of $T(\hat{F}_n)$ can be clumsy, the bootstrap approach is to draw a sample instead. If there are no repeated values in the original sample, it can be shown that the number of different bootstrap samples is $\binom{2n-1}{n}$. This number becomes quite large, even for small sample sizes. For example, $n = 15$ gives 77,558,760 possible unique samples. This explains why it is more practical, instead of using all samples, to randomly select a small number B, e.g. $B = 200$. Note that the samples are drawn from $\hat{F}_n()$ and need to be *i.i.d.*. This is achieved if we draw *with* replacement from our observed data $y_1, \ldots, y_n$. In other words, drawing n samples with replacement from $y_1, \ldots, y_n$ is equivalent to

$$y_i^* \sim \hat{F}_n(), \ i.i.d. \text{ for } i = 1, \ldots, n.$$

Example 31 Consider a toy example with the following ($n = 3$) sample: $y = (20, 25, 40)$. There are 10 possible bootstrap samples y^* when sampling with replacement:

$$(20, 20, 20); (25, 25, 25); (40, 40, 40); (20, 20, 25); (20, 20, 40);$$

$$(25, 25, 40); (20, 25, 25); (20, 40, 40); (25, 40, 40); (20, 25, 40).$$

It is important to note that the probabilities of the samples are not equal. While (20,20,20) occurs only once, (20,20,25) results from the three possible samples $(20, 20, 25)$, $(20, 25, 20)$, and $(25, 20, 20)$ and is therefore more likely, but is listed only once above. For very small n, the multinomial distribution can be used to calculate bootstrap estimates based on the ideal bootstrap distribution. Using the fact that we sample with replacement, we get

$$P\left(y^* = (20, 20, 20)\right) = \frac{3!}{3!0!0!}\left(\frac{1}{3}\right)^3\left(\frac{1}{3}\right)^0\left(\frac{1}{3}\right)^0 = \frac{1}{27}$$

$$P\left(y^* = (20, 20, 25)\right) = \frac{3!}{2!1!0!}\left(\frac{1}{3}\right)^2\left(\frac{1}{3}\right)^1\left(\frac{1}{3}\right)^0 = \frac{3}{27}$$

$$P\left(y^* = (20, 25, 40)\right) = \frac{3!}{1!1!1!}\left(\frac{1}{3}\right)^1\left(\frac{1}{3}\right)^1\left(\frac{1}{3}\right)^1 = \frac{6}{27}$$

with, for example, $P\left(y^* = (20, 20, 25)\right) = P\left(y^* = (25, 25, 40)\right)$, etc. Let us focus on the median as the statistic of interest. The median of the original sample is 25. To get a bootstrap estimate of the variance of the median, we need to calculate the median of each of the above bootstrap samples, compute the arithmetic mean of all medians, and finally compute the variance using (8.1.2). Instead of enumerating all samples, we can use the multinomial probabilities to simplify the computation. In the end, this gives

$$\bar{t}_{\text{Boot}} = \frac{1}{27}(20 + 25 + 40) + \frac{3}{27}(20 + 20 + 25 + 25 + 40 + 40) + \frac{6}{27} \cdot 25 = 27.59259$$

$$\text{Var}_{\text{Boot}}(t(y)) = \frac{1}{27}\left((20 - \bar{t}_{\text{Boot}})^2 + \ldots + 6 \cdot (25 - \bar{t}_{\text{Boot}})^2\right) = 58.09328 \ .$$

▷

To further motivate the plug-in principle, we consider two "worlds": the "real world" and the "bootstrap world". In the real world, the sample y is drawn from the unknown population distribution $F(.)$ and we calculate the statistic $t(y)$. In the bootstrap world, the bootstrap samples y^{*b} are drawn from the known empirical distribution $\hat{F}_n(.)$ and the statistic $t(y^{*b})$ is calculated B times. Therefore, we are trying to gain information about $F(.)$ and the properties of the statistic $t(.)$ through repeated sampling from the empirical distribution of the data $\hat{F}_n(.)$, which then gives an empirical distribution of our statistic. At first glance, it may appear unreasonable that re-use of the original sample y allows us to gain further information, which we were not able to extract from y alone. However, theoretical and empirical results show that the bootstrap works well with various kinds of data. This is a direct result of the plug-in principle. The bootstrap procedure as introduced above is also called the *nonparametric bootstrap*, as no parametric distributional assumptions are made. Later, we will also introduce the *parametric bootstrap* in Sect. 8.2, which streamlines the process when parametric assumptions are appropriate.

8.1.3 Bootstrap Estimate of a Standard Error

We have already introduced the procedure to estimate a variance using bootstrap samples and the estimation of a standard error follows the same principle. We now focus on a parameter of interest, ξ, which we estimate with $\hat{\xi} = t(y)$. The task is then to find a bootstrap approximation of the standard error of $\hat{\xi}$, which we calculate using B estimates $\hat{\xi}^{*b} = t(y^{*b})$ derived from the bootstrap samples. The bootstrap now works as follows

1. Generate B bootstrap samples $y^{*1}, \ldots, y^{*B}$.
2. Calculate $\hat{\xi}^{*b}$, $b = 1, \ldots, B$.
3. Estimate the standard error $\mathrm{se}_F(\hat{\xi}) = \sqrt{\mathrm{Var}_F(\hat{\xi})}$ with

$$\widehat{\mathrm{se}}_B = \left\{ \frac{1}{B-1} \sum_{b=1}^{B} \left[\hat{\xi}^{*b} - \bar{\hat{\xi}}^* \right]^2 \right\}^{\frac{1}{2}} \quad \text{with} \quad \bar{\hat{\xi}}^* = \frac{1}{B} \sum_{b=1}^{B} \hat{\xi}^{*b}.$$

Consequently, the bootstrap estimate for the standard error of an estimator $\hat{\xi}$ (with data from F) is the standard deviation of the bootstrap estimates $\hat{\xi}^{*b}$. These estimates are based on random samples y^{*b}, drawn with replacement from $\hat{F}_n$. It can be shown that

$$\lim_{B \to \infty} \widehat{\mathrm{se}}_B = \mathrm{se}_{\hat{F}_n}(\hat{\xi}^*), \tag{8.1.3}$$

where $\mathrm{se}_{\hat{F}_n}(\hat{\xi}^*)$ is the estimate given all possible samples from the empirical distribution. The approximation $\widehat{\mathrm{se}}_B$ is often called the *nonparametric bootstrap estimate* of the standard error. The number of bootstrap samples B is, in general, governed by practical considerations. On the one hand, if the computation of the estimate $\hat{\xi}$ is complex and time consuming, this lends itself to a lower number of samples. On the other hand, the bootstrap variance estimate itself has its own variance, which is higher when B is small (due to the random process of drawing bootstrap samples). Therefore, B should be high enough to get a stable estimate of the variance. Usually, $B = 200$ is used in practice for variance estimates, but higher values of B may be useful, especially when one wants to estimate the bias of an estimator or construct confidence intervals.

Example 32 We consider a slightly more complex situation in which we have data that is possibly drawn from two different distributions and we are interested in the difference between the two distributions, e.g. in the difference of their mean values. To begin, let

$$\left. \begin{array}{ll} Y_1, \ldots, Y_n \sim F & i.i.d. \\ Z_1, \ldots, Z_m \sim G & i.i.d. \end{array} \right\} \text{ independent.}$$

Our target parameter is the standard error of the difference $\xi = E(Y_i) - E(Z_i) = \mu_Y - \mu_Z$, which we estimate with $\hat{\xi} = \bar{y} - \bar{z}$. For the b-th bootstrap sample

$$y^{*b} = (y_1^{*b}, \ldots, y_n^{*b}) \text{ with replacement from } \hat{F}_n$$

$$z^{*b} = (z_1^{*b}, \ldots, z_m^{*b}) \text{ with replacement from } \hat{G}_m. \qquad (8.1.4)$$

The bootstrap estimate is then:

$$\widehat{se}_B = \left\{ \frac{1}{B-1} \sum_{b=1}^{B} \left[\hat{\xi}^{*b} - \bar{\hat{\xi}}^* \right]^2 \right\}^{\frac{1}{2}}$$

with

$$\hat{\xi}^{*b} = \bar{y}^{*b} - \bar{z}^{*b} = \frac{1}{n} \sum_{i=1}^{n} y_i^{*b} - \frac{1}{m} \sum_{i=1}^{m} z_i^{*b}$$

and

$$\bar{\hat{\xi}}^* = \frac{1}{B} \sum_{b=1}^{B} (\bar{y}^{*b} - \bar{z}^{*b}) = \frac{1}{B} \sum_{b=1}^{B} \hat{\xi}^{*b} .$$

▷

The variance of the mean difference can also be estimated by standard methods using the sampling variances. However, the variance of the ratio or other nonlinear functions of the two means cannot be easily calculated by standard methods, while the bootstrap estimate can be directly transformed to this case.

8.1.4 Bootstrap Estimate of a Bias

As demonstrated above, bootstrapping allows us to estimate the variance of a statistic $t(y)$ or more specifically an estimate $\hat{\xi}$. As we learned in Chap. 3, the Mean Squared Error (MSE) of an estimator is given by the variance plus the squared bias. This raises the question of whether it is possible to determine bias of an estimate with bootstrapping. The answer is yes, which we will now demonstrate. Let our samples again be $Y_1, \ldots, Y_n \sim F$ i.i.d. from an unknown distribution function F and ξ be some parameter of interest. We can express the parameter with the functional $\xi = T(F)$. The plug-in principle suggests that the corresponding estimate is $\hat{\xi} = T(\hat{F}_n)$. The bias of the estimator $\hat{\xi}$ is given by

$$\text{bias}_F(\hat{\xi}, \xi) = E_F(\hat{\xi}) - \xi = E_F(\hat{\xi}) - T(F) . \qquad (8.1.5)$$

It is clear that the bias cannot be estimated from a single sample y directly, as it depends upon the unknown parameter ξ. Instead, we want to calculate its bootstrap estimate by following the plug-in principle and substituting unknown components in (8.1.5) with their empirical counterparts. To do so, we substitute F with $\hat{F}_n$, $\hat{\xi}$ with $\hat{\xi}^*$ and finally ξ with $\hat{\xi} = T(\hat{F}_n)$. This gives the bootstrap estimate of the bias

$$\widehat{\text{bias}}_F(\hat{\xi}, \xi) = \text{bias}_{\hat{F}_n}(\hat{\xi}^*, \hat{\xi}) = \text{E}_{\hat{F}_n}[\hat{\xi}^*] - T(\hat{F}_n) \, .$$

In practice, the ideal bootstrap bias estimate is approximated by simulation. Consequently, let $y^{*1}, \ldots, y^{*B}$ be independent bootstrap samples, which give bootstrap estimates $\hat{\xi}^{*(b)}$, $b = 1, \ldots, B$. Taking

$$\bar{\hat{\xi}}^* = \frac{1}{B} \sum_{b=1}^{B} \hat{\xi}^{*(b)}$$

the bias can be estimated with

$$\widehat{\text{bias}}_B = \bar{\hat{\xi}}^* - \underbrace{T(\hat{F}_n)}_{\hat{\xi}} \, .$$

In principle, we now correct for the bias to construct the bias corrected estimate

$$\hat{\hat{\xi}} = \hat{\xi} - \widehat{\text{bias}}_B = \hat{\xi} - [\bar{\hat{\xi}}^* - \hat{\xi}] = 2\hat{\xi} - \bar{\hat{\xi}}^* \, .$$

It should be noted that bias correction in this form is not always recommendable, as correcting the bias typically leads to the bias corrected estimate having a higher variance. Nonetheless, bootstrapping does allow the quantification and estimation of the bias, which can often be helpful.

8.2 Parametric Bootstrap

The bootstrap approach discussed so far was built upon sampling from the empirical distribution function. Even though this approach is quite broadly applicable, it is not always appropriate, in particular, if data are not $i.i.d.$. We therefore propose the parametric bootstrap as an alternative approach to resampling. First of all, we assume that the unknown distribution function $F(.)$ depends on some parameter θ, which we denote with $F(y) = F(y; \theta)$. It is assumed that θ uniquely determines the distribution function and that it is estimated by some statistical method, e.g. Maximum Likelihood. Note that θ, the parameter of the distribution model and the parameter of interest ξ can be different. For instance, θ can be the parameter λ in a Poisson model, while the parameter of interest ξ is a quantile of the distribution.

As a special case we can take $\theta = \xi$. The parametric bootstrap procedure then uses samples taken from the distribution $F(.; \hat{\theta})$, where $\hat{\theta}$ is the estimate of the true parameter θ. An important consequence of this approach is that our bootstrap samples can be different from our original data, which is even guaranteed in the case of a continuous distribution. In this case, the parametric bootstrap samples can take any values. However, this comes at the cost of making stricter parametric assumptions about the distribution of Y. To demonstrate the parametric bootstrap, let us look at the subsequent two examples.

Example 33 We will first demonstrate the parametric bootstrap of the median of a Poisson distribution. Let $Y_1, \ldots, Y_n$ be an *i.i.d.* sample from a Poisson distribution, $Po(\lambda)$, where $\lambda = E(Y_i)$, $i = 1, \ldots, n$. We take the median as the parameter of interest ξ, for which we want to compute a bootstrap standard error estimate. The parametric bootstrap now proceeds as follows.

1. Compute the maximum likelihood estimate of λ which is $\hat{\lambda} = \frac{1}{n} \sum_{i=1}^{n} y_i$.
2. For $b = 1, \ldots, B$:

   (a) Generate bootstrap samples $y^{*b} = (y_1^{*b}, \ldots, y_n^{*b})$ by generating random numbers

   $$y_i^{*b} \sim Po(\hat{\lambda}) .$$

   (b) Compute the median $\hat{\xi}^{*b}$ of sample y^{*b}.

3. Compute the standard error estimate

$$\hat{se}_B = \left\{ \frac{1}{B-1} \sum_{b=1}^{B} \left[\hat{\xi}^{*b} - \bar{\hat{\xi}}^{*} \right]^2 \right\}^{\frac{1}{2}} \quad \text{with} \quad \bar{\hat{\xi}}^{*} = \frac{1}{B} \sum_{b=1}^{B} \hat{\xi}^{*b}$$

   exactly as for the nonparametric bootstrap.

   ▷

Example 34 (Bivariate Normal Distribution) With the next example we want to calculate the correlation coefficient for a bivariate normal distribution. Let $((Y_1, Z_1)', \ldots, (Y_n, Z_n)')$ with

$$\begin{pmatrix} Y_i \\ Z_i \end{pmatrix} \sim F_{Y,Z}(.) \quad i.i.d..$$

Assume, that $F_{Y,Z}(.)$ is a bivariate normal distribution with parameter vector $\theta = (\mu, \Sigma)$, where

$$\mu = \begin{pmatrix} E(Y_i) \\ E(Z_i) \end{pmatrix} \qquad \Sigma = \begin{pmatrix} Var(Y_i) & Cov(Y_i, Z_i) \\ Cov(Y_i, Z_i) & Var(Z_i) \end{pmatrix} .$$

Assume that the parameter of interest is the Pearson correlation coefficient ϱ between Y and Z and that we want to compute an estimate for the standard error of its empirical counterpart.

The Maximum Likelihood estimator $\hat{\theta}$ is

$$\hat{\mu} = \begin{pmatrix} \bar{y} \\ \bar{z} \end{pmatrix},$$

$$\hat{\Sigma} = \frac{1}{n} \begin{pmatrix} \sum_{i=1}^{n}(y_i - \bar{y})^2 & \sum_{i=1}^{n}(y_i - \bar{y})(z_i - \bar{z}) \\ \sum_{i=1}^{n}(y_i - \bar{y})(z_i - \bar{z}) & \sum_{i=1}^{n}(z_i - \bar{z})^2 \end{pmatrix}.$$

The distribution function $F(; \hat{\mu}, \hat{\Sigma}) = N(\hat{\mu}, \hat{\Sigma})$ is therefore used to generate the bivariate bootstrap samples. For each bootstrap sample the Pearson correlation coefficient is computed, i.e.

$$\hat{\varrho}^{*b} = \frac{\sum_{i=1}^{n}(y_i^{*b} - \bar{y}^{*b})(z_i^{*b} - \bar{z}^{*b})}{\sqrt{\sum_{i=1}^{n}(y_i^{*b} - \bar{y}^{*b})^2 \sum_{j=1}^{n}(z_j^{*b} - \bar{z}^{*b})}}$$

and the bootstrap estimated standard error is the usual standard deviation of the values $\hat{\varrho}^{*1}, \ldots, \hat{\varrho}^{*B}$. $\triangleright$

8.3   Bootstrap in Regression Models

Let us now look at the linear model discussed in Sect. 7.1 and assume that the data are given as pairs of a scalar response variable and a vector of covariates: (y_i, x_i), $i = 1, \ldots, n$ where x_i denotes a p-dimensional row vector whose first entry is 1 to simplify the calculation of the intercept. We assume the linear model

$$y_i = x_i \beta + \varepsilon_i,$$

where $\varepsilon_i \sim F(.)$ i.i.d. and $\mathrm{E}(\varepsilon_i) = 0$ and $Var(\varepsilon_i) = \sigma^2$ for $i = 1, \ldots, n$. Suppose that we want to determine the variance of the Ordinary Least Squares (OLS) estimator $\hat{\beta} = (X^{\mathrm{T}}X)^{-1}X^{\mathrm{T}}y$. We propose four approaches to bootstrapping in this scenario, each of which has its advantages and disadvantages.

Approach 1: Residual-Based Bootstrap
The first approach uses a nonparametric bootstrap of the residuals. The true distributional model depends on β, the vector of regression coefficients, and the distribution function $F(.)$ of the residuals. We notate this with $F(.|\beta)$, which we call the "real world", in contrast to the "bootstrap world" that was sketched in Fig. 8.3. Bootstrapping now proceeds as follows.

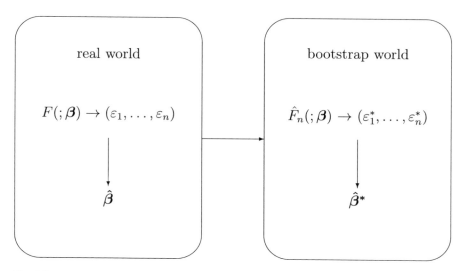

Fig. 8.3 "Real world" and "bootstrap world" for the residual bootstrap of a linear model

First we estimate $\hat{\boldsymbol{\beta}}$ using ordinary least squares (OLS) $\hat{\boldsymbol{\beta}} = (X^T X)^{-1} X^T y$ where, as in Sect. 7.1, matrix X has rows $x_i, i = 1, \ldots, n$ and $y = (y_1, \ldots, y_n)^T$. This gives the fitted residuals $\hat{\boldsymbol{\varepsilon}} = (I - X(X^T X)^{-1} X^T) y = y - X\hat{\boldsymbol{\beta}}$, giving $\hat{\varepsilon}_1, \ldots, \hat{\varepsilon}_n$ and its empirical distribution function $\hat{F}_n(.)$. This empirical distribution function $\hat{F}_n$ is now used for bootstrapping by sampling with an equal probability for each residual $\hat{\varepsilon}_i, i = 1, \ldots, n$. A single bootstrap now takes place as follows:

1. Draw a sample $\boldsymbol{\varepsilon}^* = (\varepsilon_1^*, \ldots, \varepsilon_n^*)$ from $\hat{F}_n(.)$ with replacement.
2. Set new bootstrap response values as $y_i^* = x_i \hat{\boldsymbol{\beta}} + \varepsilon_i^*$ for $i = 1, \ldots, n$, which can be written in matrix form as $y^* = X\hat{\boldsymbol{\beta}} + \boldsymbol{\varepsilon}^*$.
3. Calculate the bootstrap OLS estimate $\hat{\boldsymbol{\beta}}^* = (X^T X)^{-1} X^T y^*$.

An important result for the residual bootstrap is that a simulation is in principle unnecessary, as the bootstrap variance estimate is equal to the standard OLS estimate of the variance of the error terms. This follows because

$$\text{Var}_{\hat{F}_n}(\hat{\boldsymbol{\beta}}^*) = (X^T X)^{-1} X^T \text{Var}_{\hat{F}_n}(y^*) X (X^T X)^{-1} = \hat{\sigma}_F^2 (X^T X)^{-1},$$

where $\text{Var}_{\hat{F}_n}(y^*) = \text{Var}_{\hat{F}_n}(\boldsymbol{\varepsilon}^*) = \hat{\sigma}_F^2 I$ with $\hat{\sigma}_F^2 = \hat{\boldsymbol{\varepsilon}}^T \hat{\boldsymbol{\varepsilon}}/n$.

Approach 2: Model-Based Bootstrap

In the second approach, we use a parametric bootstrap, modelling our residuals with a normal distribution. Assuming $\varepsilon_i \sim N(0, \sigma^2)$ *i.i.d*, the strategy for a single draw is as follows:

1. Draw an *i.i.d.* sample $\boldsymbol{\varepsilon}^* = (\varepsilon_1^*, \ldots, \varepsilon_n^*)$ from the normal distribution $N(0, \hat{\sigma}^2)$, where the variance $\hat{\sigma}^2$ is estimated with Maximum Likelihood.
2. Set new bootstrap response values as $y_i^* = \boldsymbol{x}_i \hat{\boldsymbol{\beta}} + \varepsilon_i^*$ for $i = 1, \ldots, n$, which can be written in matrix form as $\boldsymbol{y}^* = \boldsymbol{X}\hat{\boldsymbol{\beta}} + \boldsymbol{\varepsilon}^*$.
3. Calculate the OLS estimate with $\hat{\boldsymbol{\beta}}^* = (\boldsymbol{X}^T \boldsymbol{X})^{-1} \boldsymbol{X}^T \boldsymbol{y}^*$.

As in Approach 1, it can be shown that no simulations are necessary, as the bootstrap variance of $\hat{\boldsymbol{\beta}}^*$ converges for $n \to \infty$ to the Fisher information. In other words, both the residual and the model-based bootstrap provide the same variance estimate as the one calculated in Chap. 7 with Maximum Likelihood.

Approach 3: Pairwise Bootstrap
Thus far, we have treated the covariates $\boldsymbol{x}_i$ as a given. That is, the matrix $\boldsymbol{X}$ remained unchanged. Approach 3 now nonparametrically treats the pairs $(y_1, \boldsymbol{x}_1), \ldots, (y_n, \boldsymbol{x}_n)$ as a sample from which bootstrap samples can be directly drawn. Hence we proceed as follows:

1. Draw $(y_i^*, \boldsymbol{x}_i^*)$ with replacement from the original sample, $i = 1, \ldots, n$.
2. Calculate

$$\hat{\boldsymbol{\beta}}^* = (\boldsymbol{X}^{*T} \boldsymbol{X}^*)^{-1} \boldsymbol{X}^{*T} \boldsymbol{y}^*,$$

where $\boldsymbol{X}^*$ is the bootstrapped design matrix with rows $\boldsymbol{x}_i^*$ and $\boldsymbol{y}^* = (y_1^*, \ldots, y_n^*)$.

This approach is less sensitive to the strict *i.i.d.* assumption applied in the previous two approaches to ensure the exchangeability of the residuals. This means that the pairwise bootstrap can better cope with model violations, for instance, due to variance heterogeneity. This will be explored in an example a bit later.

Approach 4: Wild Bootstrap
Note that in pairwise bootstrapping we not only bootstrap observations y_i^* but also the covariates $\boldsymbol{x}_i^*$. This can be questionable if the covariates are not random, but fixed by design, for example, if they were taken at regular timepoints. In this case the bootstrap sample $(y_i^*, \boldsymbol{x}_i^*)$ has, in fact, a different empirical distribution to the observed values $\boldsymbol{x}_i$. To accommodate the restriction that the distribution of $\boldsymbol{x}_i$ should not change in the bootstrap, the wild bootstrap was proposed by Wu (1986). The idea is to bootstrap residuals $\hat{\varepsilon}_i^* = V_i^* \hat{\varepsilon}_i$, with $\hat{\varepsilon}_i$ as fitted residual and V_i^* being drawn *i.i.d.* from the two point distribution

$$P(V_i^* = \frac{\sqrt{5}+1}{2}) = \frac{\sqrt{5}-1}{2\sqrt{5}}$$

$$P(V_i^* = -\frac{\sqrt{5}-1}{2}) = \frac{\sqrt{5}+1}{2\sqrt{5}}.$$

These numbers may look arbitrary but they are well chosen, as

$$E(V_i^*) = \frac{\sqrt{5}+1}{2}\frac{\sqrt{5}-1}{2\sqrt{5}} - \frac{\sqrt{5}-1}{2}\frac{\sqrt{5}+1}{2\sqrt{5}} = \frac{5-1}{4\sqrt{5}} - \frac{5-1}{4\sqrt{5}} = 0$$

$$\text{Var}(V_i^*) = \frac{5+2\sqrt{5}+1}{4}\frac{\sqrt{5}-1}{2\sqrt{5}} + \frac{5-2\sqrt{5}+1}{2\sqrt{5}}\frac{\sqrt{5}+1}{2\sqrt{5}} = 1.$$

It can also be shown that $E[(V_i^*)^3] = 1$, such that the first three moments of $\hat{\varepsilon}_i^*$ mimic the empirical estimates. The wild bootstrap proceeds now as follows:

1. Draw a sample $\boldsymbol{\varepsilon}^* = (\varepsilon_1^*, \ldots \varepsilon_n^*)$ using the two point distribution given above.
2. Set new bootstrap response values as $y_i^* = x_i\hat{\boldsymbol{\beta}} + \varepsilon_i^*$ for $i = 1, \ldots, n$, which can be written in matrix form as $\boldsymbol{y}^* = X\hat{\boldsymbol{\beta}} + \boldsymbol{\varepsilon}^*$.
3. Calculate the bootstrap OLS estimate $\hat{\boldsymbol{\beta}}^* = (X^TX)^{-1}X^Ty^*$.

The wild bootstrap is simple to apply and proves to be quite flexible, as the following example shows.

Example 35 We demonstrate the different bootstrap strategies with the rent data from Chap. 7. The variable Y_i represents the rent of an apartment with a given floor space x_i. Figure 8.4 visualises how the different bootstrap approaches affect the determined relationship between these values. In the top left, we have plotted the data and the resulting least squares fit. Note that variance heterogeneity is clearly present, as larger apartments show more variability in their rents. If we apply a residual-based or a model-based bootstrap (middle row), we ignore this fact, as, e.g. for a model-based bootstrap we simulate "new" residuals from the variance homogeneous normal distribution $N(0, \hat{\sigma}_\varepsilon^2)$. In contrast to the original observations, the bootstrapped data now have a homogeneous variance. This is an unavoidable consequence of the bootstrap design and clearly demonstrates what follows when bootstrapped data do not mimic the original data. The least squares slope estimates of the $B = 200$ bootstrap samples are shown in grey, with the original fit shown with the solid line. Both approaches are not suitable to uncover variance heterogeneity. Note also that the bootstrapped rents y_i^* can also be negative, which clearly makes no sense. Hence the bootstrap can produce invalid samples that do not fulfil the positivity requirement. The pairwise bootstrap, seen in the bottom left, circumvents this problem as (y_i^*, x_i^*) always corresponds to an observed pair and hence y_i^* is positive. We see this approach more accurately reflects the variance heterogeneity of the original data. When interpreting the plot, note that points may represent multiple data entries, as we draw the data pairs (y_i, x_i) with replacement. Moreover, the distribution of the bootstrapped values of x_i^* does not match the empirical distribution of original x_i. Finally, we apply the wild bootstrap which is shown in the bottom right plot, where variance heterogeneity is also accounted for. We can still get negative values y_i^*, but with smaller probability than the model-based and residual-based bootstrap. A comparison of the bootstrapped slope parameters $\hat{\beta}_x$ for

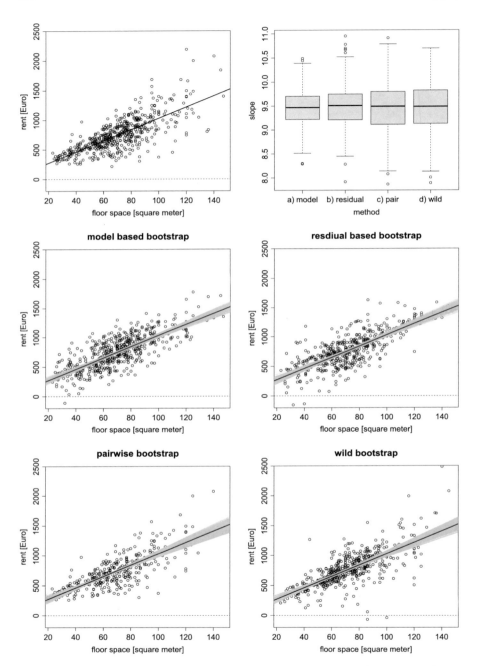

Fig. 8.4 Rental data and bootstrap samples. The top left plot shows the original data and least squares fit and the top right the distribution of bootstrapped slope estimates for each method. The remaining plots show, for each bootstrapping method, a single bootstrap sample, the original least squares slope estimate in black and the bootstrap slope estimates in grey

the four different bootstrap strategies $Y_i = \beta_0 + x_i \beta_x$ is shown in the top right plot. Clearly, incorrectly assuming variance homogeneity underestimates the variance of the slope estimate of $\hat{\beta}_x$, which is accounted for by the pairwise and wild bootstrap.

▷

8.4 Theory and Extension of Bootstrapping

8.4.1 Theory of the Bootstrap

The bootstrap aims to estimate the distribution of a statistic or some characteristic measures of this distribution, such as the expected value, median, variance or quantiles. In this section, we aim to elucidate some of the theory behind this process. Consistency of the bootstrap in estimating the asymptotic distribution of a statistic or estimator can be proven under certain conditions and in simple settings. We think that these proofs can be instructive and will explore some of them here. We will also give counterexamples where the bootstrap is inconsistent to demonstrate its limitations. Take note that, even though the use of bootstrapping is intended as a finite sample approximation of the distribution, asymptotic arguments can come into play. The following results are largely based on Horowitz (2001).

We first introduce some additional notation. Let $Y = (Y_1, \ldots, Y_n)$ be a random sample from an unknown distribution $F(.)$. With $\hat{F}_n$ we denote the empirical distribution $\hat{F}_n(y) = \frac{1}{n} \sum_{i=1}^{n} 1_{\{Y_i \leq y\}}$ or some other (e.g. smooth, parametric) estimate of $F(.)$. Additionally, we take $T_n = \tilde{t}(Y_1, \ldots, Y_n)$ as some statistic of the data (with finite moments). Further, let $G_n(t, F) = P(T_n \leq t)$ be the *exact finite sample* distribution of T_n if the true distribution function of the sample Y is $F(.)$. Note that if $G_n(t, F)$ does not depend on $F(.)$, then T_n is called a pivotal statistic, see Definition 3.15. The bootstrap can be seen as a method for the estimation of $G_n(t, F)$, or some of its characteristic measures, by plugging in $\hat{F}_n$ into G_n, i.e. by using the approximation

$$G_n(t, \hat{F}_n) = P(t(Y^*) \leq t | \hat{F}_n),$$

where $Y^* = (Y_1^*, \ldots, Y_n^*)$ is a bootstrap sample drawn from $\hat{F}_n(.)$. Loosely speaking, the main idea is that for large n

$$G_n(\cdot, \hat{F}_n) \approx G_\infty(\cdot, \hat{F}_n) \approx G_\infty(\cdot, F) \approx G_n(\cdot, F).$$

In the following, we give conditions for the consistency of a bootstrap estimator.

Definition 8.1 The bootstrap estimator $G_n(\cdot, \hat{F}_n)$ is consistent if

$$\lim_{n \to \infty} P_n \left\{ \sup_t |G_n(t, \hat{F}_n) - G_\infty(t, F)| > \epsilon \right\} = 0$$

for each $\epsilon > 0$. P_n is the joint probability distribution of the sample $(Y_1, \ldots, Y_n)$.

Clearly, $\sup_t |F(t) - G(t)|$ is only one possible measure to define the distance between distributions. Bickel and Freedman (1981) introduced a different metric to derive alternative conditions for consistency. They applied what is called the Wasserstein-Mallows metric (the special case of the Mallows metric for $p = 2$), which is defined as follows.

Definition 8.2 Set $p \geq 1$. Let $\mathcal{F}_p$ denote the set of all distribution functions F for which $\int_{-\infty}^{\infty} |t|^p dF(t) < \infty$. For $F, G \in \mathcal{F}_p$, the **Mallows metric** is defined as

$$\rho_p(F, G) = \inf_{\mathcal{T}_{XY}} \left\{ E|X - Y|^p \right\}^{\frac{1}{p}},$$

where $\mathcal{T}_{XY}$ is set of all joint distributions of pairs of two random variables (X, Y) whose marginal distributions of X and Y are F and G, respectively. It implicitly assumes that the moments up to order p exist.

The idea is that it may be easier to show the convergence of $\rho_p(F, G)$ to 0 than that the equation in Definition 8.1 holds. For the special case $p = 2$, we get $\rho_2(F, G) = \inf_{\mathcal{T}_{XY}} \left\{ E|X - Y|^2 \right\}^{\frac{1}{2}}$. The metric is useful because of the following property: $\rho_2(F, G) \to 0$ holds if and only if F converges in distribution to G and $E_F(X^k) \to E_G(X^k)$, $k = 1, 2$. Because the bootstrap is most often used to estimate the distribution function, mean and variance of a statistic, consistency in ρ_2 is well suited to this problem. The following theorem is fundamental to understanding the bootstrap and gives sufficient conditions for its consistency.

Theorem 1 (Beran and Ducharme 1991) $G_n(\cdot, \hat{F}_n)$ *is consistent, i.e.* $G_n(t, \hat{F}_n) \to G_\infty(t, F)$, *if, for any* $\varepsilon > 0$ *and* F, *the following three conditions hold:*

(i) $\lim_{n \to \infty} P_n(\rho_2(\hat{F}_n, F) > \varepsilon) = 0$,
(ii) $G_\infty(t, F)$ *is a continuous function of* t,
(iii) *for any* t *and any sequence* $\{H_n\}$ *such that* $\lim_{n \to \infty} \rho(H_n, F) = 0$:

$$G_n(t, H_n) \to G_\infty(t, F).$$

This theorem holds, for example, with the sample average $\bar{Y}$ and the $p = 2$ Mallows metric, which implies convergence of the corresponding sequences of first and second moments.

Example 36 Assume that Y_i has finite variance and define $T_n = \sqrt{n}(\bar{Y} - \mu)$ with $\mu = E(Y_i), i = 1, \ldots, n$. Then

$$G_n(t, F) = P_n \left\{ \sqrt{n}(\bar{Y} - \mu) \leq t \right\} .$$

Let $\hat{F}_n$ be the empirical distribution function of the data and $T_n^* = \sqrt{n}(\bar{Y}^* - \bar{Y})$ be the bootstrap analogue to T_n. Then

$$G_n(t, \hat{F}_n) = P_n^* \left\{ \sqrt{n}(\bar{Y}^* - \bar{Y}) \leq t \right\} .$$

$\triangleright$

We also want to show a counterexample to demonstrate where the bootstrap can fail. Let $Y = (Y_1, \ldots, Y_n)$ with $Y_i \sim \text{Uniform}(0, \theta)$ *i.i.d.*. The sample maximum is the Maximum Likelihood estimator and is denoted by $\hat{\theta}_{ML} = Y_{(n)} = \max\{Y_i\}$. The probability that $Y_{(n)}$ is not in the bootstrap sample is $\left(1 - \frac{1}{n}\right)^n$. The probability that $Y_{(n)}$ *is* in the bootstrap sample is therefore

$$1 - \left(1 - \frac{1}{n}\right)^n \rightarrow 1 - e^{-1} \approx 0.632 \quad \text{for } n \rightarrow \infty .$$

Thus, $P(\hat{\theta}^* = \hat{\theta}_{ML}) \approx 0.632$ for $n \rightarrow \infty$ and the distribution of $\hat{\theta}^*$ has a point mass of 0.632, even asymptotically, on the Maximum Likelihood estimator. One can show that the distribution of $\hat{\theta}_{ML}$ is not asymptotically normal and the bootstrap is not consistent. In fact it holds that

$$P(n(\theta - Y_{(n)}) \leq x) = 1 - \left(1 - \frac{y}{\theta n y}\right)^n \rightarrow 1 - e^{-\theta y} ,$$

i.e. the distribution of $n(\theta - Y_{(n)})$ converges to an exponential distribution with parameter θ, a continuous distribution with point mass zero everywhere. Therefore, the bootstrap distribution does not converge to the target distribution.

There are a number of other important exceptions. For example, the bootstrap is inconsistent for the maximum of a sample. The bootstrap is also inconsistent for the distribution of an estimator, when the true parameter is on the boundary of the parameter space. A simple example is the square $\bar{Y}^2$ of the sample average of *i.i.d.* variables with mean μ and variance σ^2 if $\mu = 0$. Further examples are given in Horowitz (2001).

8.4.2 Extensions of the Bootstrap

There are several extensions to the bootstrap for alternative data types, e.g. time series, panel data, longitudinal data and stochastic processes. These situations are characterised by the fact that the observations can be dependent upon each other. Recall that bootstrap essentially assumes *i.i.d.* observations (univariate and multivariate) or, as in residual sampling in regression, *i.i.d.* errors. This is not the case with dependent observations or heteroscedastic errors, which we saw could be managed with the wild and pairwise bootstrap.

A more general approach, however, is the Bayesian bootstrap proposed by Rubin (1981), which is able to directly address these more complex systems. It generates new data by weighting the original sample of size n and calculating weighted statistics. To begin with, we need a random weight for each sample, such that all weights add to 1 and $E(G_i) = 1/n$, i.e. the expected proportion of a sample's inclusion in the nonparametric bootstrap. One simple approach is to draw $n - 1$ random numbers $u_1, \ldots, u_{n-1}$ from a standard uniform distribution on $(0, 1)$ and order them as $u_{(0)}, u_{(1)}, \ldots, u_{(n-1)}, u_{(n)}$ with $u_{(0)} = 0$ and $u_{(n)} = 1$. Calculate now the n gaps $g_i = u_{(i)} - u_{(i-1)}, i = 1, \ldots, n$ leading to $g = (g_1, \ldots, g_n)$ as the vector of probabilities used for weighting the original sample.

8.4.3 Subsampling

In the bootstrap approaches above we sample single variables. For time series data, Politis et al. (1999) have proposed different possibilities for bootstrapping, including block bootstrapping, where instead of a single variable, entire blocks of observations are sampled from the observed data. This in turns maintains serial dependence during the bootstrapping process. Politis and Romano (1994) lists several ideas and bootstrap extensions for bootstrapping in the context of prediction.

Subsampling generally takes a sample of size m from the original sample of size $n > m$. This is also called the m-out-of-n bootstrap. In principle, we have two options: sampling with replacement (as in the usual bootstrap), called **replacement subsampling** or sampling without replacement, called **non-replacement subsampling**.

Sampling without replacement gives $\binom{n}{m}$ possible sub-samples and is conceptually different from the standard bootstrap, as the sub-sample of size $m < n$ is then theoretically a sample from the unknown F and not from $\hat{F}_n$. Both methods give consistent estimates under more general conditions than the standard bootstrap, as long as m and n tend to infinity and m/n tends to zero. It was shown by Politis and Romano (1994) and Politis et al. (1999) that non-replacement subsampling works under much weaker conditions than bootstrapping. The disadvantage is that one has to choose the sample size m, which is a tuning parameter that influences the performance of subsampling, especially in real applications where samples are finite.

8.5   Bootstrapping the Prediction Error

8.5.1   Prediction Error

In predictive modelling, the most interesting target is the prediction error and its minimisation. Predictive models relate a response variable Y to a vector of explanatory variables x, also called predictor variables. We estimate a structural relationship or regression

$$E(Y|x) = m(x),$$

where $m(.)$ is the prediction model. This can be done with linear regression models, i.e. $m(x) = \beta_0 + x\beta_x$, but also with more complex models, such as regression trees or even neural networks. For a metric response variable Y, the interesting quantity is often the prediction error measured by the mean squared error of prediction (MSEP)

$$E(Y - \hat{Y})^2 ,$$

where $\hat{Y} = \hat{m}(x)$ is a prediction given by a fitted model (or any other prediction algorithm). Note that

$$E\{(Y - \hat{Y})^2\} = E\left\{(Y - m(x) + m(x) - \hat{Y})^2\right\} = \mathrm{Var}(Y) + \mathrm{MSE}(\hat{Y}).$$

If Y is categorical, then this is a classification problem and the prediction error is often measured as the probability of a false classification, i.e. $P(\hat{Y} \neq Y)$. In the following, we restrict ourselves to the estimation of the MSEP. The MSEP itself is a theoretical quantity, which therefore needs to be estimated. A naive estimator in the regression context would be

$$\widehat{\mathrm{MSEP}} = \frac{1}{n} \sum_{i=1}^{n} (y_i - \hat{y}_i)^2 \tag{8.5.1}$$

or

$$\widehat{\mathrm{MSEP}} = \frac{1}{n-p} \sum_{i=1}^{n} (y_i - \hat{y}_i)^2,$$

where $\hat{y}_i = \hat{m}(x_i)$ and p is the number of predictor variables. This implies that we estimate the prediction with the same observations that were used to fit the prediction model $m(x)$. Consider a simple linear regression model with $Y_i = \beta_0 + x_i \beta_x + \varepsilon_i$ and data (y_i, x_i), $i = 1, \ldots, n$. We fitted β_0 and β_x by minimising the squared error

$$(\hat{\beta}_0, \hat{\beta}_x) = \arg \min \sum_{i=1}^{n} (y_i - \beta_0 - x_i \beta_x)^2$$

which also minimises the MSEP in (8.5.1). However, because we used the same data for building the model and for assessing its predictive quality, this "in-sample" estimate is too optimistic, i.e. it underestimates the error. In an ideal situation, new independent data are available and the MSEP can be estimated with the following procedure:

- Use the **training data** $y_1, \ldots, y_n$ to build a model $\hat{m}\,(\cdot)$.
- Use the fitted model to predict new data y_j^0 with $\hat{y}_j^0$, for $j = 1, \ldots, m$. The new data pairs (y_j^0, x_j^0) are different than to the data used for fitting the model and are therefore usually called the **test data**.
- Estimate the prediction error with

$$\widehat{\text{MSEP}} = \frac{1}{m} \sum_{i=1}^{m} (y_i^0 - \hat{y}_i^0)^2.$$

In practice, new data are often not directly available or only available at a later date upon collection of more data. Cross validation is one approach to estimating the MSEP with only the current dataset.

8.5.2 Cross Validation Estimate of the Prediction Error

The most common strategy for cross validation is called k-fold cross validation and involves randomly splitting the data into K sections of roughly the same size. For each section $k = 1, \ldots, K$, a model is fitted with the data from $K - 1$ sections while prediction quality is assessed with data from the remaining section. In Fig. 8.5, we visualise the process for $K = 6$. Let $k(i)$ be the section which contains the i-th observation y_i. For example, in Fig. 8.5 we have $k(m) = 2$.

Furthermore, let $\hat{y}_i^{-k(i)}$ denote the prediction for y_i, calculated without the section $k(i)$, i.e. without the section which contains y_i. The estimation of the MSEP using cross validation is then given by

$$\widehat{\text{MSEP}} = \frac{1}{n} \sum_{i=1}^{n} \left(y_i - \hat{y}_i^{-k(i)} \right)^2 . \tag{8.5.2}$$

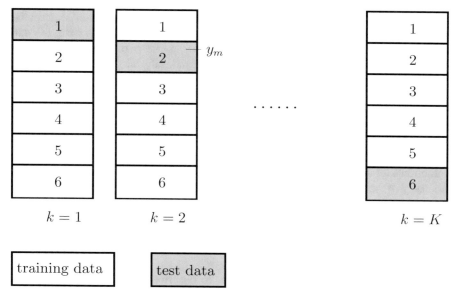

Fig. 8.5 Sketch of cross validation process

The major advantage of cross validation is, that y_i is independent of $\hat{y}^{-k(i)}$. Replacing the observations with random variables gives

$$E\left(\frac{1}{n}\sum_{i=1}^{n}(Y_i - \hat{Y}_i^{-k(i)})^2\right) = \frac{1}{n}\sum_{i=1}^{n}E\left\{(Y_i - \hat{Y}_i^{-k(i)})^2\right\}$$

$$= \frac{1}{n}\sum_{i=1}^{n}E\left\{(Y_i - m(x_i) + m(x_i) - \hat{Y}_i^{-k(i)})^2\right\}$$

$$= \frac{1}{n}\sum_{i=1}^{n}E\left\{(Y_i - m(x_i))^2\right\} + \frac{1}{n}\sum_{i=1}^{n}E\{(m(x_i) - \hat{Y}_i^{-k(i)})^2\}.$$

$$(8.5.3)$$

Clearly, as a result of independence this decomposes into the variance of Y (or the average variance in case of variance heterogeneity) and the MSE of the estimate $\hat{Y}_i^{-k(i)} = \hat{m}^{-k(i)}(x_i)$. When $K = n$, this is called leave-one-out cross validation, which has low variance but is less useful for large n and complex regression methods due to the high computational cost.

Note that our prediction error estimates are conditional on the configuration of the predictor variables, i.e. the estimates depend on the distribution of x. If this distribution differs between the training and test data, e.g. in future data, the true prediction error may be different and our MSEP estimate will not be correct. Therefore, our estimates are only **conditional error estimates**. Alternative strategies can be used if a large amount of data is available. In this case, the data can be randomly split into two parts, simply using one part for training and the other for testing. One may even split the data randomly into three parts: training data, validation data and test data. This is recommended when the training data is used heavily for model fitting. The validation data then works as temporary test data until a final model is decided upon, which is then finally evaluated on the test data. Simple random sampling can also be replaced with more complex approaches, such as stratified sampling. The proportions in the splits can also vary. For example, with the simple train-test split approach, data can be split into two equal sized parts or sometimes a ratio of two thirds training data to one third test data is recommended. A survey of various cross validation strategies is given by Arlot and Celisse (2010).

8.5.3 Bootstrapping the Prediction Error

Let us use regression again as an example to describe a bootstrap procedure for estimating the prediction error. Given paired data $(y_i, x_i), i = 1, \ldots, n$, the goal is to predict a new observation Y for a given x, both drawn from a population distribution F. Let z denote the data $(y_i, x_i), i = 1, \ldots, n$. Note that the data z depend on the sample size n, which is, however, suppressed in the subsequent notation. The prediction is subsequently denoted as $\hat{m}(x)$, which clearly depends on the data z used for fitting. The prediction error for $\hat{m}(x)$ is then given by

$$\text{err}(z, F) \equiv E_F \left\{ (Y - \hat{m}(x))^2 \right\}.$$

Note that the data z used to fit the prediction model is also drawn from $F(.)$. In other words, we are testing how our model, fitted with the data z, performs on the new datapoint (Y, x), which is also drawn randomly from F.

A naive approach to estimating this theoretical quantity is the **sample error**, which is defined as

$$\text{err}(z, \hat{F}_n) = E_{\hat{F}_n} \left\{ (Y - \hat{m}(x))^2 \right\} = \frac{1}{n} \sum_{i=1}^{n} (y_i - \hat{m}(x_i))^2,$$

where we replace the expectation with the average error over the data. We have already mentioned that, because it uses the same data for both fitting and evaluating the model, this quantity is too optimistic and hence gives a biased estimate of $\text{err}(z, F)$. Note that attempting to bootstrap the prediction error with the existing model $\text{err}(z, \hat{F}_n)$ is not a plug-in version of $\text{err}(z, F)$, because the training data z is not drawn from $\hat{F}_n(.)$. In other words, we need to be more careful when constructing a plug-in version of $err(z, F)$. To apply the plug-in principle rigorously we draw the B bootstrap samples (y_i^{*b}, x_i^{*b}) with $i = 1, \ldots, n$ and $b = 1, \ldots, B$. Then, for any b,

$$\text{err}(z^{*b}, \hat{F}_n) = \frac{1}{n} \sum_{i=1}^{n} (y_i - \hat{m}^{*b}(x_i))^2$$

results as a plug-in estimate of $\text{err}(z, F)$, where y_i and x_i are from the original sample but $\hat{m}^{*b}(.)$ is the fitted predictor using the b-th bootstrapped data. Hence, the prediction model is estimated with bootstrapped data, while the prediction error is calculated from the original sample. The prediction error still depends on the data and we could now question how to calculate the expected prediction error, i.e. the average of $\text{err}(z, F)$. We will stop here, because this is getting a little bit clumsy and, while bootstrapping is easy and recommended for quantifying uncertainty, methods like cross validation are easier and more appropriate when calculating prediction error in practice.

8.6   Bootstrap Confidence Intervals and Hypothesis Testing

8.6.1   Bootstrap Confidence Intervals

Thus far, we have used bootstrapping to derive properties of estimates $\hat{\xi} = t(Y)$. However, the bootstrap distribution of the estimate that we calculate also allows us to directly obtain confidence intervals. Let, as above, $t(Y)$ be a statistic that estimates a parameter ξ. The confidence interval is defined as $[t_l(Y), t_r(Y)]$, such that (as in Definition 3.15)

$$P(\hat{\xi} \in [t_l(Y), t_r(Y)]) \geq 1 - \alpha.$$

Using the bootstrap principle, we replace the above expression with

$$P(\hat{\xi} \in [t_l(y^*), t_r(y^*)]) \geq 1 - \alpha,$$

where $y^* = (y_1^*, \ldots, y_n^*)$ is drawn from the empirical distribution $\hat{F}_n(.)$. In practice, one takes $t_l(.)$ and $t_r(.)$ to be the $\alpha/2$ and $1 - \alpha/2$ quantile of the bootstrap samples $\hat{\xi}^{*b}$. That is, we draw B bootstrap samples $y_1^{*b}, \ldots, y_n^{*b}$ with $b = 1, \ldots, B$ and derive the bootstrap estimates $\hat{\xi}^{*b}$. Note that B needs to be sufficiently large, as the $\alpha/2$ quantile of B bootstrap replicates relies on only $(\alpha B/2)$ observations. For $\alpha = 0.05$ and $B = 200$ this means that we have only 5 bootstrap samples below the 2.5% quantile, too few to provide reliable estimates of the confidence interval boundaries. Hence, if confidence intervals are derived from quantiles of bootstrap samples one should work with a relatively large B.

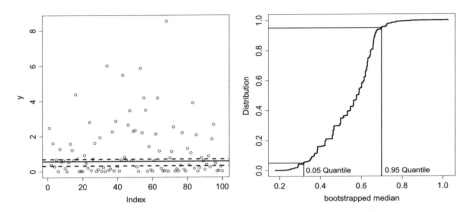

Fig. 8.6 Chi-squared distributed data (left) and the fitted median (solid line) with bootstrapped confidence intervals (dashed lines). The right-hand plot shows bootstrap distribution

Example 37 We simulate 100 data-points from a chi-squared distribution with one degree of freedom and consider the median as the parameter of interest. The data is shown on the left of Fig. 8.6, with the median delineated by the solid line and the 90% bootstrapped median values by the dashed lines. We can see a clear asymmetry of the confidence interval. On the right-hand side is the entire distribution of the bootstrapped median values. We see that in fact quite a lot of information is available and the entire bootstrap distribution can be used to derive properties of the estimate.

$\triangleright$

8.6.2 Testing

Bootstrap hypothesis testing is similar to permutation testing, in that both are usually implemented using a Monte-Carlo procedure. However, for permutation testing one draws without replacement and for bootstrap with replacement. We will sketch both ideas in the coming section. In order to keep things simple, we limit ourselves here to the two-sample hypothesis. We assume

$$Y_i \sim F(.) \quad i.i.d., i = 1, \ldots, n$$

$$Z_j \sim G(.) \quad i.i.d., j = 1, \ldots, m$$

and question the hypothesis

$$H_0 : F = G.$$

The data are denoted as $y = (y_1, \ldots, y_n)$ and $z = (z_1, \ldots, z_m)$ and the combined sample is written as $x = (y, z)$. Note that both the individual samples and the combined sample are drawn from the same distribution under H_0.

The first step for testing is to choose a test statistic $t(x)$ which is sensitive to differences between F and G, that is, one that will also guarantee the power of the test. Once we have defined $t(x)$, we need to simulate the distribution of $t(x)$ under the null hypothesis. In principle, there are several possible test statistics for this situation, but for demonstration purposes let us just compare the arithmetic means. Let $t(x) = \bar{y} - \bar{z}$ where $\bar{y} = \frac{1}{n} \sum_{i=1}^{n} y_i$ and $\bar{z} = \frac{1}{m} \sum_{j=1}^{m} z_j$. Then under $H_0 : F = G = F_0$ we can draw bootstrap samples of length $n + m$ from the combined sample x. The procedure is then as follows:

1. Draw B bootstrap samples x^{*b} of size $n + m$ with replacement from x.
2. Consider x^{*b} as (y^{*b}, z^{*b}) and use the first n observations to calculate $\bar{y}^{*b}$ and the remaining m observations to calculate $\bar{z}^{*b}$ and evaluate $t(x^{*b}) = \bar{y}^{*b} - \bar{z}^{*b}$ for $b = 1, \ldots, B$.
3. The two-sided bootstrapped p-value of the test is then given by

$$p\text{-value}_{\text{boot}} = \frac{1}{B} \sum_{b=1}^{B} 1\{|t(x^{*b})| > |t(x)|\}.$$

If only a few bootstrap samples generate higher absolute values than the observed test statistic $|t(x)|$, then the bootstrapped p-value is low. In this case, $t(x)$ is rather extreme under the null hypothesis H_0, which therefore should be rejected. Note that the procedure is different from the one pursued in (8.1.4), where we generated a bootstrap estimate of the variance of $t(x)$ and the bootstrap samples were generated separately for each of the two groups.

A permutation test proceeds very similarly. Again, we pool the data in x but instead of drawing a sample from x with replacement, we draw a sample from x without replacement, i.e. we permute the order of the entries of x. The remaining calculation of the p-value remains unchanged. Usually, using a studentised statistic increases the accuracy of testing and is therefore preferable. For example, we could use

$$t(x) = \frac{\bar{y} - \bar{z}}{s}$$

with $s^2 = \frac{1}{n+m-2}(\sum_{i=1}^{n}(y_i - \bar{y})^2 + \sum_{j=1}^{m}(z_j - \bar{z})^2)$. This is the two-sample t statistic under the assumptions of equal variances in the two groups. One could also use

$$t(x) = \frac{\bar{y} - \bar{z}}{s_y^2/n + s_z^2/m}$$

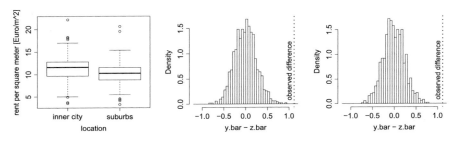

Fig. 8.7 Difference on rent per square metre for inner city apartments and apartments in the suburbs (left plot). Simulated differences using permutation (middle plot) and bootstrapping (right-hand plot)

with $s_y^2 = \frac{1}{n-1} \sum_{i=1}^{n} (y_i - \bar{y})^2$ and $s_z^2 = \frac{1}{m-1} \sum_{j=1}^{m} (z_j - \bar{z})^2$ and the bootstrap analogues for testing.

We have used a mean difference for testing H_0, but, of course, other statistics are also possible. This choice of statistic also affects the power to detect differences between the distributions. It is difficult to generally recommend a particular statistic as we can only approximate the distribution of $t(x)$ under H_0 but not under unspecified alternatives. The power is therefore dependent on $H_1 : F \neq G$ and cannot be calculated without further parametric assumptions on F and G. A comprehensive overview of bootstrap and permutation testing can be found in Good (2005). Let us round off this section with a few examples that hopefully demonstrate the ease and flexibility with which bootstrapping can be applied to testing.

Example 38 Let us once again take a look at the Munich rental data and question whether the location of the apartment has an influence on the rent. We therefore take the rent per square metre as response variable and make two location categories: *inner city* and *suburbs*. The rent per square metre is shown in Fig. 8.7. We denote the rent of apartments in the city centre with $y = (y_1, \ldots, y_n)$ and the rent of apartments in the suburbs with $z = (z_1, \ldots, z_m)$, where $n=278$ and $m=222$. We take the test statistic $t(x) = t(y, z) = \bar{y} - \bar{z}$ and apply both the permutation and bootstrap tests. The resulting simulated differences are shown in the middle plot in Fig. 8.7 for the permutation test and on the right for the bootstrap test. Clearly, there is a significant difference and apartments in the city centre are significantly more expensive than those in the suburbs.   ▷

Example 39 Let $Y_i, \ldots, Y_n \sim F(.)$ i.i.d. and let us test whether $F(.)$ is symmetric around a known c. Note that $F(.)$ is symmetric around c if and only if the distribution of Y and $c - Y$ is the same. In other words, we have the hypothesis

$$H_0 : F(y - c) = 1 - F(c - y)$$

for all $y \in \mathbb{R}$. A number of test statistics and approaches have been proposed for this problem and we refer to Zheng and Gastwirth (2010) for a recent summary. In our bootstrap approach we take

$$t(Y) = \bar{Y} - Y_{med}$$

as the test statistic. That is, we take the difference between the mean estimate $\bar{Y}$ and the median estimate Y_{med}, where Y_{med} is the empirical median of $Y_1, \ldots, Y_n$. This leads to the observed value $t(y) = \bar{y} - y_{med}$, which, if close to zero, suggests a symmetric distribution and if large suggests asymmetry. To apply a bootstrap test, we need to bootstrap (simulate) from a symmetric distribution, that is, we need to simulate under H_0. Therefore, we need some way to force $\hat{F}_n$ to be symmetric. The trick is to complement the data with "symmetrised" values $y_{n+1}, \ldots, y_{n+n}$ which are defined as

$$y_{n+i} = 2c - y_i = c + (c - y_i).$$

In other words, we mirror all observations around c. This gives an extended dataset $\tilde{y} = (y_1, \ldots, y_n, y_{n+1}, \ldots, y_{2n})$, from which we now draw, with replacement, a bootstrap sample of size n. From the bootstrap sample y^{*b} we calculate $\bar{y}^{*b}$ and y_{med}^{*b}, from which we can derive the bootstrapped $t(y^{*b})$. Taking B bootstraps allows us to compare $t(y)$ with the empirical distribution function of $t(y^{*b})$. We apply the test to the apartment size, labelled as y, also from the Munich rental data. The question is whether this is symmetric around the value 69, which is the median. We take $c = 69$ as fixed. The left of Fig. 8.8 shows the empirical distribution (solid line) and the symmetrised distribution (dotted line). Clearly, there is some level of asymmetry that can be assessed with a test. The observed value of the test statistic is $t(y) = \bar{y} - y_{med} = -1.83$. Bootstrapped values $t(y^{*b})$ are shown on the right. We can see that the observed value lies far away from the bootstrapped values, with a

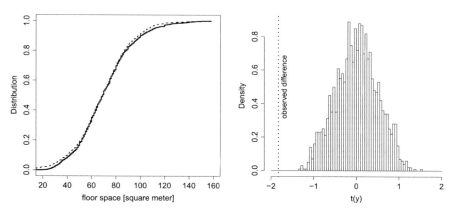

Fig. 8.8 Left: Empirical distribution of floor space and its symmetrised counterpart; Right: Bootstrapped differences between the mean and the median and the observed difference

bootstrapped p-value less than $1/B = 1/2000$, with our 2000 bootstrap replicates. This suggest clearly that the floor space is not symmetrically distributed.

▷

Example 40 In Sect. 6.5, we already discussed tests on independence. Let us take the correlation coefficient test and demonstrate its bootstrap version. Let Y_{i1} and Y_{i2} be drawn jointly from F:

$$(Y_{i1}, Y_{i2}) \sim F(y_1, y_2).$$

We want to test the hypothesis

$$H_0 : F(y_1, y_2) = F_1(y_1) F_2(y_2)$$

with $F_1(.)$ and $F_2(.)$ being the univariate marginal distributions of Y_{i1} and Y_{i2}, respectively. As a test statistic, we use the empirical correlation (even though other dependence statistics can also be used):

$$\hat{\rho}_{12} = \frac{\sum_{i=1}^{n}(Y_{i1} - \bar{Y}_1)(Y_{i2} - \bar{Y}_2)}{\sqrt{\left(\sum_{i=1}^{n}(Y_{i1} - \bar{Y}_1)^2\right)\left(\sum_{i=1}^{n}(Y_{i2} - \bar{Y}_2)^2\right)}}.$$

We are interested in the distribution of $\hat{\rho}_{12}$ in the case of H_0. This is easily bootstrapped by replacing $F_1(.)$ with its empirical counterpart $\hat{F}_{1n}(.)$ and likewise for $\hat{F}_2(.)$. In other words, we draw a univariate sample, with replacement, of the first variable $(y_{11}^*, \ldots, y_{n1}^*)$ and the second variable $(y_{12}^*, \ldots, y_{n2}^*)$. These two independent samples are put together to give the final bootstrap sample

$$(y_{i1}^*, y_{i2}^*) \text{ with } i = 1, \ldots, n.$$

From this sample we can calculate a bootstrap correlation coefficient $\hat{\rho}^*$, which is calculated under H_0. Repeating the bootstrap B times provides the reference distribution under H_0.

▷

8.7   Sampling from Data

The idea of sampling that we have thus far only applied to bootstrapping and other resampling methods can be seen from a different angle, which can also be useful in the age of big data. Assume that we need to analyse a massive database and the number of observations N is very large. While this is certainly a pleasant situation, more is not always better if the quality of the data is not sufficient, as we will discuss in more detail in Sect. 11.3. But for now let us focus on the computational effort for the data analysis, which increases with the size of available data. This increase is particularly pronounced for some analytic models and procedures and often hardware limitations can mean that the entire dataset cannot be processed at once. In this case, it is both advisable and statistically sound to draw a sample from the data and run the analysis only with this sample. We define with n the sample size. As we have seen, the variance of estimates decreases with increasing sample size n, which also applies in this case. Hence, instead of analysing the entire database of N entries, we allow for a random error and draw n observations from the N data. If all observations get the same probability of inclusion, this is called **simple random sampling**. In fact, one usually draws without replacement and hence the central *i.i.d.* assumption is violated. Indeed it can be shown (see Thompson 2002) that drawing without replacement reduces the variance compared to drawing with replacement. In other words, if we treat data which are drawn without replacement as if they were drawn with replacement, we overestimate the variance of statistics calculated from the sampled data. The difference is of order $\frac{n}{N}$, meaning that the variance without replacement is smaller by a factor $(1 - \frac{n}{N})$ compared to the variance with replacement. This also means that, if we pretend that data are drawn *i.i.d.*, we still develop valid confidence intervals or tests, as we would have effectively overestimated the variance of the quantity of interest by taking a sample with replacement. The variance of any estimate decreases with n, which we sketched in Fig. 2.2 already. Hence, the standard deviation decreases with $1/\sqrt{n}$ and with increasing sample size n the accuracy of our analysis increases. Let us informally define accuracy as the reciprocal of the width of a confidence interval calculated from the data. As the standard deviation decreases with order $1/\sqrt{n}$ (in the best case), accuracy then increases with order $\sqrt{n}$. If we calculate the estimates with more data, the computational complexity increases. In the best case, this is linear in n, but most algorithms have a higher complexity, e.g. $n \log(n)$ or n^2. In Fig. 8.9, we sketch both computational complexity and achieved accuracy and their dependence on the data size. We see that to simply analyse all of the data as a rule is counterproductive, as the computational burden increases sharply while the gain in accuracy is small. This suggests that in many cases it can be useful to simply sample data and run the analysis on sampled data. We emphasise that the message mirrored in Fig. 8.9 holds in general, that is not only for statistical models but for all data analytic algorithms. Sampling from massive databases therefore appears to be a recommendable strategy if the computational effort for analysing all of the data does not justify the gains in accuracy.

So far, we recommended using simple random sampling. This means each data entry (or individual) has the same probability of selection and inclusion in the data analysis of the sampled data. While this sounds sensible, it is not necessarily the most efficient way to draw a sample. In some situations, it is advisable to draw units with unequal probability, sometimes called oversampling of units. The resulting sampled dataset is clearly biased and subsequent data analysis needs to take this into account, which can be done through a weighted analysis. We already treated sampling weights and a corresponding weighted data analysis in regression already in Sect. 7.2. Here, we just want to mention and emphasise that the idea of drawing

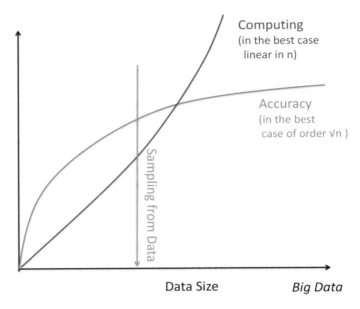

Fig. 8.9 Trade-off between computational effort and accuracy of data analysis for increasing sample size

a representative sample from a database does not require that the resulting sampled data are representative. We refer to Thompson (2002) for further details.

8.8   Exercises

Exercise 1 (Use R Statistical Software)
We consider n $i.i.d.$ realisations $(y_1, \ldots, y_n)$ of a normal random variable $Y \sim N(\mu, \sigma^2)$ with unknown μ and $\sigma^2 > 0$.

1. Consider the Maximum Likelihood estimates

$$\hat{\mu} = \bar{y} = \frac{1}{n} \sum_{i=1}^{n} y_i \qquad \hat{\sigma}^2 = \frac{1}{n} \sum_{i=1}^{n} (y_i - \hat{\mu})^2.$$

   Try to calculate *estimates* for the variances of $\hat{\mu}$ and $\hat{\sigma}^2$, $\widehat{Var}(\hat{\mu})$ and $\widehat{Var}(\hat{\sigma}^2)$ and the corresponding standard errors $\sqrt{\widehat{Var}(\hat{\mu})}$ and $\sqrt{\widehat{Var}(\hat{\sigma}^2)}$.
2. Write pseudo code for the estimation of these standard errors with a nonparametric bootstrap.

3. Simulate with varying sample sizes $n = 10, 50, 100, 500$ samples from a normal distribution with $\mu = 10$ and $\sigma^2 = 25$. Implement your pseudo code and test your function with $B = 500$ replications. Compare your results by using the function boot of the R package boot.

4. Compare your results additionally to the estimates in (1).

Exercise 2

The spatial data in the R package bootstrap contains the outcomes of two tests A and B on the spatial perception of 26 neurologically impaired children. We want to construct a 95% confidence interval for the correlation $\rho = Cor(A, B)$ between the two test results.

1. Construct a confidence interval using the formula $\hat\rho \pm 1.96\hat{se}$ where $\hat\rho$ is the usual Pearson correlation and $\hat{se}$ is a nonparametric bootstrap estimate of the standard error of $\hat\rho$.

2. Construct a confidence interval using the Fisher Z transformation

$$\theta = \frac{1}{2} \log \left(\frac{1 + \rho}{1 - \rho} \right) .$$

Hint: Apply the inverse transformation on the endpoints of the interval to get an interval for ρ once you have an interval for θ.

3. Construct an interval using a parametric bootstrap assuming a bivariate normal distribution for the two outcomes.

4. Construct a nonparametric bootstrap interval for the Spearman rank correlation.

5. Compare the results of all the above methods.

Chapter 9
Model Selection and Model Averaging

In Chaps. 4 and 5 we explored Maximum Likelihood estimation and Bayesian statistics and, given a particular model, used our data to estimate the unknown parameter θ. The validity of the model itself was not questioned, except for a brief detour into the Bayes factor. In this chapter we will delve a little deeper into this idea and explore common routines for selecting the most appropriate model for the data. Before we start, let us make the goal of model selection a little more explicit. We defined with

$$l(\theta; y_1, \ldots, y_n) = \sum_{i=1}^{n} \log f(y_i; \theta)$$

the log-likelihood. This likelihood of the data clearly depends on the probability model $f(.; \theta)$, which we thus far have simply assumed to be true and have not specified in the notation. However, the specification and selection of the probability model is already an important step in statistical modelling and should absolutely be explicitly notated. In principle, the likelihood should then be denoted with $l(\theta; y_1, \ldots, y_n | f)$. Hence, a likelihood can only be derived if we condition upon the model, that is, we assume that $f(.; \theta)$ holds. The phrase "model" is loosely defined here and simply states that we condition on a class of probability models by assuming

$$Y_i \sim f(y; \theta) \quad i.i.d. \tag{9.0.1}$$

This "specification" step (9.0.1) is what we will focus on for the moment. Note that all derivations of Chap. 4 for the Maximum Likelihood estimate were based on the validity of (9.0.1). But what happens if the data come from a different distribution? For instance, let

$$Y_i \sim g(y) \quad i.i.d., \tag{9.0.2}$$

where $g(.)$ is the unknown true distribution.

G. Kauermann et al., *Statistical Foundations, Reasoning and Inference*,
Springer Series in Statistics, https://doi.org/10.1007/978-3-030-69827-0_9

If we do not know $g(.)$, the question arises whether it is possible to learn anything about it at all. More importantly, if we have falsely taken $f(.; \theta)$ as the true distribution, what is the meaning and nature of the Maximum Likelihood estimate $\hat{\theta}$ under this misspecification? That is, what is the use of (9.0.1) if $f(.; \theta) \neq g(.)$? Should we be warned? Is it wrong to use the Maximum Likelihood estimate when the model is incorrect? After all, as data in general are complex and their distributions usually unknown, we have no a priori insight into their true distribution $g(.)$. This clearly implies that most (if not all) models are wrong. But are they still useful?

At the very least, we want to guarantee that we are not acting like the proverbial drunkard, looking for his keys under a lamppost. A policeman arrives, helps him with his search and, after a while, asks the man where exactly he dropped the keys. The drunkard replies that he dropped them on the other side of the street, but it was dark over there, so he decided to instead search under the lamppost where there was light. That is, we need to know if, when we search for the parameters of a model, we are searching under the lamppost or where we lost the key, i.e. where it is convenient or from where the data actually stem. We will fortunately see that the Maximum Likelihood approach makes sense, even if the true model is not within the assumed model class (9.0.1) but some unknown model (9.0.2). Hence, the likelihood approach sheds light on the true model $g(.)$, even if we do not know its structure.

Sometimes it happens that we have two models that describe the data equally well. This begs the question, why we should explicitly select one of the two, instead of using both models. This perspective will lead us to model averaging, where we fit multiple models and, instead of discarding all but the best model, we apply them together as a group. But first we will begin with Akaike Information Criterion—an essential component of model selection.

9.1   Akaike Information Criterion

9.1.1   Maximum Likelihood in Misspecified Models

A key component of both model selection and model averaging is the Akaike Information Criterion (AIC), which we already briefly introduced in Chap. 7. We will explore the theory behind the AIC in detail in the following section. Our exploration begins with the Kullback–Leibler (KL) divergence, defined in Sect. 3.2.5. Recall that for two densities $f(.)$ and $g(.)$, the KL divergence is defined as

$$KL(g, f) = \int \log\left(\frac{g(y)}{f(y)}\right) g(y) dy$$

$$= \int \log(g(y)) g(y) dy - \int \log(f(y)) g(y) dy. \qquad (9.1.1)$$

We now consider $g(.)$ to be the true, but unknown, distribution, while $f(y) = f(y; \theta)$ is our distributional model. If $g(.)$ belongs to the model class, i.e. there exists a true parameter value θ_0 such that $g(y) = f(y; \theta_0)$ for all y, then the Kullback–Leibler divergence takes the value zero for $\theta = \theta_0$. Generally, however, the Kullback–Leibler divergence is positive, meaning that $f(y; \theta) \neq g(y)$, for some y, no matter what value θ takes. A sensible strategy would be to choose a θ, so as to minimise the Kullback–Leibler divergence between $f(y; \theta)$ and the unknown true distribution $g(.)$. We denote this parameter value with θ_0 such that

$$\theta_0 = \arg \min_{\theta \in \Theta} KL(g, f). \qquad (9.1.2)$$

Looking at (9.1.1), it is clear that the first component does not depend on $f(.; \theta)$, but only on the true but unknown distribution $g(.)$. From here on, this component can therefore be ignored, as it is effectively a constant of no particular interest. So equivalently, we can instead maximise

$$\int \log(f(y; \theta))g(y)dy$$

with respect to θ. Taking the derivative gives $\int s(\theta; y)g(y)dy$, where $s(\theta; y) = \partial \log f(y; \theta)/\partial \theta$ is the score function. The best parameter θ_0 apparently needs to fulfil

$$\int s(\theta_0; y)g(y)dy = E_g(s(\theta_0; Y)) = 0, \qquad (9.1.3)$$

where the expectation is calculated using the true but unknown distribution $g(.)$. In other words, the best parameter θ_0 gives a vanishing expectation of the score function. This property holds regardless of the true model.

Given this result and observed data $y_1, \ldots, y_n$ drawn from (9.0.2), let us now attempt to estimate θ_0. This is done by setting the empirical score to zero, i.e.

$$\sum_{i=1}^{n} \frac{\partial \log f(y_i; \hat{\theta})}{\partial \theta} = \sum_{i=1}^{n} s_i(\hat{\theta}; y_i) = 0. \qquad (9.1.4)$$

Clearly, this is identical to Maximum Likelihood estimation, i.e. the Maximum Likelihood approach provides an estimate for θ_0. This finding serves an important purpose, in that it allows us to interpret the Maximum Likelihood estimate from a different perspective. Setting θ to θ_0 minimises the difference between $g(.)$ and $f(y; \theta)$, and hence $f(., \theta_0)$ is the closest density to $g(.)$ as measured by the Kullback–Leibler divergence.

We can go even further and derive, as in Chap. 4, asymptotic properties of $\hat{\theta}$.

Property 9.1 For data $Y_1, \ldots, Y_n$ drawn *i.i.d.* from $g(.)$, the Maximum Likelihood estimate $\hat{\theta}$ for model $f(.; \theta)$ fulfils

$$\hat{\theta} - \theta_0 \overset{a}{\sim} N(0, I^{-1}(\theta_0)V(\theta_0)I^{-1}(\theta_0)),$$

where θ_0 is defined as per (9.1.3), $I(\theta_0)$ is the Fisher matrix and $V(\theta_0)$ is the variance of the score function.

Proof To derive the above property, we expand (9.1.4) around θ_0. We begin with some notation. Firstly, let us explicitly allow the parameter θ to be multidimensional, giving $\theta = (\theta_1, \ldots, \theta_p)^T$. Secondly, let us define the score as

$$s(\theta; y_1, \ldots, y_n) = \sum_{i=1}^{n} s_i(\theta; y_i).$$

As defined in (3.3.3), the observed Fisher information is given by

$$J(\theta; y_1, \ldots, y_n) = -\sum_{i=1}^{n} \frac{\partial s_i(\theta; y_i)}{\partial \theta^T} = -\frac{\partial s(\theta; y_1, \ldots, y_n)}{\partial \theta^T},$$

which is a $p \times p$ dimensional matrix. The corresponding Fisher matrix is defined as

$$I(\theta) = E_g \left(-\frac{\partial s(\theta; Y_1, \ldots, Y_n)}{\partial \theta^T} \right) = \frac{\partial^2}{\partial \theta \partial \theta^T} \sum_{i=1}^{n} \int \log f(y_i; \theta) g(y_i) dy_i.$$

$$(9.1.5)$$

Note that the expectation is carried out with respect to the true but unknown density $g(.)$, not with the model density $f(.; \theta)$ as in Chap. 4. The first order Taylor expansion of (9.1.4) gives

$$0 = s(\hat{\theta}; y_1, \ldots, y_n) \tag{9.1.6}$$

$$\approx s(\theta_0; y_1, \ldots, y_n) + J(\theta_0; y_1, \ldots, y_n)(\hat{\theta} - \theta_0) \tag{9.1.7}$$

$$\Leftrightarrow (\hat{\theta} - \theta_0) \approx J^{-1}(\theta_0; y_1, \ldots, y_n) s(\theta_0; y_1, \ldots, y_n). \tag{9.1.8}$$

Note that $J(\theta; y_1, \ldots, y_n)$ is the observed version of $I(\theta)$, which allows the simplification of (9.1.8) to

$$(\hat{\theta} - \theta_0) \approx I^{-1}(\theta_0) s(\theta_0; y_1, \ldots, y_n). \tag{9.1.9}$$

With standard arguments based on the central limit theorem (as in Chap. 4), this gives

$$(\hat{\theta} - \theta_0) \overset{a}{\sim} N(0, I^{-1}(\theta_0)V(\theta_0)I^{-1}(\theta_0)), \tag{9.1.10}$$

where

$$V(\theta_0) = Var(s(\theta_0; Y_1, \ldots, Y_n)).$$

If the true model is of the form $f(y; \theta)$, then the variance simplifies further, as $Var(s(\theta_0; Y_1, \ldots, Y_n))$ is equal to the Fisher information $I(\theta)$. For misspecified models, however, this property does not apply and we get

$$Var(s(\theta_0; Y_1, \ldots, Y_n)) = \int s(\theta_0; y_1, \ldots, y_n)s^T(\theta_0; y_1, \ldots, y_n) \prod_{i=1}^{n} g(y_i)dy_i$$

$$= \sum_{i=1}^{n} \int s_i(\theta_0; y_i)s_i^T(\theta_0; y_i)g(y_i)dy_i, \tag{9.1.11}$$

where $y = (y_1, \ldots, y_n)$. This is not necessarily equal to the Fisher information. However, the empirical version of (9.1.11) can be used for estimating the variance.

□

9.1.2 Derivation of AIC

We have shown so far that the Maximum Likelihood approach gives the best possible estimate $\hat{\theta}$ for θ_0, as defined in (9.1.2). The next task is to evaluate how close our best model $f(y; \hat{\theta})$ is to $g(y)$. Note that y here represents any possible value of a random variable Y, while $\hat{\theta}$ is the Maximum Likelihood estimate derived from the data $y_1, \ldots, y_n$. Instead of looking at specific values, we take the expectation over all possible values of Y and look at the Kullback–Leibler divergence between $g(.)$ and $f(.; \hat{\theta})$. This is given by

$$KL\left(g(.); f(.; \hat{\theta})\right) = \int \log\left(\frac{g(y)}{f(y; \hat{\theta})}\right)g(y)dy \tag{9.1.12}$$

$$= \int \log(g(y))g(y)dy - \int \log(f(y; \hat{\theta}))g(y)dy. \tag{9.1.13}$$

Again, the first component is a constant and we can simply focus on the second. Clearly, because $g(.)$ is unknown, the integral cannot be directly calculated. Moreover, the estimate $\hat{\theta}$, and hence, the Kullback–Leibler divergence (9.1.13), depends on the observed data $y_1, \ldots, y_n$. To obtain a general measure that does not depend on the observed values, it seems sensible to take the expectation over the sample $Y_1, \ldots, Y_n$. This gives the expected Kullback–Leibler divergence

$$E_{Y_1,\ldots,Y_n} \left\{ KL\big(g(.), f(.; \hat{\theta}(Y_1, \ldots, Y_n))\big) \right\}$$

$$= \text{const} - E_{Y_1,\ldots,Y_n} \left\{ \int \log f\big(y; \hat{\theta}(Y_1, \ldots, Y_n)\big) g(y) dy \right\}, \qquad (9.1.14)$$

where we made the dependence of our estimate $\hat{\theta}$ on the data $y_1, \ldots, y_n$ explicit. The constant term can be ignored in the following. Note that we have two integrals in the second term in (9.1.14): the inner integral over y in the Kullback–Leibler divergence and the outer integral resulting from the expectation with respect to $Y_1, \ldots, Y_n$. To be explicit, we can write the expectation in the second component of (9.1.14) as

$$\int \left[\int \log \big(f(y; \hat{\theta}(y_1, \ldots, y_n)) \big) g(y) dy \right] g(y_1, \ldots, y_n) dy_1 \ldots dy_n, \qquad (9.1.15)$$

where $g(y_1, \ldots, y_n) = \prod_i g(y_i)$. This integral may look rather clumsy, but we will nevertheless attempt to approximate it. This will lead us to the famous Akaike Information Criterion.

Definition 9.1 The Akaike Information Criterion (AIC) is defined as

$$\text{AIC} = -2 \sum_{i=1}^{n} \log f(y_i; \hat{\theta}) + 2p. \qquad (9.1.16)$$

The multiplication with 2 above has no specific meaning and was suggested by Akaike himself. The AIC is very broadly applicable and, as we will show, is much more than just balance between goodness of fit (the first term) and complexity (the second term in (9.1.16)). The deeper meaning of the AIC becomes clear through its derivation.

Derivation We will now show how the AIC can be explicitly derived. We begin by disentangling the double integral and approximating $f(y; \hat{\theta})$ with a second order Taylor series expansion around θ_0. To simplify notation, from now on we write $\hat{\theta}(y_1, \ldots, y_n)$ as $\hat{\theta}$, but it should be kept in mind that $\hat{\theta}$ depends on $y_1, \ldots, y_n$ and not on y. This gives the approximation of (9.1.15) as

$$\int \int \left\{ \log f(y; \theta_0) + \frac{\partial \log f(y; \theta_0)}{\partial \theta^T}(\hat{\theta} - \theta_0) - \frac{1}{2}(\hat{\theta} - \theta_0)^T J(\theta_0)(\hat{\theta} - \theta_0) \right\}$$

$$\times g(y) dy g(y_1, \ldots, y_n) dy_1 \ldots dy_n$$

$$\approx \int \log \big(f(y; \theta_0) \big) g(y) dy - \frac{1}{2} \int (\hat{\theta} - \theta_0)^T I(\theta_0)(\hat{\theta} - \theta_0) g(y_1, \ldots, y_n) dy_1 \ldots dy_n$$

$$(9.1.17)$$

using (9.1.10) such that the second component vanishes. Let us first look at the final component in (9.1.17). Making use of (9.1.10), this can be asymptotically approximated by $tr(I^{-1}(\theta_0)V(\theta_0))$. Note that neither the integral in the first component in (9.1.17) nor $V(\theta_0)$ and $I(\theta_0)$ can be calculated explicitly, as they depend on the unknown $g(.)$. Our aim is therefore to estimate the above quantity based on the data $y_1, \ldots, y_n$. We do know that the data $y_1, \ldots, y_n$ are drawn from $g(.)$. Therefore, we replace the integral in the first component with its arithmetic mean. To do so, we look at the empirical version of (9.1.15) but leave $\hat{\theta}$ fixed for now. That is, we replace the expectation over $g(y)$ with the arithmetic mean using $y_1, \ldots, y_n$. To be specific, we calculate

$$\frac{1}{n} \sum_{i=1}^{n} \log f(y_i; \hat{\theta}). \tag{9.1.18}$$

This is clearly just the likelihood function at its maximum divided by the sample size. We again expand (9.1.18) around θ_0 and obtain an approximation of (9.1.18) with

$$\frac{1}{n} \sum_{i=1}^{n} \left\{ \log f(y; \theta_0) + s_i(\theta_0; y_i)^T (\hat{\theta} - \theta_0) - \frac{1}{2}(\hat{\theta} - \theta_0)^T J_i(\theta_0; y_i)(\hat{\theta} - \theta_0) \right\},$$

$$\tag{9.1.19}$$

where $J_i(\theta_0, y_i) = -\partial^2 \log f(y_i; \theta_0)/\partial\theta\partial\theta^T$. Note that the first component in (9.1.19) can approximate the first component in (9.1.17). This means

$$\frac{1}{n} \sum_{i=1}^{n} \log f(y_i; \theta_0) \quad \rightarrow \quad \int \log f(y; \theta_0)g(y)dy \tag{9.1.20}$$

for increasing n, which can in fact be proven to be a consistent estimate. Moreover, because with increasing sample size n

$$\frac{1}{n} \sum_{i} J_i(\theta_0; y) \quad \rightarrow \quad I(\theta_0), \tag{9.1.21}$$

we can argue that the integrand of the third component in (9.1.19) converges to

$$\frac{1}{2}(\hat{\theta} - \theta_0)^T I^{-1}(\theta_0)(\hat{\theta} - \theta_0).$$

Taking the expectation with respect to $Y_1, \ldots, Y_n$ and using the asymptotic distribution (9.1.10), we see that

$$E_{Y_1,\ldots,Y_n}\left(\frac{1}{2}(\hat{\theta} - \theta_0)^T I^{-1}(\theta_0)(\hat{\theta} - \theta_0)\right) = tr(I^{-1}(\theta_0)V(\theta_0)).$$

What remains is the second component in (9.1.19). If we subtract this and take (9.1.20), (9.1.21) and the approximation (9.1.17), we get

$$\frac{1}{n}\sum_{i=1}^{n}\log f(y_i; \hat{\theta}) - \frac{1}{n}\sum_{i=1}^{n}s_i^T(\theta_0; y_i)(\hat{\theta} - \theta_0) \quad \rightarrow \quad \int\int \log f(y; \hat{\theta})g(y)dy\, g(y_1, \ldots, y_n)dy_1 \ldots dy_n.$$

This suggests the use of $\frac{1}{n}\sum_{i=1}^{n}s_i^T(\theta_0; y_i)(\hat{\theta} - \theta_0)$ as a bias correction, which we now try to simplify further. Using (9.1.10) gives

$$\frac{1}{n}\sum_i s_i^T(\theta_0; y_i)(\hat{\theta} - \theta_0) \approx \frac{1}{n}\sum_i s_i^T(\theta_0; y_i)I^{-1}(\theta_0)\sum_j s_j(\theta_0; y_j).$$

Bearing in mind that $E(s_i(\theta_0; y_i)) = 0$, if we take the expectation with respect to $Y_1, \ldots, Y_n$, we get

$$E_{Y_1,\ldots,Y_n}\left(\frac{1}{n}\sum s_i(\theta_0; y_i)^T(\hat{\theta} - \theta_0)\right) \approx \frac{1}{n}tr\left(I^{-1}(\theta_0)V(\theta_0)\right). \tag{9.1.22}$$

As both $I(\theta_0)$ and $V(\theta_0)$ depend on the unknown distribution $g(.)$, we cannot calculate their values. Instead, we replace the matrices with their empirical counterparts, which gives

$$\hat{I}(\theta) = -\frac{1}{n}\sum_{i=1}^{n}\frac{\partial^2 \log f(y_i; \theta)}{\partial\theta\partial\theta^T} \quad \text{and}$$

$$\hat{V}(\theta) = \frac{1}{n}\sum_{i=1}^{n}s_i^T(\theta; y_i)s_i(\theta; y_i).$$

This approximation was proposed by Takeuchi (1979), see also Shibata (1989) or Konishi and Kitagawa (1996). It is, however, more common and much more stable for small n to approximate the term (9.1.22) even further. To do so, assume that $g(.) = f(.; \theta_0)$. Then $V(\theta_0) = I(\theta_0)$ such that

$$\frac{1}{n}tr(I^{-1}(\theta_0)V(\theta_0)) = \frac{1}{n}p,$$

where p is the dimension of the parameter. This approximation is attractively simple and was proposed by Akaike (1973). Combining the above derivations gives the famous **Akaike Information Criterion (AIC)** .   □

We can conclude from the above derivation that the AIC can be interpreted more deeply than simply as a balance between goodness of fit and complexity. This deeper meaning can be drawn from the Kullback–Leibler divergence, which we write explicitly in the following important property:

Property 9.2   The AIC in (9.1.16) serves as estimate for

$$2E_{Y_1,\ldots,Y_n}\{KL(g(.),f(.;\hat{\theta}))\} - 2\int \log\big(g(y)\big)g(y)dy. \tag{9.1.23}$$

9.1.3   AIC for Model Comparison

We will now discuss how the AIC can be applied to model selection. Looking at (9.1.23), we see that the latter component is unknown, and hence, the absolute value of the AIC is not informative for us. Consequently, we are not able to explicitly estimate a value for the expected Kullback–Leibler divergence $E_{Y_1,\ldots,Y_n}\{KL(g(.),f(.;\hat{\theta}))\}$. In other words, we can never evaluate how close $f(;\theta)$ is to $g(.)$. However, the AIC is eminently useful for *relative* comparisons. Let us demonstrate the concept with two candidate models. Assume that we have the two models $f_1(.;\theta_1)$ and $f_2(.;\theta_2)$. These could be two completely different distributions or the same distribution with two different parameterisations. A common setting is $f_1(.) = f_2(.) = f(.)$, meaning they belong to the same distributional model, but θ_1 can be derived from θ_2 by setting some values of θ_2 to zero. This is also called **nested model selection**, as $f_1(.)$ is a special case of $f_2(.)$. We can now calculate the AIC values for the two models and compare them. Assume that θ_2 is p_2 dimensional and θ_1 is p_1 dimensional with $p_2 > p_1$. Then

$$AIC(1) = -2\sum_{i=1}^{n} \log\big(f(y_i;\hat{\theta}_1)\big) + 2p_1$$

$$AIC(2) = -2\sum_{i=1}^{n} \log\big(f(y_i;\hat{\theta}_2)\big) + 2p_2.$$

In this case, let us assume that AIC(1) < AIC(2). At first glance, this is a surprising result. This means that we are better able to model, as measured by our approximation of the expected Kullback–Leibler divergence, the true unknown density $g(.)$ with the smaller model $f_1(.)$ than with more complex model $f_2(.)$. In the above case, $f_1(.)$ is just a special case of $f_2(.)$ with some parameters set to zero, and clearly we are able to get as close to $g(.)$ with $f_2(.)$ as we can with $f_1(.)$. With a higher dimensional model, we achieve more flexibility, and hence the smallest Kullback–Leibler divergence between $f_2(.)$ and $g(.)$ is always smaller than that of $f_1(.)$ and $g(.)$. However, the AIC does not measure how close we can get, but how close we are on average, after estimating the unknown parameters, i.e. we take the expectation with respect to $Y_1, \ldots, Y_n$. This takes into account the fact that additional parameters are estimated in $f_2(.)$, which are set to zero in $f_1(.)$ and this implies that we suffer from more estimation variability with $f_2(.)$ compared to $f_1(.)$, which in turn makes the expected Kullback–Leibler divergence of $f_2(.)$ larger than that of $f_1(.)$. To sum up, we emphasise that the AIC compares both how close we can get with the model to the unknown distribution $g(.)$ and the variability of parameter estimation.

It also helps to simply understand the AIC as a balance between the fit and complexity of a model, which is best explained in a regression context. To demonstrate, let

$$Y_i = \beta_0 + x_{1i}\beta_1 + x_{2i}\beta_2 + \varepsilon_i,$$

where x_{i1} and x_{i2} are known covariates. For the residuals, we assume normality $\varepsilon_i \sim N(0, \sigma^2)$ and independence. The log-likelihood is given by

$$l(\beta, \sigma) = \sum_{i=1}^{n} -\log(\sigma) - \frac{1}{2}\frac{(y_i - x_i\beta)^2}{\sigma^2},$$

where $x_i = (1, x_{i1}, x_{i2})$ and $\beta = (\beta_0, \beta_1, \beta_2)^T$. The Maximum Likelihood estimate is then

$$\hat{\beta} = (X^T X)^{-1} X^T y$$

$$\hat{\sigma}^2 = \sum_{i=1}^{n}(y_i - x_i\hat{\beta})^2/n,$$

where X is the design matrix and y is the vector of observations as defined in (7.1.5). The AIC is then given by ($p = 2$ in this example)

$$AIC = n\log(\hat{\sigma}^2) + 2(p + 2).$$

Note that if we include more covariates, the estimated variance $\hat{\sigma}^2$ becomes smaller, but the number of parameters grows. The term $\log(\hat{\sigma}^2) = \log\left(\sum_{i=1}^{n}(y_i - x_i\hat{\beta})^2/n\right)$ measures the goodness of fit, i.e. how close the predicted value $x_i\hat{\beta}$ is to the observed value y_i. The term $2(p + 2)$ measures the complexity of the model, in this case how many parameters have been fitted. These two aspects, goodness of fit and the number of parameters, are balanced by the AIC.

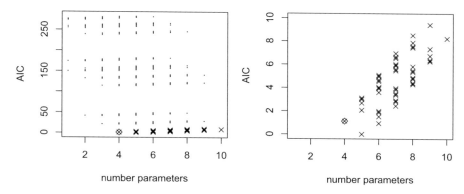

Fig. 9.1 AIC values for all 1024 possible models. Models including the true 4 covariates are indicated as crosses. The true model is shown as diamond. The right-hand plot is zoomed in on the true model, just showing the models with the smallest AIC

Example 41 Let us demonstrate the AIC with a simulated example. We generate 10 independent covariates $x_{ij} \sim N(0, 1)$, where $i = 1, \ldots, 200$ and $j = 1, \ldots, 10$, and simulate

$$Y_i \sim N(0.25x_{i1} + 0.25x_{i2} + 0.1x_{i3} + 0.1x_{i4}, 1).$$

Hence, the response variable depends on the first 4 covariates, while the remaining 6 are spurious. We fit $2^{10} = 1024$ possible regression models, with each covariate either absent or present in the model. Given the AIC value of the 1024 possible models, we want to select the best model. The results are visualised in Fig. 9.1, where we plot the number of parameters included in the model against the resulting AIC value. We indicate models that correctly contain the first 4 covariates with crosses and the true model also with a diamond. It appears that models that include the true covariates clearly reduce the AIC, but the best model selected by the AIC contains 5 parameters. This means that, in this particular example, we falsely select one of the spurious variables, which by chance explains some of the residual error, but correctly include the four relevant variables in the model. This reflects a known property of the AIC: that it tends to select overly complex models, i.e. models with too many parameters.

$\triangleright$

9.1.4 Extensions and Modifications

Bias-Corrected AIC

Akaike originally proposed (9.1.16) as an approximation of the expected Kullback–Leibler divergence (9.1.15). Despite the potential violation of this elegant interpretation, alternatives to the multiplicative factor 2 of the second term have nevertheless been proposed. The most prominent suggestion that still conforms to Akaike's framework is a bias-corrected version of the AIC, proposed by Hurvich and Tsai (1989)

$$AIC_C = -2l(\hat{\theta}) + 2p\left(\frac{n}{n-p-1}\right).$$

As n increases, the correction term $n/(n-p-1)$ converges to 1, but for small samples the bias-corrected version is generally advised. A rule of thumb given by Burnham and Anderson (2002) states that the corrected AIC is preferred if n/p is less than 40.

The Bayesian Information Criterion

Very similar to the AIC is the **Bayesian Information Criterion (BIC)**. To motivate this, let us consider model selection from a Bayesian perspective. Assume we have a set of models $M_1, \ldots, M_K$, each of which corresponds to a distributional model, such that

$$M_k \Leftrightarrow Y_i \sim f_k(y; \theta_k) \quad i.i.d.$$

Model selection can now be interpreted as a decision problem and a plausible strategy is to select the most likely model or, in Bayesian terminology, the model with the highest posterior probability. To this end, we need to calculate the posterior probability of each model. For model M_k, this is given by

$$P(M_k|y) = \frac{f(y|M_k)P(M_k)}{f(y)} = \frac{\int f_k(y; \vartheta_k)f_{\theta_k}(\vartheta_k)d\vartheta_k}{f(y)}P(M_k), \qquad (9.1.24)$$

where $P(M_k)$ is the prior belief in model M_k. The denominator $f(y)$ is calculated by summing over all models

$$f(y) = \sum_k \int f_k(y; \vartheta_k)f_{\theta_k}(\vartheta_k)d\vartheta_k.$$

Although (9.1.24) allows us to quantify the posterior model probability, it is often complicated, or even infeasible, to calculate. As we already discussed in Chap. 5, this model needs to be approximated, possibly with MCMC or other alternatives. However, this problem is even more complex here, as we need to approximate the integral in each of K models separately. We therefore pursue a simplifying approach and apply a Laplace approximation to the above integral using a slightly modified version of our original approach in Sect. 5.3.2. To begin, let

$$l_k(\theta_k) = \sum_{i=1}^{n} \log f_k(y_i; \theta_k)$$

be the log-likelihood for model M_k and define with $\hat{\theta}_k$ the corresponding Maximum Likelihood estimate. With the Laplace approximation, the integral component in (9.1.24) is given by

$$\int \exp\left(l_k(\vartheta_k)\right) f_{\theta_k}(\vartheta_k) d\vartheta_k \approx \left(\frac{2\pi}{n}\right)^{\frac{p_k}{2}} \exp\left(l_k(\hat{\theta}_k)\right) f_{\theta_k}(\hat{\theta}_k) \left|\frac{1}{n} I_k(\hat{\theta}_k)\right|^{-\frac{1}{2}},$$

where p_k is the dimension of θ_k. Taking the log of the right-hand side gives

$$\frac{p_k}{2} \log(2\pi) - \frac{p_k}{2} \log(n) + l_k(\hat{\theta}_k) - \frac{1}{2} \log \left|\frac{1}{n} I_k(\hat{\theta}_k)\right| + \log f_{\theta_k}(\hat{\theta}_k).$$

If we collect all the components that grow with order n, i.e. the second and third, and multiply them by -2, we obtain the **Bayesian Information Criterion** (BIC)

$$BIC_k = -2l_k(\hat{\theta}_k) + \log(n) p_k.$$

Maximising the posterior probability of a model therefore corresponds, at least approximately, to minimising the BIC.

Definition 9.2 The Bayesian Information Criterion (BIC) is defined as

$$BIC = -2l(\hat{\theta}) + \log(n) p.$$

The BIC clearly appears very similar to the AIC. The main difference is that the multiplication by 2 in the AIC is replaced by a $\log(n)$ in the BIC. As $\log(n) > 2$ for $n > 7$, for any reasonably sized sample, the BIC prefers models with fewer parameters, as compared to the AIC. It should also be noted that the BIC essentially shows the diminishing influence of the prior distribution with increasing n, which asymptotically goes to 0. This holds for both priors: the prior on the parameter $f_k(\theta_k)$ and the prior on the model $P(M_k)$.

Example 42 We continue with the previous example, but now calculate the BIC for each of the 1024 possible models, which can be seen in Fig. 9.2. The models containing the true parameter are shown as crosses and the true model as a diamond. The right-hand plot is an enlarged version of the left-hand plot, focusing on the area

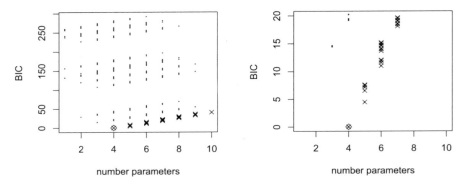

Fig. 9.2 BIC values for all 1024 possible models. Models including the true 4 covariates are indicated with crosses and the true model with a diamond. The right-hand plot is zoomed in on the true model

around the true model. Comparing the plot with Fig. 9.1, we see that the BIC tends to prefer simpler models with fewer parameters, with the true model having the lowest BIC score in this case.

▷

Deviance Information Criterion

The **Deviance Information Criterion** (DIC) was proposed by Spiegelhalter et al. (2002) and can be seen as a Bayesian version of the AIC. We do not give a detailed discussion here but only motivate its construction. The term deviance usually refers to how much the maximised likelihood for a particular model differs from that of the full model, which is sometimes called the saturated model. For this section, it is sufficient to define the deviance $D(y; \theta)$ as the log-likelihood multiplied by negative 2,

$$D(y; \theta) := -2l(\theta).$$

If θ_0 is the true parameter, then the difference in the deviance is

$$\Delta D(y; \theta_0, \hat{\theta}) := D(y; \theta_0) - D(y; \hat{\theta})$$
$$= 2\{l(\hat{\theta}) - l(\theta_0)\}.$$

As $\hat{\theta}$ is the Maximum Likelihood estimate, we can clearly see that $\Delta D(y; \theta, \hat{\theta}) \geq 0$. In Sect. 4.4, we investigated the likelihood-ratio and derived the asymptotic distribution (4.4.3)

$$2\{l(\hat{\theta}) - l(\theta)\} \overset{a}{\sim} \mathcal{X}_p^2,$$

and in particular we get

$$E\left(2\{l(\hat{\theta}) - l(\theta)\}\right) \approx p$$

with p as the dimension of θ. Taking a Bayesian view, we can consider θ as random and $\hat{\theta}$ as fixed, conditional on the data. We continue with this line of reasoning and replace $\hat{\theta}$ with the posterior mean estimate $\hat{\theta}_{postmean}$, defined in (3.2.2). This leads us to the deviance difference

$$\Delta D(y; \theta, \hat{\theta}_{postmean}) = 2\left\{l(\hat{\theta}_{postmean}) - l(\theta)\right\}.$$

Spiegelhalter et al. (2002) use this quantity to derive a deviance-based approximation for the parameter dimension

$$p_D := E(\Delta D(y; \theta, \hat{\theta}_{postmean})|y) = \int \Delta D(y; \vartheta, \theta_{postmean}) f_\theta(\vartheta|y) d\vartheta$$

$$= \int D(y, \vartheta) f(\vartheta|y) d\vartheta - D(y, \hat{\theta}_{postmean}).$$

The first integral can be approximated using MCMC as in Chap. 5. The **Deviance Information Criterion** (DIC) is now defined by

$$DIC = D(y, \theta_{postmean}) + 2p_D = \int D(y, \vartheta) f_\theta(\vartheta|y) d\vartheta + p_D.$$

The DIC had a wide-ranging influence on model selection in Bayesian statistics, even though, when strictly following the Bayesian paradigm, the selection of a single model is of little interest. Instead, the calculation of its posterior probability is of central importance, which we will explore later in this chapter.

Cross Validation

Recall that the AIC aims to estimate the Kullback–Leibler divergence

$$E_{Y_1,\dots,Y_n}(KL(g(.); f(,; \hat{\theta}))) = E_{Y_1,\dots,Y_n}\left\{E_Y\left(\log \frac{g(Y)}{f(Y|\hat{\theta}(Y_1,\dots,Y_n))}\right)\right\}$$

as motivated in (9.1.16). If we now replace the Kullback–Leibler divergence with a squared distance, we can motivate the AIC as a theoretical counterpart to cross validation, as discussed in Sect. 8.5. Assume that we are interested in the mean of Y, which we denote with

$$\mu = E(Y) = \int y g(y) dy.$$

To predict the mean, we use the model $f(.; \theta)$, such that the prediction is given by

$$\hat{\mu} = \int yf(y; \hat{\theta})dy,$$

where $\hat{\theta}$ depends on the data $y_1, \ldots, y_n$. The mean squared prediction error is then given by

$$E_Y\left((Y - \hat{\mu})^2\right) = \int \left\{y - \hat{\mu}(y_1, \ldots, y_n)\right\}^2 g(y)dy.$$

This replaces the Kullback–Leibler divergence in (9.1.14). We intend to select a model that minimises the mean squared error of prediction (MSEP), i.e.

$$E_{Y_1, \ldots, Y_n}\left\{E_Y[Y - \hat{\mu}(Y_1, \ldots, Y_n)]^2\right\} = \int \{\int \{y - \hat{\mu}(y_1, \ldots, y_n)\}^2 g(y)dy\}g(y_1, \ldots, y_n)dy_1 \ldots dy_n.$$

(9.1.25)

Note that (9.1.25) also shows a double integral, namely over the observations and over a new value Y. The major difference is that the Kullback–Leibler divergence is replaced with a squared distance.

9.2 AIC/BIC Model Averaging

So far we have used the AIC and some alternatives to select a single best model. Let us change our focus a little and look at multiple different models simultaneously, weighted by their relative suitability in explaining the data. In order to do so, we need to derive weights for the different models that represent their suitability or validity. The AIC values of different models provide one such measure. To begin with, let $M_1, \ldots, M_K$ be a set of models and define with

$$\Delta AIC_k = AIC_k - \min_k AIC_k,$$

where AIC_k is the Akaike Information Criterion calculated for model k and $\min_k AIC_k$ is the minimum of all AIC values. Clearly, if $\Delta AIC_k = 0$, model k is the best model (possibly one of many) and would be chosen by AIC model selection. Moreover, if ΔAIC_k is large, then model k appears to be unsuitable for modelling the data. Burnham and Anderson (2002) propose that models with $\Delta AIC < 2$ are eminently suitable for the data, while models with $\Delta AIC > 10$ are essentially unsuitable.

The value of ΔAIC can be transformed into a weight, with Akaike (1983) proposing the simple value

$$\exp(-\frac{1}{2}\Delta AIC_k).$$

Instead of weighting, it seems more natural to construct model probabilities out of the weights

$$P(M_k|y) := \frac{\exp(-\frac{1}{2}\Delta AIC_k)}{\sum_{k=1}^{K} \exp(-\frac{1}{2}\Delta AIC_k)},$$

which is, however, not a true Bayesian approach, because prior probabilities for the different models are not incorporated. That being said, calculating probabilities from weights is nevertheless intuitive and the approach is somewhat self-explanatory. Instead of using the AIC, one can also use the BIC by simply replacing $\Delta_k AIC$ with

$$\Delta_k BIC = BIC_k - \min_k(BIC_k),$$

with obvious definitions for BIC_k and $\min_k BIC_k$. This in turn gives the model probabilities

$$P(M_k|y) := \frac{\exp(-\frac{1}{2}\Delta BIC_k)}{\sum_{k'=1}^{K} \exp(-\frac{1}{2}\Delta BIC_{k'})}.$$

Example 43 Let us return once again to the original regression example and calculate the model probabilities $P(M_k|y)$ with both the AIC and the BIC. These can now be used to determine the probability that a particular covariate is included in the model. This probability is a measure of the importance of the variable, which we define as follows. Let $\mathcal{I}_1$ be the index set of all models that contain the first covariate. Note that with 10 potential covariates, we have $|\mathcal{I}_1| = 2^9 = 512$ models that include covariate x_1. We then define the probability that covariate 1 is in the model with

$$P(\text{covariate 1 in model }|y) = \sum_{k \in \mathcal{I}_1} P(M_k|y).$$

Similarly, we can calculate the probabilities for all other covariates. We show the resulting probabilities for the 10 covariates in Table 9.1, for both the AIC and the BIC. We can see that the first four covariates, which are the covariates of the true model, are always included. The remaining spurious variables have rather high posterior probabilities for the AIC, with a probability of approximately 15% that the

Table 9.1 Probability in % that the covariate is included in the model

	x_1	x_2	x_3	x_4	x_5	x_6	x_7	x_8	x_9	x_{10}
AIC	100.0	100.0	100.0	100.0	13.3	12.0	14.6	15.6	16.2	14.0
BIC	100.0	100.0	100.0	100.0	0.1	0.1	0.1	0.3	0.1	0.1

variables are included in the model. For the BIC, however, they have practically no chance of inclusion.

$\triangleright$

9.3   Inference After Model Selection

Inference for Maximum Likelihood estimation as derived in Chap. 4 was based on the assumption that the data are drawn from $Y \sim f(.;\theta)$ *i.i.d.*. That is, we considered the model as given. We showed at the beginning of this chapter that the Maximum Likelihood approach remains reasonable for the estimation of θ_0, returning the closest approximation of model $f(y;\theta)$ as measured by the Kullback–Leibler divergence. The inference and asymptotic behaviour of the variance of $\hat{\theta}$, however, had a more complicated structure. These properties hold if we just consider a single model $f(.;\theta)$. This begs the question: what happens if we first employ model selection to obtain a model before estimating our parameters. To be specific, let us assume we have K models

$$M_k \Leftrightarrow Y_i \sim f_k(;\theta_k)   \quad i.i.d.,$$

where $k = 1, \ldots, K$, from which we select one model, e.g. the model that minimises the AIC. Let $\hat{k}$ be the corresponding model index, e.g.

$$\hat{k} = \arg\min_k \mathrm{AIC}_k.$$

The corresponding estimate is denoted with $\hat{\theta}_{\hat{k}}$, whose properties we now want to determine. In Chap. 4, we explored the behaviour of $\hat{\theta}_k$ for increasing n, but now the choice of model k is also informed by the data. Clearly, $\theta_1, \ldots, \theta_K$ may differ in size, i.e. the parameters in the different models $f_k(.;\theta_k)$ may be of a different dimensionality. For ease of explanation, let us limit ourselves to nested model classes. This means that we assume an overall model,

$$Y \sim f(.;\theta),$$

where $\theta_1, \ldots, \theta_K$ is given by θ with certain components set to 0. To make this more explicit, we need a slight change in notation and subsequently denote the parameter in the k-th model with $\theta_{(k)}$, that is, we set the model index in brackets. The probability model itself remains unchanged, but we simply set components of the parameter θ to zero, i.e. $f_k(y;) = f(y;\theta_{(k)})$. If $\theta = (\theta_1, \ldots, \theta_p)$ is a p-dimensional parameter and $\mathcal{I}_k$ is the index set of parameter components set to zero in the k-th model, i.e. $\mathcal{I}_k \subset \{1, \ldots, p\}$, then $\theta_{(k)}$ can be generated from θ by setting $\theta_j = 0$ for $j \in \mathcal{I}_k$. Assume now that we are interested in drawing inference about a single parameter $\vartheta = \theta_j$ for some fixed $j \in \{1, \dots, p\}$. Note that ϑ may be set to zero in some models and needs to be estimated in other models. Applying model selection as proposed above gives the estimate

$$\hat{\vartheta} = \hat{\theta}_{(\hat{k})j},$$

where $\hat{\theta}_{(\hat{k})}$ denotes the Maximum Likelihood estimate in the selected model and $\hat{\theta}_{(\hat{k})j}$ the corresponding estimate of component θ_j. This can be either zero if $j \in \mathcal{I}_{\hat{k}}$ or the estimate of the component if $j \notin \mathcal{I}_{\hat{k}}$. Hence, the estimate $\hat{\vartheta}$ can be fixed to zero but can also take values in $\mathbb{R}$, depending on the model that is selected. This property already shows that inference statements as derived in Chap. 4 do not directly apply and that we need deeper insight to understand how to derive inference after model selection.

Even though this is a relevant and very practical problem, it is surprisingly not often discussed in depth in statistics. Breiman (1992) called this the "quiet scandal", complaining that inference after model selection is not treated rigorously enough in statistics. With recent work for example by Leeb and Pötscher (2005), Berk et al. (2013) and Leeb et al. (2015), new results have been proposed, see also Claekens and Hjort (2008) or Symonds and Moussalli (2011) for a more comprehensive discussion of the topic. It appears that proper inference after model selection is methodologically demanding and rather complicated, which makes it difficult to use in an applied context. We will motivate these problems and make use of the model averaging approach that was explored in the last subsection.

Assume we select model k with probability π_k, where $\sum_k \pi_k = 1$. The probability may depend on the data, but let us for the moment assume that π_k is fixed and, somewhat unrealistically, known in advance. In practice, we would replace π_k with the model weights defined in the previous section, that is

$$\pi_k = \frac{\exp(-\frac{1}{2}\Delta AIC_k)}{\sum_{k'=1}^{K} \exp(-\frac{1}{2}\Delta AIC_{k'})},$$

which clearly depends upon the data. However, to keep the mathematics simple, we assume, as said, that π_k is known for all k. Let $\hat{\theta}_{(k)}$ be the estimate if we select model k, and define with $\hat{\vartheta}_k$ the component estimate in model k. That is, $\hat{\vartheta}_k$ equals 0 if the component is not in the model, i.e. if index $j \in \mathcal{I}_k$ where $\vartheta = \theta_j$. Given a model k, we saw in Sect. 9.1.1 that $\hat{\vartheta}_k$ converges to ϑ_k, where ϑ_k is the closest parameter to the true unknown model, as measured by the Kullback–Leibler divergence between the distributions. If the component is not in the model, then $\hat{\vartheta}_k$ is set to zero and clearly $\vartheta_k = 0$. Now let

$$\bar{\vartheta} = \sum_{k=1}^{K} \pi_k \vartheta_k$$

be the average of the parameter ϑ in the different models. This gives the expectation

$$E(\hat{\vartheta}) = \sum_{k=1}^{K} \pi_k E(\hat{\vartheta}|\text{model } k) = \sum_{k=1}^{K} \pi_k \vartheta_k = \bar{\vartheta},$$

where the conditional expectation $E(.|\text{model } k)$ denotes the expected value if we consider model k to be the true model. The variance is calculated using Property 2.3, that is

$$Var(\hat{\vartheta}) = E_{\text{model}}(Var(\hat{\vartheta}|\text{model})) + Var_{\text{model}}(E(\hat{\vartheta}|\text{model}))$$

$$= \sum_{k=1}^{K} \pi_k Var_k(\hat{\vartheta}_k) + \sum_{k=1}^{K} \pi_k (\vartheta_k - \bar{\vartheta})^2, \tag{9.3.1}$$

where $Var_k(\hat{\vartheta}_k)$ denotes the variance if model k is considered true. Estimating the variance in (9.3.1) is somewhat clumsy. The ad hoc approach would be to replace the unknown quantities in (9.3.1) with their estimates. This is not a problem for the first component, i.e. we can replace $Var_k(\hat{\vartheta}_k)$ with its estimate $\widehat{Var_k}(\hat{\vartheta}_k)$. The variance is either zero, if ϑ_k is equal to zero in model k, or the inverse Fisher information of model k. To replace the second component in (9.3.1), we first need the averaged estimate

$$\hat{\bar{\vartheta}} = \sum_{k=1}^{K} \pi_k \hat{\vartheta}_k.$$

If we now replace the latter sum in (9.3.1) with

$$\sum_{k=1}^{K} \pi_k (\hat{\vartheta}_k - \hat{\bar{\vartheta}})^2, \tag{9.3.2}$$

we can see that the estimates are correlated, as they are all derived from the same data. That is, $Cor(\hat{\vartheta}_k, \hat{\vartheta}_{k'}) \neq 0$ and in fact the correlation will be high, possibly even close to 1. This induces a bias and precludes the use of (9.3.2) as an estimate for the second component in (9.3.1). Instead, Burnham and Anderson (2002) proposed the estimation of the variance (9.3.1) with

$$\widehat{Var}(\hat{\vartheta}) = \left\{ \sum_{k=1}^{K} \pi_k \sqrt{\widehat{Var_k}(\hat{\vartheta}_k) + (\hat{\vartheta}_k - \hat{\bar{\vartheta}})^2} \right\}^2.$$

Given their results, this variance estimate can also be used as an estimate for the variance of the averaged estimator $Var(\hat{\bar{\vartheta}})$. Recently, Kabaila et al. (2016) came to the conclusion that "it seems difficult to find model-averaged confidence intervals that compete successfully with the standard confidence intervals based on the full model". In other words, it appears most useful to just fit the full model with no parameter component set to zero and take the variance estimate from this model to quantify the variability of $\hat{\vartheta}$ or $\hat{\bar{\vartheta}}$. However, one may also follow the recommendation of Burnham and Anderson (2002) who state on Page 202 "Do not include (...) p-values when using the information theoretic approach as this inappropriately mixes different analysis paradigms". That is to say, one either uses model selection to get $\hat{\vartheta}$, which is either the Maximum Likelihood estimate if it is in the selected model or zero if not, or one tests whether $\vartheta \neq 0$, but never both. An alternative may be to follow a bootstrap strategy and run both the model selection and the estimation to obtain bootstrap confidence intervals, which is certainly recommended if sufficient computing resources are available. The bootstrap approach also allows us to accommodate the uncertainty from the estimation of the model probabilities π_k using AIC weights, whose derivation would be even more complicated. In other words, inference after model selection is complex and clumsy, and if a simple and practical approach is desired, then bootstrapping appears as suitable method, at the cost of heavy computation.

9.4   Model Selection with Lasso

A conceptually attractive alternative to model selection is the Least Absolute Shrinkage and Selection Operator (Lasso) approach. The Lasso allows us to pursue both model selection and estimation in a single step. The method was proposed by Tibshirani (1996). Here we provide an overview using a penalised likelihood approach. A comprehensive discussion of the method is provided in Hastie et al. (2015). Let $l(\theta)$ be the likelihood of a statistical model, and let $\theta = (\theta_1, \ldots, \theta_p)$ be the parameter. We assume that p is large and our aim is, as in the previous section, to select a model by setting numerous components of θ to zero. Let $\mathcal{I}$ be the index of components of θ set to zero through model selection, where $\mathcal{I} \subset \{1, \ldots, p\}$. The Lasso approach aims to find the maximum of the extended, or penalised, log-likelihood

$$l_p(\theta, \lambda) = l(\theta) - \lambda \sum_{j \in \mathcal{I}} |\theta_j| \qquad (9.4.1)$$

with λ as the tuning parameter. Clearly, setting $\lambda = 0$ gives the normal likelihood, while setting $\lambda \to \infty$ implies that $\theta_j \equiv 0$ for all $j \in \mathcal{I}$. Hence, the term λ plays the role of a model selection parameter.

The penalised likelihood (9.4.1) can be solved with iterative quadratic programming. To do so, let $\hat{\theta}_{(0)}$ be a starting value, for instance, the parameter estimate with $\theta_j = 0$ for $j \in \mathcal{I}$, and set $\theta_{(t)} = \hat{\theta}_{(0)}$. We approximate the log-likelihood with

$$l(\theta) \approx l(\theta_{(t)}) + s(\theta_{(t)})(\theta_{(t)} - \theta) - \frac{1}{2}(\theta_{(t)} - \theta)I(\theta_{(t)})(\theta_{(t)} - \theta) =: Q(\theta_{(t)}, \theta).$$

The right-hand side is a quadratic function in θ. This allows us to approximate the Lasso likelihood (9.4.1) with

$$Q(\theta_{(t)}, \theta) - \lambda \sum_{j \in \mathcal{I}} |\theta_j| \to \max.$$

It can be shown that the above maximisation problem is equivalent to

$$Q(\theta_{(t)}, \theta) \to \max \text{ subject to } \sum_{j \in \mathcal{I}} |\theta_j| \leq c \qquad (9.4.2)$$

for some c, which is clearly related to our penalty term λ. Equation (9.4.2) gives a quadratic optimisation problem, which leads us to a new estimate $\theta_{(t)}$. We set $\theta_{(t+1)} = \hat{\theta}_{(t)}$ and this new value is in turn used to derive $Q(\theta_{(t+1)}, \theta)$, which gives the next iteration.

The Lasso approach will result in an estimate where numerous parameter values of the index set $\mathcal{I}$ are zero. This can be nicely visualised, as shown in Fig. 9.3. We show the log-likelihood for a two-parameter model with $\hat{\theta} = (\hat{\theta}_1, \hat{\theta}_2)$ as Maximum Likelihood estimate, i.e. when λ is (9.4.1) is set to zero. The grey shaded area shows the parameter values with $|\theta_1| + |\theta_2| \leq 3.7$, i.e. $c = 3.7$ in (9.4.1). Clearly, the constrained likelihood in (9.4.2) is maximised if θ_1 is set to zero.

The Lasso penalisation can also be taken from a Bayesian perspective. In this case, we interpret the penalty as the logarithm of a prior distribution over the coefficients θ_j, $j \in \mathcal{I}$. That is, we assume

$$f_\theta(\theta_j, j \in \mathcal{I}) \propto \exp(-\sum_{j \in \mathcal{I}} |\theta_j|)$$

or, due to its factorisation,

$$f_{\theta_j}(\theta_j) \propto \exp(-|\theta_j|) \quad i.i.d. \text{ for } j \in \mathcal{I}.$$

The resulting distribution is a Laplace prior (see Park and Casella 2008), i.e.

$$(\theta_j; j \in \mathcal{I}) \sim \prod_{j \in \mathcal{I}} \frac{1}{\sigma} \exp\left(-\frac{|\theta_j|}{\sigma}\right),$$

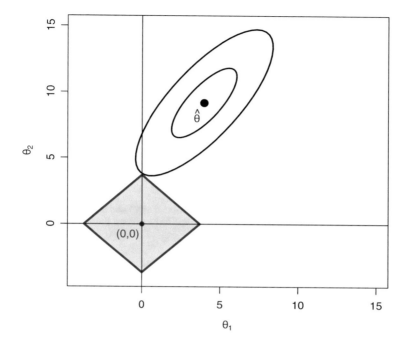

Fig. 9.3 Representation of Lasso penalty

where σ is a scaling parameter. In this case, σ plays the role of a hyperparameter, which needs to be set by the user or estimated from the data as per empirical Bayes.

Variance estimates for the fitted parameters after Lasso estimation have been derived by Lockhart et al. (2014). As motivated in the previous section, inference after model selection is clumsy, so we refer to the cited article for details, see also Hastie et al. (2015).

Example 44 Let us continue with the data from the previous examples and apply the Lasso estimation approach. The likelihood in this case results in a normal distribution, such that the Lasso estimate minimises

$$\frac{1}{2n}\sum_{i=1}^{n}(y_i - \beta_0 - \sum_{j=1}^{p} x_{ij}\beta_j)^2 + \lambda \sum_{j=1}^{p}|\beta_j| \rightarrow \min_{\beta}.$$

The resulting estimates for different values of λ are shown in Fig. 9.4. The left-hand plot shows the estimates β_j plotted against $\log \lambda$. It is more convenient to show the estimates plotted against $\sum_{j=1}^{p}|\hat{\beta}_j|$, which is shown in the right-hand plot. For clarification we include the true values of the parameters as horizontal grey lines, which are 0.25 for the β_1 and β_2 and 0.1 for β_3 and β_4. We see that the Lasso

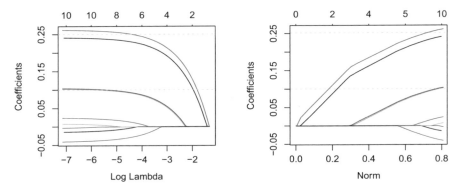

Fig. 9.4 Lasso estimates for different values of λ (left-hand side) and the norm $\sum_{j=1}^{p} |\hat{\beta}_j|$ (right-hand side). The true parameter values are shown as dashed horizontal lines

approach picks the correct values of β and sets the other coefficients to zero as long as $\lambda > \exp(-3)$.

▷

9.5   The Bayesian Model Selection

We already briefly addressed the subject of Bayesian model selection in Sect. 6.7, where we introduced the Bayes factor. We will extend this here a little and sketch the fundamentals of how the Bayesian paradigm can allow us to validate and select a model. We refer, for example, to Ando (2010) for a deeper discussion of the field. Let $M_1, \ldots, M_K$ be a set of K candidate models, each parameterised by θ_k. For each model, we define with $f_{\theta_k}(.)$ the prior distribution of the parameter, such that

$$f(y|M_k) = \int f_k(y; \theta_k) f_{\theta_k}(\theta_k) d\theta_k$$

is the distribution of the data in the k-th model, where $f_k(.)$ denotes the distribution if the k-th model holds. We further include the model prior $P(M_k)$ for $k = 1, \ldots, M$, where we may allow simpler models to have a higher prior probability. This, by extension, is equivalent to the penalisation of complex models. The posterior model probability is then

$$P(M_k|y) = \frac{f(y|M_k) P(M_k)}{f(y)},$$

where $f(y) = \sum_{k=1}^{K} f(y|M_k)P(M_k)$. Model selection is now simply choosing the model with the largest posterior probability. We may also use the posterior model probability to pursue model averaging, taking $P(M_k|y)$ as the weight. This setting can be further extended by sampling directly from the space of all possible models. Generally, the Bayesian view is quite suitable for model selection, and in fact the idea of model averaging discussed in Sect. 9.2 is very much in this vein. We do not go deeper into the field here but refer to Berger and Pericchi (2001), Cui and George (2008) or Hoeting et al. (1999) for further reading.

9.6 Exercises

Exercise 1 (Use R Statistical Software)
Let us once again consider Example 41. We generate 10 independent covariates $x_{ij} \sim N(0, 1)$, where $i = 1, \ldots, n$ and $j = 1, \ldots, 10$, and simulate

$$Y_i \sim N(10 + 0.25x_{i1} + 0.25x_{i2} + 0.1x_{i3} + 0.1x_{i4}, 1) .$$

Note that we include an additional intercept term $\beta_0 = 10$. Therefore, the response variable depends only on the first 4 covariates with true coefficients $\beta = (10, 0.25, 0.25, 0.1, 0.1)$, while the remaining 6 are spurious. Perform the following with different sample sizes $n = 50, 100, 150, 200, 500, 1000, 5000$ and with $S = 500$ repetitions.

1. For every of the $S = 500$ experiments, find the best model (of all possible models) using the AIC criterion. Calculate the percentage of experiments in which the selected model (1) contains exactly four correct predictors and (2) contains at least four correct predictors. Calculate (using the best model) the average value of all 10 coefficients over the $S = 500$ experiments.
2. Instead of selecting the best model, use the function stepAIC of the R package MASS to select a final model and repeat the experiments again $S = 500$ times. Evaluate the coverage probabilities, i.e. the percentage of cases where the confidence interval contains the true coefficient (coefficient-wise, assume a value of zero and a variance of zero for a coefficient not included in the final model). Comment on your results.
3. Evaluate how the results change, when you increase or decrease the absolute size of the coefficients or the number of coefficients which are different from zero. As an example, consider $\beta = (10, 5, 5, 1, 1)$ and $\beta = (10, 0.05, 0.05, 0.01, 0.01)$ and $\beta = (10, 1, 1, 1, 1, 1, 0.1, 0.1, 0.1, 0.1, 0.1)$.

Chapter 10
Multivariate and Extreme Value Distributions

Up to this point, we have mainly focused our efforts on univariate distributions. This was mostly just to keep the notation simple. Multivariate data, however, appear often in practice and multivariate distributions are eminently useful and important. It is time now to formalise multivariate distributions and explicitly discuss models for multivariate observations. The workhorse in this field is certainly the multivariate normal distribution, which we will explore in depth in Sect. 10.1. Beyond the normal distribution, copula-based distributions have been a hot topic in recent years. Copulas allow for complex dependence structures and this chapter provides a short introduction to the basic ideas of copulas in Sect. 10.2. Besides modelling multivariate data, it often occurs that extreme events are of interest, e.g. a maximal loss or a minimal supply level. This is addressed in Sect. 10.3, where we introduce extreme value distributions.

10.1 Multivariate Normal Distribution

10.1.1 Parameterisation

In Sect. 2.1 we introduced many properties of multivariate random variables and defined the multivariate normal distribution. When modelling multivariate data, if the individual covariates are all independent, they can simply be addressed with individual univariate models. Therefore, the principal objective in multivariate statistics is to investigate and model dependencies between the variables. Throughout the following chapter, we assume that

$$Y = (Y_1, \ldots, Y_q)^T$$

© The Author(s), under exclusive license to Springer Nature Switzerland AG 2021
G. Kauermann et al., *Statistical Foundations, Reasoning and Inference*,
Springer Series in Statistics, https://doi.org/10.1007/978-3-030-69827-0_10

is a q-dimensional random variable, which is drawn from a multivariate distribution. Hence, we observe

$$y_i = (y_{i1}, \ldots, y_{iq})^T,$$

where y_i is a realisation of Y_i. We assume that Y_i is drawn from

$$Y_i \sim f(; \theta) \quad i.i.d.$$

with the parameter vector θ representing all parameters of the distribution. For example, for a multivariate normal distribution, $Y_i \sim N(\mu, \Sigma)$. This means that θ consists of the mean vector $\mu = (\mu_1, \ldots, \mu_q)^T$ and the covariance matrix Σ. If we let $f(.; \theta)$ denote the density of a q-dimensional normal distribution, it was shown in Chap. 2 that

$$f(y; \theta) = \frac{1}{(2\pi)^{\frac{q}{2}}} |\Sigma|^{-\frac{1}{2}} \exp\left\{-\frac{1}{2}(y - \mu)^T \Sigma^{-1}(y - \mu)\right\}.$$

Given the data $y_1, \ldots, y_n$ and with the results of Chap. 4, we obtain the log-likelihood

$$l(\mu, \Sigma; y_1, \ldots, y_n) = -\frac{n}{2} \log |\Sigma| - \frac{1}{2} \sum_{i=1}^{n} (y_i - \mu)^T \Sigma^{-1}(y_i - \mu). \qquad (10.1.1)$$

With a little calculation, the resulting Maximum Likelihood estimates are given by

$$\hat{\mu} = \frac{1}{n} \sum_{i=1}^{n} y_i = \left(\frac{1}{n} \sum_{i=1}^{n} y_{i1}, \ldots, \frac{1}{n} \sum_{i=1}^{n} y_{iq}\right)^T$$

$$\hat{\Sigma} = \frac{1}{n} \sum_{i=1}^{n} (y_i - \hat{\mu})(y_i - \hat{\mu})^T.$$

These estimates are easy to derive by differentiating (10.1.1) and setting all components of the score vector to zero.

Let us now investigate whether this multivariate normal model can be simplified, i.e. if we are able to reduce the dimensionality of the parameters. To do so, we focus primarily on the covariance matrix Σ and mostly ignore the role of the mean components in μ. In practice, we can centre the variables by using $Y_{ij} - \mu_j$ for $j = 1, \ldots, q$ or, empirically, with $y_{ij} - \bar{y}_j$, where $\bar{y}_j = \frac{1}{n} \sum_{i=1}^{n} y_{ij}$. Note that Σ is symmetric and hence has $q(q + 1)/2$ parameters. We also know that Σ is positive definite, meaning that it comes from the set of positive definite $q \times q$ matrices

$$\Sigma \in \{A : A \text{ is symmetric and } a^T A a > 0 \text{ for all } a \in \mathbb{R}^q\}.$$

Our aim is now to reduce the above parameter set, that is, to constrain the set of possible variance matrices. This can be split into two main objectives. Firstly, we want to reduce the number of parameters. We have extensively discussed this point in Chap. 9, where we applied model selection by setting certain parameters to zero. Secondly, this is somewhat more relevant for practical purposes, we want to restrict the parameter space such that we can obtain models that allow for interpretations in terms of independences and conditional independences. This is shown subsequently by making use of different multivariate models.

10.1.2   Graphical Models

The simplest approach to reducing the parameter space would be to check whether particular components of Σ can be set to zero. For the normal distribution, this implies independence of the form

$$\Sigma_{jk} = 0 \quad \Leftrightarrow \quad Y_{ij} \text{ and } Y_{ik} \text{ are independent}$$

for $i = 1, \ldots, n$. To simplify notation, we will subsequently drop the observation index and simply write $Y_{.j}$ and $Y_{.k}$. Such independence is called marginal independence, as it holds when looking only at the variables in question. In other words, we completely ignore all of the other variables Y_l for $l = \{1, \ldots, q\}/\{j, k\}$.

An alternative form of independence is called conditional independence, which always considers all observed data. Conditional independence corresponds to zero entries in the **concentration matrix** $\Omega = \Sigma^{-1}$. Let us more thoroughly explain and demonstrate this concept. To begin with, let $Y_{.\overline{\{j,k\}}}$ be defined as $\{Y_l, l = 1, \ldots, q \text{ and } l \neq j \text{ and } l \neq k\}$. Hence, $Y_{.\overline{\{j,k\}}}$ is the random vector without components $Y_{.j}$ and $Y_{.k}$. Note that $Y_{.j}$ and $Y_{.k}$ are conditionally independent given $Y_{.\overline{\{j,k\}}}$ if and only if

$$f(y_{.j}, y_{.k} | y_{.\overline{\{j,k\}}}) = f(y_{.j} | y_{.\overline{\{j,k\}}}) f(y_{.k} | y_{.\overline{\{j,k\}}}). \tag{10.1.2}$$

For simplicity, we have omitted the parameters in the previous densities and will do so in subsequent notation. To be more specific, conditional independence occurs if and only if the density factorises to

$$f(y) \propto h(y_{.j}, y_{.\overline{\{j,k\}}}) g(y_{.k}, y_{.\overline{\{j,k\}}}). \tag{10.1.3}$$

This is easily seen because

$$f(y_{.j}, y_{.k} | y_{.\overline{\{j,k\}}}) = \frac{f(y)}{f_{.\overline{\{j,k\}}}},$$

which factorises as in (10.1.2) if and only if $f(y)$ decomposes as in (10.1.3). In the multivariate normal distribution, we have

$$f(y_{.j}, y_{.k} | y_{.\overline{\{j,k\}}}) \propto f(y)$$

$$\propto \exp\{-\frac{1}{2}y^T \Omega y\}$$

$$\propto \exp\left(-\frac{1}{2}\sum_{l=1}^{q}\sum_{m=1}^{q} y_{.l}y_{.m}\Omega_{lm}\right),$$

where Ω_{lm} refers to the (l, m) elements of $\Omega = \Sigma^{-1}$. We see that conditional independence in the multivariate normal distribution occurs if there is no component in the density that contains both $y_{.j}$, $y_{.k}$ and $y_{.\overline{\{j,k\}}}$. Consequently, if the (j, k) component of Ω (and apparently the (k, j) component, as Ω is symmetric) is equal to zero, we obtain a factorisation that matches (10.1.2). We can therefore conclude that $Y_{.j}$ and $Y_{.k}$ are conditionally independent, given the rest of the variables, if and only if $\Omega_{jk} = (\Sigma^{-1})_{jk} = 0$. This shows that the inverse of the covariance matrix plays an interesting role, in that it reflects conditional dependence between the variables.

This property of conditional independence has led to the development of an entire model class, today known as **graphical models**. The name is derived from the fact that these conditional independences can simply and elegantly be visualised in a graph. The term graph here refers to its mathematical definition $G = (V, E)$, which is defined by a set of nodes V and edges between the nodes E. We define the q variables as nodes, i.e. $V = \{1, \ldots, q\}$ and the existing dependencies as edges $E \subset V \times V$. A graphical model is defined such that $(j, k) \notin E$ and $(k, j) \notin E$, if and only if $\Omega_{jk} = 0$ and $\Omega_{kj} = 0$. We can easily visualise the conditional dependence structure of a q-dimensional random vector with a graph, where a missing edge indicates a conditional independence. The model class is extensively described in Whittaker (1989) and Lauritzen (1996) or more recently in Højsgaard et al. (2012). For a general overview and discussion on equivalent conditional independence statements, we refer to Wermuth and Cox (1996). Let us demonstrate the idea with an example.

Example 45 For this example, we make use of data that describe the crime rate for different criminal offences in the USA. The rates are given per state, with each state considered an independent observation. We look at the following eight criminal acts: damage to *property, murder, rape, robbery, assault, burglary, larceny* and *car theft*. We assume that the rates follow a multivariate normal distribution and our aim is to simplify the concentration matrix Ω by setting as many entries to zero as possible. We select a model using the AIC, as discussed in the previous chapter. This gives a sparse concentration matrix, which is represented by the resulting graph, shown in Fig. 10.1. If there is no edge between variables, which are represented as nodes, we can infer conditional independence. For instance, *car theft* and *rape* are not connected, implying that the criminal rates of car thefts and rape are conditionally

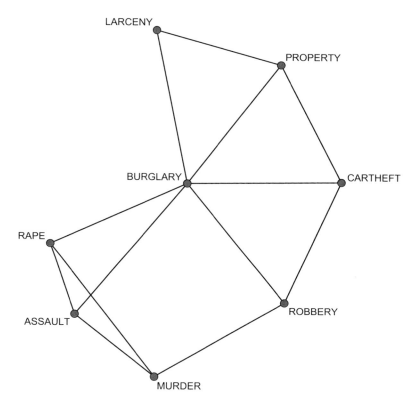

Fig. 10.1 Graph model

independent, given the values of the remaining variables. Looking at Fig. 10.1, we
see numerous missing edges. To explore the consequences of this sparseness, we
need to specify further independence properties, which can be easily read from the
graph. For instance, all paths from the node *car theft* to the node *rape* go via *burglary*
or *robbery*. This implies that the criminal rates of *car theft* are independent of *rape*,
conditional on the values of *burglary* and *robbery*. In other words, we do not need
to condition on all variables, but only on the intersecting set in the graph. This
independence interpretation is also valid and shows that the particular pattern of
zeros in $\Omega = \Sigma^{-1}$ allows for interpretation beyond simple pairwise independence.

$\triangleright$

  Independence is generally notated with the following symbol $\perp\!\!\!\perp$, which mirrors
that independence can be understood as a type of orthogonality in multivariate data.
In other words, the conditional independence, as given in (10.1.2), could be notated
with

$$i \perp\!\!\!\perp j \mid (V \setminus \{i, j\}),$$

Fig. 10.2 Graphical model

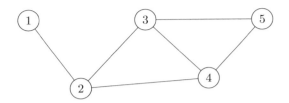

where $V = \{1, \ldots, q\}$ is the index set of all variables. This statement is also called the **pairwise Markov property** (see e.g. Lauritzen (1996)). The graph representation easily allows us to derive induced independences, which we can demonstrate with an example. Let us look at the graphical model shown in Fig. 10.2. We see that Node 1 is not directly connected to Node 5, which we denote with $1 \perp\!\!\!\perp 5|\{2, 3, 4\}$. In fact, there are several edges missing, which lead to the following list of conditional independence statements:

- $1 \perp\!\!\!\perp 3|\{2, 4, 5\}$
- $1 \perp\!\!\!\perp 4|\{2, 3, 5\}$
- $1 \perp\!\!\!\perp 5|\{2, 3, 4\}$
- $2 \perp\!\!\!\perp 5|\{1, 3, 4\}$

Note that in all statements above, we always condition on all variables apart from the pair being considered. These four conditional independences allow the derivation of further statements. For example, we can see that all paths from Node 1 to Nodes 3,4 or 5 go via Node 2. This allows us to state

$$1 \perp\!\!\!\perp \{3, 4, 5\}|2.$$

This is called the **global Markov property**. The same reasoning allows the statement

$$\{1, 2\} \perp\!\!\!\perp 5|\{3, 4\},$$

as all paths from $\{1, 2\}$ to 5 go via $\{3, 4\}$. A third Markov property results if we condition on the neighbours. For instance, Node 5 has neighbours 3 and 4. If we condition on these, then the remaining nodes are conditionally independent; that is,

$$5 \perp\!\!\!\perp \{1, 2\}|\{3, 4\}.$$

This property is called the **local Markov property**. Similarly, Node 3 has neighbours $\{2, 4, 5\}$, such that

$$3 \perp\!\!\!\perp 1|\{2, 4, 5\}.$$

These properties are defined as follows.

Definition 10.1 Let $G = (V, E)$ be an undirected graph expressing conditional independence. G satisfies the following Markov properties:

- Pairwise Markov Property: For two non-adjacent nodes i and j, we have

$$i \perp\!\!\!\perp j | V \setminus \{i, j\}.$$

- Local Markov Property: Defining with $N(i)$ the neighbours of node i, we get

$$i \perp\!\!\!\perp V \setminus \{N(i), i\} | N(i).$$

- Global Markov Property: For two subsets $A, B \subset V$ with $A \cap B = \emptyset$, assume that there is no direct edge from A to B and every path from A and B goes via $C \subset V$, and then

$$A \perp\!\!\!\perp B | C.$$

It can be shown that these three Markov properties are equivalent. This demonstrates the benefit of graphical models, as simple and visible properties of the graph allow for clear interpretation with respect to variable dependencies. Graphical models are also extensively used in spatial data analysis. The idea is that, when observing data on a grid, it is reasonable to assume that the observation at one grid point may depend on the neighbouring grid points, but not on any others. This leads us to Gaussian–Markov Random Fields, which are discussed, for instance, in Rue and Held (2005). We assume in this case that spatial observations are connected in a grid form, as visualised in Fig. 10.3. The plot was produced using the R package `bamlss` (2017) and shows the districts of London. We induce a Markov structure, such that each measurement taken on a district level is conditionally independent of the remaining measurements when conditioned on its direct neighbours. This is represented by the overlaid network structure. Note that this setting also allows for additional modifications to the dependence structure. For instance, in the graph in Fig. 10.3, we took the river Themes into account and did not connect districts separated by the river. This is reasonable for measurements where the spatial dependence relies upon accessibility of neighbouring districts, i.e. transport. However, if we consider airborne measurements, then neighbouring districts should nevertheless be connected. We do not want to go any deeper here and close by emphasising that graphical models allow for various extensions and applications which, at their core, are simply modelling the inverse covariance matrix Σ^{-1}.

10.1.3 Principal Component Analysis

As the covariate dimension q grows large, the dimension of the corresponding covariance matrix Σ may explode intractably. A commonly used strategy for

Fig. 10.3 Graphical Model for spatial dependence showing the city of London

simplifying Σ is to make use of techniques from linear algebra. Note that with spectral decomposition, we get

$$\Sigma = U \Lambda U^T,$$

where $U \in \mathbb{R}^{q \times q}$ is the matrix of orthonormal eigenvectors of Σ and $\Lambda = diag(\lambda_1, \ldots, \lambda_q)$ is the diagonal matrix of eigenvalues. This is clearly just a different representation of Σ at the moment and not a simplification. The matrix can, however, be simplified if we order the eigenvalues $\lambda_1 \geq \lambda_2 \geq \ldots \geq \lambda_q$ and prune them. That is, we set small values of λ to zero, such that Λ is replaced by $\tilde{\Lambda} = diag(\lambda_1, \ldots, \lambda_r, 0, \ldots, 0)$ with $r << q$. This in turn gives

$$\tilde{\Sigma} = U \tilde{\Lambda} U^T = \tilde{U} diag(\lambda_1, \ldots, \lambda_r)\tilde{U}^T,$$

where $\tilde{U}$ is the matrix built from the first r columns of U. Typically, r is chosen to be rather small and by doing so we can simplify the model significantly. The resulting distribution can be rewritten using the **Karhunen–Loève** expansion. To demonstrate, let Z be a q-dimensional standard normal distributed random vector. Hence,

$$Z \sim N \left(0, \begin{pmatrix} 1 & & \\ & \ddots & \\ & & 1 \end{pmatrix} \right). \tag{10.1.4}$$

If we multiply Z by the $q \times q$ dimensional matrix $U \Lambda^{1/2}$, it follows the distribution

$$U \Lambda^{1/2} Z \sim N(0, U \Lambda^{1/2} \Lambda^{1/2} U) = N(0, \Sigma).$$

Therefore, we obtain the distributional model for variables Y by multiplying independent standard normal variables Z by $U \Lambda^{1/2}$. We simplify the model if we let $\tilde{Z} = (z_1, \ldots, z_r)$, where $r << q$, and define $\tilde{Y}$ as

$$\tilde{Y} = \tilde{U} \tilde{\Lambda}^{1/2} \tilde{Z},$$

where $\tilde{Y}$ is q-dimensional, while $\tilde{Z}$ is only r-dimensional. Such model simplification can be obtained with **principal component analysis (PCA)**, where the covariance matrix is simplified by a smaller number of principal components. To demonstrate, let $Y \in \mathbb{R}^{n \times q}$ be the entire data matrix, where the rows of Y consist of observations $(y_{i1}, \ldots, y_{iq})$ and $n >> q$. We assume that the columns of Y (variables) are centred. Then, with spectral decomposition, we can write Y as

$$Y = V \Lambda^{1/2} U^T,$$

where $V \in \mathbb{R}^{n \times q}$ and $\Lambda^{1/2} \in \mathbb{R}^{q \times q}$ and $U \in \mathbb{R}^{q \times q}$. Matrix V contains the eigenvectors of $Y Y^T$ and U contains the eigenvectors of $Y^T Y$ with Λ as a diagonal matrix of the corresponding eigenvalues. Model simplification with principal components allows us to now simplify the data matrix Y with

$$\tilde{Y} = \tilde{V} \tilde{\Lambda}^{1/2} \tilde{U}^T,$$

where $\tilde{V}$ is the matrix built from the first r columns of V, corresponding to the r largest eigenvalues and, as above, $\tilde{U}$ are the first r columns of U.

Example 46 We continue with the crime rate example from above and simplify the model with the first r principle components. Note that the selection of r cannot directly be made with tools like the AIC, because the reduced covariance matrix $\tilde{\Sigma}$ does not have full rank and hence $|\tilde{\Sigma}| = 0$ and the inverse of $\tilde{\Sigma}$ does not exist. One may take a generalised inverse instead, like in formula (7.4.4). However, instead of the AIC, a more common metric is the explained variance over the entire variance. This is given by the quantity

$$\text{explained variance} = \frac{\sum_{l=1}^{r} \lambda_r}{\sum_{m=1}^{q} \lambda_m}.$$

Table 10.1 The total percentage explained variance for an increasing number of included principle components

r	1	2	3	4	5
Explained variance	92%	98 %	99%	99.5%	≈100%

Table 10.2 The first two eigenvectors of the covariance matrix for the crime data

Variable	First eigenvector	Second eigenvector
property	−0.83	−0.21
murder	0.00	−0.01
rape	−0.01	−0.01
robbery	−0.05	−0.19
assault	−0.04	−0.13
burglary	−0.27	−0.54
larceny	−0.44	0.73
car theft	−0.08	−0.26

We calculate this value for the crime rate data, where the eigenvalues are those of the empirical covariance matrix, shown as follows. Table 10.1 shows that only two principle components explain 98% of the variation, which suggests that the model can be simplified. Let us look at the first two eigenvectors, given in Table 10.2. These results can be interpreted as follows: the crimes *property, burglary* and *larceny* contribute most to the variance of crime rates. Moreover, *larceny* contributes differently to the variance as compared to *burglary, car theft, property* and *robbery*, which is mirrored in the larger values of the second eigenvector. Hence, if *larceny* increases (positive values of the second eigenvector U_2 in (10.1.4)), then *burglary, car theft, property* and *robbery* decrease.

▷

The decomposition and simplification of the covariance matrix of multivariate normally distributed variables have been the inspiration for many models in statistics. The field is called multivariate statistical analysis and we have only touched here on two possible model classes. Classical textbooks that show the full scope of possibilities are Mardia et al. (1979) or Anderson (2003). More recent descriptions of the field are provided for example by Härdle and Simar (2012) or Zelterman (2015).

10.2 Copulas

10.2.1 Copula Construction

The shape of the normal distribution is based on linear dependencies that are defined by the correlation matrix Σ. This allows us to model elliptic densities and hence elliptic dependencies. A completely different approach to modelling dependencies is available with **copulas**. Originally proposed by Sklar (1959), the model class has recently seen a boom in interest and has been widely implemented and extended in the last two decades. We refer to Nelsen (2006) or Krupskii and Joe (2015) for a comprehensive and thorough mathematical introduction. We also recommend Härdle and Okhrin (2010). The central idea is to define a unique function (called a copula), which describes the dependence between the variables. To formalise this concept, assume a multivariate random vector $Y. = (Y_{.1}, \ldots, Y_{.q})$ with a continuously differentiable distribution function

$$F(y_1, \ldots, y_q) = P(Y_{.1} \leq y_1, \ldots, Y_{.q} \leq y_q).$$

Sklar's theorem states that $F(.)$ can be uniquely formulated with a copula $C(.)$, such that

$$F(y_1, \ldots, y_q) = C(F_1(y_1), \ldots, F_q(y_q)), \tag{10.2.1}$$

where

$$C : [0, 1]^q \rightarrow [0, 1]$$

and $F_j(y_j) = P(Y_{.j} \leq y_j)$ is the marginal, univariate cumulative distribution function of variable $Y_{.j}$. The term "copula" emphasises that the function $C(.)$ couples the univariate distributions together to form a true multivariate cumulative distribution function. The copula $C(.)$ has some important properties. Firstly, $C(.)$ must be monotonically increasing, which is clear because the marginal distributions $F_j(.)$ are themselves monotonically increasing. Secondly, the copula $C(.)$ must be a valid cumulative distribution function on $[0, 1]^q$, which naturally implies again that $C(.)$ is monotonically increasing. If we now take the univariate margins, we get for the first component (and analogously for all other components)

$$
\begin{aligned}
C(F_1(y_1), 1, \ldots, 1) &= C(F_1(y_1), F_2(\infty), \ldots, F_q(\infty)) \\
&= F(y_1, \infty, \ldots, \infty) \\
&= P(Y_{.1} \leq y_1, Y_{.2} \leq \infty, \ldots, Y_{.q} \leq \infty) \\
&= P(Y_{.1} \leq y_1) = F_1(y_1).
\end{aligned}
$$

Hence, the univariate margins of $C(.)$ are all uniform distributions on $[0, 1]$, i.e. $C(u_1, 1, \ldots, 1) = u_1$ for $u_1 \in [0, 1]$. This shows that the univariate margins of $C(.)$ do not carry any information about the marginal distribution of $Y_{.j}$. This mirrors the property of interest, namely that copulas allow us to decompose the modelling of a multivariate distribution into two steps:

1. *Marginal Distribution:* Find a suitable marginal distribution $F_j(\cdot)$ for $Y_{.j}$ for $j = 1, \ldots, q$.

2. *Dependence Structure:* Find a suitable copula for $F_1(y_1), \ldots, F_q(y_q)$, which carries the dependence structure among the components $(Y_{.1}, \ldots, Y_{.q})$.

Let us explore this observation in a bit more depth. Assume that we have data $y_i = (y_{i1}, \ldots, y_{iq})$ for $i = 1, \ldots, n$. Then, instead of finding a joint q-dimensional distribution $F(.)$ such that

$$Y_i \sim F(.) \quad i.i.d.,$$

we simplify the task by modelling the univariate marginal distributions and the dependence structure separately. For the first task, the entire machinery of Chaps. 3 and 4, where we focused primarily on univariate distributions, can be applied. To demonstrate, let $\hat{F}_j(.)$ be an estimate of $F_j(.)$. This can either be a fitted parametric distribution, i.e. $\hat{F}_j(.) = F(; \hat{\theta}_j)$, or simply the empirical cumulative distribution, i.e. $\hat{F}_j(y_j) = \frac{1}{n} \sum_{i=1}^n 1_{\{y_{ij} \le y_j\}}$. To estimate the copula, we now calculate

$$\hat{u}_{ij} := \hat{F}_j(y_{ij}),$$

where $\hat{u}_{ij} \in [0, 1]$ for $i = 1, \ldots, n$ and $j = 1, \ldots, q$. That is, we transform our observation, such that $\hat{u}_i$ lies in the q-dimensional cube $[0, 1]^q$. In the second step, we aim to find a suitable copula for the transformed data. Hence, we consider $\hat{u}_i = (\hat{u}_{i1}, \ldots, \hat{u}_{iq})$ to be sampled *i.i.d.* from some copula $C(.)$, which now needs to be fitted.

As stated above, the copula is a cumulative distribution function $C(.) : [0, 1]^q \to [0, 1]$. If the univariate margins are continuous, as we have assumed, we can derive a density for the copula, which is often more intuitive and allows for better visualisation. The copula density is given by

$$c(u_1, \ldots, u_q) = \frac{\partial^q C(u_1, \ldots, u_q)}{\partial u_1 \ldots \partial u_q}.$$

Note that $c(u_1, \ldots, u_q)$ is a density on $[0, 1]^q$, with the side constraint that its univariate margins are uniform distributions on the interval $[0, 1]$. We will later visualise this density for a number of copulas. Typically, the copula and density depend on parameters, e.g. θ, that we suppress in the notation for simplicity. If we take $\hat{u}_i$ to be our observation, then we obtain the log-likelihood

$$l(\theta) = \sum_{i=1}^n \log c(\hat{u}_{i1}, \ldots, \hat{u}_{iq}),$$

which needs to be maximised to yield an estimate for θ.

The set of available copula models is extensive and will not be discussed exhaustively here. Instead, we present a few copulas and sketch the idea behind pair copula construction. Before doing so, we will quickly demonstrate the derivation of the multivariate density $f(y_1, \ldots, y_n)$ from the copula density. With (10.2.1), we obtain

$$f(y_1, \ldots, y_q) = \frac{\partial^q}{\partial y_1 \ldots \partial y_n} F(y_1, \ldots, y_q) = \frac{\partial C(F_1(y_1), \ldots, F_q(y_q))}{\partial y_1 \ldots \partial y_q}$$

$$= c(F_1(y_1), \ldots, F_q(y_q)) \prod_{j=1}^{q} f_j(y_j).$$

Hence, the density decomposes to the copula density multiplied by the product of the marginal densities. This again demonstrates that the model can be decomposed into the dependence structure and univariate marginal densities.

10.2.2 Common Copula Models

Gaussian and Elliptical Copulas

Let us begin with the Gaussian copula, which is based upon the normal distribution. It is defined by

$$C(u_1, \ldots, u_q) = \Phi(Z_1 \leq \Phi_0^{-1}(u_1), \ldots, Z_q \leq \Phi_0^{-1}(u_q); R),$$

where $\Phi(.; R)$ is the cumulative distribution function of an $N(0, R)$ distribution, with an $R \in \mathbb{R}^{q \times q}$ correlation matrix. Moreover, $\Phi_0(.)$ here denotes the univariate distribution function of an $N(0, 1)$ distribution. The copula density is given by

$$c(u_1, \ldots, u_q) = \frac{1}{|R|^{1/2}} \exp(-\frac{1}{2} z^T R^{-1} z), \tag{10.2.2}$$

where $z = (\Phi_0^{-1}(u_1), \ldots, \Phi_0^{-1}(u_q))$. In fact, assuming $Y_i \sim N(\mu, \Sigma)$, we can rewrite the density of Y_i to make the role of the copula more explicit. Note that for $Y_{ij} \sim N(\mu_j, \sigma_j^2)$, we obtain for the density

$$f(y_{ij}; \mu_j, \sigma_j^2) = \frac{1}{\sqrt{2\pi}\sigma_j} \exp(-\frac{1}{2}(y_{ij} - \mu_j)^2/\sigma_j^2) = \frac{1}{\sigma_j} \varphi_0(z_{ij}),$$

where φ_0 is the density of an $N(0, 1)$ distribution and $z_{ij} = (y_{ij} - \mu_j)/\sigma_j$ is the standardised version of y_{ij}. Note that z_{ij} depends on the parameters μ_j and σ_j, which clearly only relate to the corresponding univariate marginal distribution. From z_{ij}, we can easily calculate $u_{ij} = \Phi_0^{-1}(z_{ij})$, where Φ_0 is, as above, the $N(0, 1)$ distribution function. We can make use of (10.2.2) to derive an estimate for the correlation matrix R. We visualise this in Fig. 10.4. The left-hand plot shows data points, and their corresponding isolines, sampled from a bivariate normal

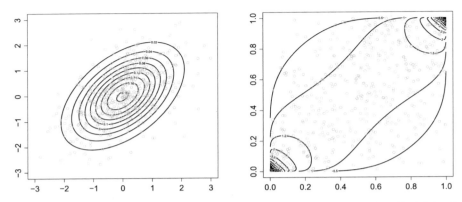

Fig. 10.4 Gaussian distribution with correlation 0.5 and standard $N(0, 1)$ margins. The left-hand plot shows simulated data and isolines of the density. The right-hand plot shows the isolines of the corresponding copula with the transformed data points

distribution, with variance 1 and correlation 0.5. The right-hand plot shows the resulting Gaussian copula, with the data given by $\Phi_0(y_{ij})$ for $i = 1, \ldots, n$ and $j = 1, 2$. It appears that the probability mass is shifted to the corners of the unit square. The Gaussian copula can be extended towards elliptic copulas, which are derived from elliptic distributions, whose density functions have elliptically shaped isolines, like those of the normal distribution.

Archimedean Copula

Archimedean copulas have the advantage that they only require few parameters, even when modelling high dimensional data. Unlike the Gaussian copula, where each pairwise dependence is modelled with a coefficient in the covariance matrix, Archimedean copulas model the entire dependence structure with one (or only a few) parameter(s). The principal idea is to make use of a parametric generator function

$$\psi(\,;\theta) : [0, 1] \to [0, \infty),$$

where $\psi(\,;\,)$ must be continuous, strictly decreasing and convex with the side constraint $\psi(1; \theta) = 0$ for all θ. Some examples are shown in Fig. 10.5. The corresponding copula is then defined by

$$C(u_1, \ldots, u_q; \theta) = \psi^{[-1]}(\psi(u_1; \theta) + \ldots + \psi(u_q; \theta); \theta),$$

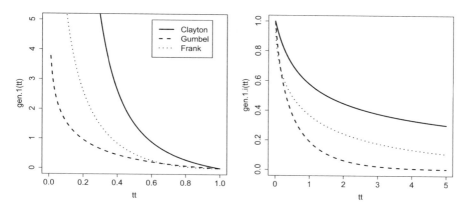

Fig. 10.5 Generator functions (left) and their inverse $\psi^{[-1]}$ (right) for Clayton, Frank and Gumbel copula with $\theta = 2$

where $\psi^{[-1]}(.)$ refers to the inverse of ψ for the appropriate range, that is,

$$\psi^{[-1]}(t; \theta) = \begin{cases} \psi^{-1}(t; \theta) & \text{for } 0 \le t \le \psi(0; \theta) \\ 0 & \text{otherwise.} \end{cases}$$

A few frequently used Archimedean Copulas are as follows:

- Clayton Copula:
  $C(u_1, \ldots, u_q; \theta) = \max(0, u_1^{-\theta} + \ldots + u_q^{-\theta} - q + 1)^{-\frac{1}{\theta}}$
  with $\theta > 0$ (or more precisely $\theta \ge -1/(q - 1)$), such that
  $\psi(t; \theta) = \frac{1}{\theta}(t^{-\theta} - 1)$.
- Frank Copula:
  $$C(u_1, \ldots, u_q; \theta) = -\frac{1}{\theta} \log \left\{ 1 + \frac{\prod\limits_{j=1}^{q} \{\exp(-\theta u_j) - 1\}}{\exp(-\theta) - 1} \right\}$$

  with $\theta \ge 1$ and corresponding generation function
  $\psi(t, \theta) = -\log\left(\frac{\exp(-\theta t) - 1}{\exp(-\theta) - 1}\right)$.
- Gumbel Copula:
  $$C(u_1, \ldots, u_q; \theta) = \exp\left\{ -\left(\sum_{j=1}^{q} -(\log(u_j))^{\theta} \right)^{\frac{1}{\theta}} \right\}$$

  for $\theta \ge 0$ and corresponding generator function
  $\psi(t; \theta) = (-\log(t))^{\theta}$.

We show the generator functions ψ and $\psi^{[-1]}$ for the above Archimedean copulas in Fig. 10.5. The resulting copulas are visualised in Fig. 10.6, where we consider, as above, the univariate marginal distributions as $N(0, 1)$ and set the parameter $\theta = 2$. We see that the approach allows us to model dense, non-elliptic dependencies that are strong in the lower left or the upper right corner. This shows not only the flexibility of copulas but also the need for model selection. Note that each of the copulas shown in Fig. 10.6 has a parameter which needs to be estimated from the data. Clearly, taking $u_{i.} = (u_{i1}, \ldots, u_{iq})$ as the data, we can write the likelihood as

$$l(\theta) = \sum_{i=1}^{n} \log c(u_{i.}, \theta),$$

where $c(u_{i.}; \theta) = c(u_{i1}, \ldots, u_{iq}; \theta)$ is the copula density of the chosen Archimedean copula $C(; \theta)$. Simple Maximum Likelihood allows us to obtain a parameter estimate for θ.

Pair Copula

Archimedean copulas are attractive because they allow us to model high dimensional dependencies with only a few parameters. On the other hand, this is also a serious constraint. A more flexible approach is to work with paired copulas. This has gained increasing interest in the last decade, see e.g. Czado (2010). We exemplify the principal idea with $q = 3$. Let $f_{123}(y_1, y_2, y_3)$ be the joint density of $Y_{.1}, Y_{.2}, Y_{.3}$. For any bivariate marginal density, we can write

$$f_{12}(y_1, y_2) = c_{12}(F_1(y_1), F_2(y_2))f_1(y_1)f_2(y_2),$$

which gives the univariate conditional distributions

$$f_{2|1}(y_2|y_1) = c_{12}(F_1(y_1), F_2(y_2))f_2(y_2).$$

The same holds for $f_{3|1}$, and with similar arguments we obtain

$$f_{3|12}(y_3|y_1, y_2) = c_{23|1}(F_{2|1}(y_2|y_1), F_{3|1}(y_3|y_1)|y_1)f_{3|1}(y_3|y_1)$$

$$= c_{23|1}(F_{2|1}(y_2|y_1); F_{3|1}(y_3|y_1)|y_1)c_{13}(F_1(y_1), F_3(y_3))f_3(y_3),$$

where $c_{23|1}(.)$ denotes the conditional pairwise copula density for $(Y_{.2}, Y_{.3})$ given $y_{.1}$. We can now factorise the three-dimensional distribution as follows:

$$f_{123}(y_1, y_2, y_3) = \qquad f_1(y_1)f_{2|1}(y_2|y_1) \qquad f_{3|2,1}(y_3|y_1, y_2)$$

$$= \quad f_1(y_1) \quad f_2(y_2) \quad f_3(y_3) \quad c_{12}(F_1(y_1), F_2(y_2))$$

$$c_{13}(F_1(y_1), F_3(y_3)) \; c_{23|1}(F_{2|1}(y_2|y_1), F_{3|1}(y_3|y_1)|y_1).$$

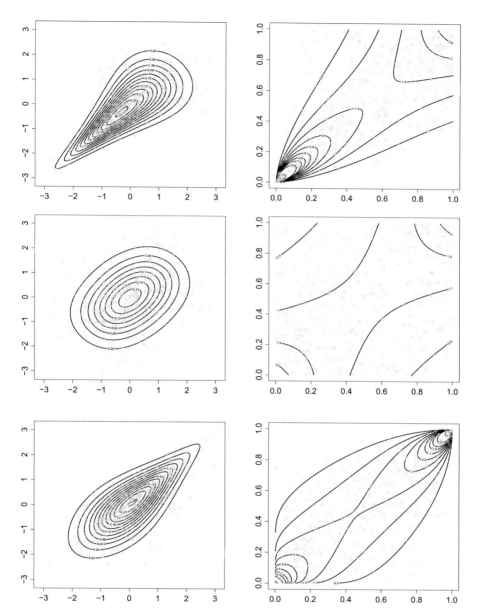

Fig. 10.6 Density and simulated data with $N(0, 1)$ margins from a Clayton Copula (1st row), Frank Copula (2nd row) and Gumbel Copula (3rd row). The left-hand column shows the data, and the right-hand column shows the copula density and the transformed data

We see that the three-dimensional copula density $c_{123}(F_1(y_1), F_2(y_2), F_3(y_3))$ decomposes to a product of three pairwise copulas. The idea of paired copulas is now to take advantage of this factorisation and assume that the conditional copula above does not depend on the value of $Y_{.1}$ in the conditional statement. This means that one assumes

$$c_{23|1}(F_{2|1}(y_2|y_1), F_{3|1}(y_3|y_1)|y_1) = c_{23|1}(F_{2|1}(y_2|y_1), F_{3|1}(y_3|y_1)).$$

In other words, the conditional copula depends on the arguments of the univariate conditional distributions, but the dependence of $Y_{.2}$ and $Y_{.3}$, which is expressed with the copula $c_{23|1}(.)$, does not depend on the value of $Y_{.1}$. We may simplify the notation above by omitting the arguments and writing

$$f_{123} = \left(\prod_{j=1}^{3} f_j \right) c_{12} c_{13} c_{23|1}.$$

With that, we obtain three pairwise copulas that need to be estimated. While estimation of c_{12} and c_{13} can be carried out by Maximum Likelihood as proposed in the above section, we need to look at $c_{23|1}(.)$. The arguments for the copula density are here $F_{2|1}(y_2|y_1)$ and $F_{3|1}(y_3|y_1)$, and hence we need estimates for the conditional distributions of $F_{2|1}$ and $F_{3|1}$. These estimates are again available from the copula itself, in this case from $C_{23|1}(F_{2|1}(y_2|y_1), F_{3|1}(y_3|y_1))$. This can be seen by noting that

$$F_{3|1}(y_3|y_1) = \int_{-\infty}^{y_3} c_{13}(F_1(y_1), F_3(v_3)) f_3(v_3) dv_3$$

$$= \int_{-\infty}^{y_3} \frac{\partial^2 C_{13}(F_1(y_1), F_3(v_3))}{\partial F_1(y_1) \partial F_3(v_3)} \frac{\partial F_3(v_3)}{\partial v_3} dv_3$$

$$= \frac{\partial}{\partial F_1(y_1)} \underbrace{\int_{-\infty}^{y_3} \frac{\partial^2 C_{13}(F_1(y_1), F_3(v_3))}{\partial F_3(v_3)} \frac{\partial F_3(v_3)}{\partial v_3} dv_3}_{\dfrac{\partial C_{13}(F_1(y_1), F_3(y_3))}{\partial v_3}}$$

$$= \frac{\partial C_{13}(F_1(y_1), F_3(y_3))}{\partial F_1(y_1)}.$$

In other words, conditional distributions can again be calculated from pairwise copulas, which in turn can be directly used in estimation. To do so, we replace $F_{3|1}(y_3|y_1)$ with $\partial \hat{C}_{13}(\hat{F}_1(y_1), \hat{F}_3(y_3))/\partial \hat{F}_1(y_1)$, where $\hat{C}_{13}$ is the fitted copula.

The key, and rather elegant, idea behind pair copulas is that the dependence structure is broken down to pairwise dependencies. Each pairwise copula itself can be chosen arbitrarily, meaning we can take elliptic, Archimedean or any other copula. The appropriate copula can be chosen using model selection approaches, as presented in Chap. 9. The procedure even scales to high dimensional models, where we need to calculate $q \cdot (q - 1)/2$ pairwise copulas to obtain a flexible joint distribution function for q variables. The approach has recently gained tremendous interest and has seen several extensions, which go well beyond the purview of this book. We refer to Joe (2014) for further details.

10.2.3   Tail Dependence

Copula models have been frequently used to model extreme events in dependent random variables. The classical example is the stock market, where the prices of stocks are clearly dependent upon one another. But what happens if the price for one (or more) stocks falls dramatically? How would this influence other stocks? This question brings us to the topic of tail dependence. Let us explain this for a bivariate setting.

Definition 10.2 Let Y_1 and Y_2 be two random variables with marginal distribution functions $F_1(.)$ and $F_2(.)$. The upper tail dependence of the variables is defined by

$$\lambda_{upper} = \lim_{u \to 1} P(Y_1 \geq F_1^{-1}(u)|Y_2 \geq F_2^{-1}(u)).$$

Accordingly, their lower tail dependence is given by

$$\lambda_{lower} = \lim_{u \to 0} P(Y_1 \leq F_1^{-1}(u)|Y_2 \leq F_2^{-1}(u)).$$

Tail dependence measures the probability of an extreme observation of one variable, if we know that the second variable has taken an extreme value. If the tail dependence is positive, we call the variables tail-dependent. If $\lambda_u = 0$ or $\lambda_l = 0$, we call them upper or lower tail-independent, respectively. For copulas, the tail dependence can be explicitly written as

$$\lambda_u = \lim_{u \to 1} \frac{1 - 2u + C(u, u)}{1 - u}$$

$$\lambda_l = \lim_{u \to 0} \frac{C(u, u)}{u}.$$

This can be easily derived by noting

$$P(Y_1 \leq F_1(u)|Y_2 \leq F_2(u)) = \frac{C(u, u)}{u}$$

because $P(Y_2 \leq F_2(u)) = u$. The above copula models have different tail dependences, listed in Table 10.3. Hence, the Clayton copula allows us to model

Table 10.3 Tail dependence
of classical copula models

Copula	λ_u	λ_l
Gaussian	0	0
Clayton	0	$2^{-1/\theta}$
Frank	0	0
Gumbel	$2-2^{1/\theta}$	0

lower tail dependence, while the Gumbel copula can model upper tail dependence. The Gaussian copula, on the other hand, cannot cope with tail dependence, which explains why Gaussian copulas are not suitable for stock market data.

10.3   Statistics of Extremes

Tail dependence allows us to model extreme events in multivariate data. But what can be said in general about extreme events? Many, if not most, statistical methods are focused on the general tendency of a variable, which is expressed by the mean or median of a distribution. Let us now shift the focus to the tails of the distribution, i.e. to extremely large or small values. Extreme values are sometimes labelled as outliers and often interfere with analysis. Methods for outlier detection and robust methods have been developed specifically to reduce this influence. However, there are many situations where the focus of the analysis *is* these outliers. For civil engineers who are building houses in areas with possible high winds, the maximum wind speed is definitely relevant. For insurance companies, extreme claims are of interest. The same is true for extreme changes in financial markets. Emil Julius Gumbel pioneered the statistics of extremes in 1958 with his first book, see Gumbel (1958). Analysing the frequency of extreme floods, which is relevant for building dams, was his original motivation. A more recent compilation of statistical extreme value theory is available in Coles (2001), see also Beirlant et al. (2004).

The typical approach to modelling extreme values is to consider the maxima of fixed numbers of random variables, sometimes called **block maxima**, for example, the monthly maxima of a river's water level or the monthly wind speed maxima. We are then interested in a distribution of random variables of the form $M_n = \max(Y_1, \ldots, Y_n)$ with *i.i.d.* random variables Y_i. Our intention is to develop a limit distribution for M_n, or, to put it differently, to investigate the behaviour of M_n for increasing n. In classical statistics, one is interested in means or sums, i.e. in random variables of the form $S_n = \sum_{i=1}^{n} Y_i$. According to the central limit theorem (see (2.4)), the random variable S_n asymptotically follows a normal distribution, such that

$$\frac{(S_n - nE(Y_1))}{\sqrt{nVar(Y_1)}}$$

converges to a standard normal distribution, if Y_i are $i.i.d.$ with mean $E(Y_1)$ and variance $Var(Y_1)$. There is an analogous theorem for block maxima. The distribution function of the maximum M_n is given by

$$F_{M_n}(y) = P(M_n \leq y) = P(Y_1 \leq y, \ldots, Y_n \leq y) = (F_Y(y))^n. \tag{10.3.1}$$

For $n \to \infty$, we see that M_n converges to a degenerate distribution

$$\lim_{n \to \infty} F_{M_n}(y) = \begin{cases} 1 & \text{if } F(y) = 1 \\ 0 & \text{if } F(y) < 1. \end{cases} \tag{10.3.2}$$

In other words, no useful limit distribution exists if we just look at the maximum. Instead, we need to standardise the maximum with

$$\frac{M_n - b_n}{a_n},$$

where (a_n, b_n) are fixed sequences depending on the distributional form of Y. Surprisingly, there are only three types of distributions, to which a standardised sequence of $(M_n - b_n)/a_n$ can converge for $n \to \infty$. This is the key result of extreme value theory and was proven by Fisher and Tippett (1928) and reformulated by de Haan (1970). This property is stated in the following theorem, whose proof we will also sketch.

Property 10.1 (Limiting Distributions)
Let $\{Y_i\}_{i=1}^{\infty}$ be $i.i.d.$ random variables and $M_n := \max(Y_1, \ldots, Y_n)$. If there are sequences of real numbers $\{a_n\}_{n=1}^{\infty}, \{b_n\}_{n=1}^{\infty}$ with $\frac{M_n - b_n}{a_n} \xrightarrow{d} Z$, as $n \to \infty$, then Z has a generalised extreme value (GEV) distribution. There are three types of GEV distributions, characterised by their distribution functions G_1, G_2 and G_3:

1. Extreme value distribution or Gumbel distribution:

$$G_1(x) = \exp(-\exp(-x)).$$

2. (Inverted) Weibull distribution:

$$G_2(x) = \begin{cases} \exp\left(-(-x)^{1/k}\right) & \text{for } x \leq 0 \\ 1 & \text{for } x > 0. \end{cases}$$

3. Fréchet Pareto distribution:

$$G_3(x) = \begin{cases} 0 & \text{for } x < 0 \\ \exp\left(-x^{-1/k}\right) & \text{for } x \geq 0. \end{cases}$$

Proof The proof makes use of the following simple property. A maximum of a matrix can be written as the maximum of the maxima of the rows of the matrix. We define

$$
\begin{pmatrix}
Y_{11} & \dots & Y_{1K} \\
Y_{21} & \dots & Y_{2K} \\
\vdots & & \\
Y_{L1} & \dots & Y_{LK}
\end{pmatrix}
$$

and set

$$
M_{LK} := \max_{1 \le l \le L} \max_{1 \le k \le K} (Y_{lk}) = \max_{1 \le k \le K} \{M^*_{(L)k}\},   \tag{10.3.3}
$$

where $M^*_{(L)k} = \max_{1 \le l \le L}(Y_{lk})$ is the maxima of the k-th column of the matrix. If the standardised sequence $(M_n - b_n)/a_n$ of the maximum of n copies of Y converges to a distribution Z, then clearly the sequence $\frac{M_{LK} - b_{LK}}{a_{LK}}$ also converges to Z, for suitable sequences (a_{LK}, b_{LK}) and $K, L \to \infty$. Furthermore, each standardised column maximum converges to the same limit, i.e. $(M^*_{(L)K} - b^*_K)/a^*_K$ converges to Z for some sequence (a^*_K, b^*_K). Using (10.3.3), we get

$$
F_{M_{LK}}(x) = F^K_{M^*_{(L)}}(x),
$$

where $F_{M_{LK}}(.)$ is the distribution function of M_{LK} and $F_{M^*_{(L)}}(.)$ is the distribution function of $M^*_{(L)k}$ for $k = 1, \dots, K$. Because both sequences converge to the same distribution function for a random variable Z after appropriate standardisation, we can derive the following property for the distribution function of variable Z:

$$
F_Z(c_K + d_K x) = [F_Z(x)]^K ,   \tag{10.3.4}
$$

for some sequences (c_K, d_K). Equation (10.3.4) is sometimes called the maximum stability equation. It can be shown that distribution functions that fulfil (10.3.4) have the form stated in Theorem 10.1.   □

The three types of distributions can be merged to the generalised extreme value distribution (GEV), which has three parameters.

Definition 10.3 The generalised extreme value distribution can be written as follows using three parameters, the location parameter μ, the scale parameter σ and the shape parameter γ. The distribution function is given by

$$
G(z) = \begin{cases} \exp\left(-(1 + \gamma z)^{-1/\gamma}\right) & \text{for } \gamma \neq 0 \\ \exp(-\exp(-z)) & \text{for } \gamma = 0 \end{cases}   \tag{10.3.5}
$$

with $z = (x - \mu)/\sigma$.

Similar to other distributions, μ is the location parameter and σ is the scale parameter. The shape parameter γ characterises the type of the distribution. For $\gamma = 0$, we get the Gumbel distribution, for $\gamma > 0$ the (inverted) Weibull distribution and for $\gamma < 0$ the Fréchet–Pareto distribution. The concept of the GEV gives us the ability to estimate distributions of maxima using only Maximum Likelihood, which removes the need to first define the specific type of extreme value distribution. This demonstrates the generality of the extreme value distribution approach and its usability, if the focus is on modelling extreme events.

Example 47 As an example, we analyse monthly maxima of nitrogen monoxide measurements at one of Munich's air pollution measurement stations (Lothstraße in the centre of Munich) for the years 2005–2014. The data follow a typical skewed distribution (see Fig. 10.7). We can estimate the parameters of the corresponding GEV with Maximum Likelihood, see Table 10.4. The distribution follows an inverted Weibull distribution. However, a Gumbel distribution also fits the data well, as the confidence interval of the shape parameter includes 0. The density function is plotted in Fig. 10.7. The estimated parametric distribution can now be used for inference about extreme quantiles. For further analysis, see Beirlant et al. (2004).

▷

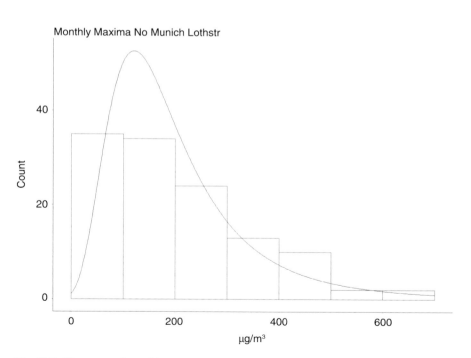

Fig. 10.7 Histogram of monthly maxima of NO taken at the measurement station Lothstraße in the centre of Munich. The solid line corresponds to the density function of the GEV estimated with Maximum Likelihood

Table 10.4 The result of Maximum Likelihood estimation of the generalised extreme value distribution parameters for the monthly maxima of NO taken at Munich Lothstraße

Parameter	EST	SE	95% CI
Location μ μ	137	9.7	[119; 157]
Scale σ	86	8.0	[71; 103]
Shape γ	0.22	0.11	[0;0.44]

10.4   Exercises

Exercise 1 (Use R Statistical Software)

Copulas are often used if one wants to model the dependency structure of high dimensional observations. In this exercise, we will take a look at the uranium data set that comes with the copula package. It contains the log concentrations of 7 chemical elements in 655 water samples. As a warm up, we will replicate the plots in the script before diving into the data.

1. Create a Gaussian copula with correlation 0.5 and one Archimedean copula (Frank, Gumble, Clayton) with parameter $\theta = 2$ (functions normalCopula, etc.). You can sample from these copulas using the rCopula function and get densities with dCopula. Finally, create bivariate distributions with the mvdc function (always assuming $N(0, 1)$ marginals).
2. Create scatter plots for both the copulas and the multivariate distributions, adding contours to visualise the densities. Another way of visualising bivariate distribution densities is 3D plots: try those out as well.
3. Take a look at the dataset. In order to visualise it, plot both the marginal and pairwise scatter plots of the data. When dealing with copulas, association measures that are often used are Spearman's Rho and Kendall's Tau as they do not rely on a linear relationship. Compute Spearman's Rho.

From now on, we will limit our analysis to Cobalt (Co) and Titanium (Ti) and model the bivariate distribution of their log concentrations.

4. Fit two normal distributions for the marginals using Maximum Likelihood. Make sure that your results describe the data reasonably well by plotting the resulting distributions over the data.
5. A priori, it is hard to choose a copula which models the dependency structure properly. In practice, model selection routines are often used to find the best fit from a pool of available copulas. Use the BiCopSelect function in the VineCopula package to fit different copulas with Maximum Likelihood, and select one using the AIC.
   Note that, according to the copula framework, the estimation of the copula does not depend on the marginals. Consequently, the BiCopSelect function expects pseudo observations (living in the unit square) computed through the pobs function.

6. Finally, visualise the data with a scatter plot, and add the multivariate distribution built from the optimal copula from Part 5. Compare the value of Spearman's Rho obtained in Part 3 with that resulting from our multivariate model. Discuss the fit.

   Note: There are several methods to more formally assess the goodness of fit for copulas which are not covered in this book.

Exercise 2 (Use R Statistical Software)

In subjects like engineering and insurance, predicting the extrema is often of great importance. Let us therefore take a look at the yearly maximum wind speed in miles per hour taken from "Extreme Value and Related Models in Engineering and Science Applications" by Castillo et al.

1. Plot the different extreme value distributions (Gumbel, (Inverted) Weibull, Fréchet–Pareto) for multiple parameters. How do they relate to the generalised extreme value distribution?
2. Read the data file evo.txt and visualise it.
3. Fit a generalised extreme value distribution to the data using Maximum Likelihood. Interpret the results. What distribution do you get?

   You may want to use the *fevd function in the extRemes package* and Part 3 to estimate the probability of a yearly maximum wind speed above 50/75/100/150 miles per hour. Interpret the results.

Chapter 11
Missing and Deficient Data

This chapter focuses on the quality and completeness of data, which is an often overlooked but essential part of data analysis. In fact, achieving the necessary quality of data and the required completeness is often more time consuming than the data analysis itself. One key aspect is missing data, referred to as "missingness" in technical literature, and the problems that can occur when this missingness is not addressed. We present statistical approaches for dealing with missing data and show how unbiased results can be obtained, even if the missingness is dependent upon the data itself. A central numerical tool in this context is the Expectation-Maximisation algorithm (or in short the EM algorithm), which make full use of available data. We also discuss the amount of information available in data, particularly with relation to the balance between quantity and quality of data. In particular, we want to demonstrate that quantity does not compensate quality. In other words, massive amounts of data do not induce massive information, true to the motto "garbage in, garbage out". Here, the question of the validity of the data is relevant. Finally, we investigate noisy measurements that is when the variables that we record do not measure what one intends to measure. We propose statistical models to correct for such errors and show how unbiased inference can be performed even from deficient data.

11.1 Missing Data

11.1.1 Missing Data Mechanisms

In practical applications, one is inevitably confronted with the problem of missing data, i.e. data entries with unobserved (or unobservable) values. In databases, this is usually coded with a unique number or character string, e.g. NA (Not Available) or MIS (MISsing). The incidence of these missing entries can follow various patterns.

© The Author(s), under exclusive license to Springer Nature Switzerland AG 2021
G. Kauermann et al., *Statistical Foundations, Reasoning and Inference*,
Springer Series in Statistics, https://doi.org/10.1007/978-3-030-69827-0_11

obs	Var 1	Var 2	Var 3	obs	R_1	R_2	R_3
1	x_{11}	NA	x_{13}	1	1	0	1
2	NA	NA	x_{23}	2	0	0	1
3	x_{31}	x_{32}	NA	3	1	1	0
4	x_{41}	x_{42}	x_{43}	4	1	1	1
5	x_{51}	NA	NA	5	1	0	0
6	NA	NA	NA	6	0	0	0

Fig. 11.1 Missing data in an example database. Observations 1, 2, 3 and 5 show item non-response, Observation 4 is complete and Observation 6 is a unit non-response. The right-hand side shows the missingness pattern with the indicator variables R_i

For example, the value of a single variable may be missing for some observations. For other observations, there may have been nothing recorded at all. In survey statistics, these are called **item non-response** and **unit non-response**, respectively. Figure 11.1 demonstrates the different response patterns. Item non-response is shown in Observations 1, 2, 3 and 5 and unit non-response in Observation 6.

The analysis of data with missing entries requires thorough investigation into patterns in their missingness and the processes that generated it. A simple, but clearly suboptimal, strategy is to completely ignore the missing data and work with a **complete case analysis**. This means dropping all observations that contain any missing entries. This heavy-handed exclusion of data is rather inefficient and we will also address later how it can lead to biased analyses. Let us demonstrate the former and assume we want to analyse a database with 1,000,000 observations, each with 100 variables. Our goal could be to investigate the dependence structure between the variables or, equivalently, to predict a single variable from the others. If each variable has approximately 1% of the entries missing and the missingness between the variables is mutually independent, then the proportion of complete observations is $0.99^{100} \approx 0.366$. Correspondingly, about 63% of the database would be excluded with a complete case analysis. Clearly, this is a waste of information and suggests the use of alternative strategies to handle missing data.

The strategy used to manage missing data depends heavily on the process that generates said missingness. In some cases, this process could be completely random. In others, it could be explainable with the non-missing values for that individual and finally, it could be dependent upon the missing variable itself. In the rest of this section, we will address these three different kinds of missingness and the effect of complete case analysis on further analyses. Let us start by formalising the notation for missing data. We define a row in the database with

$$Y_i = (Y_{i1}, \ldots, Y_{iq})$$

and assume that we are interested in the multivariate distribution

$$Y_i \sim F(y_i; \theta) = P(Y_i \le y_i; \theta).$$

For notational simplicity, from here on in, we denote the above probability with $P(Y_i)$ and drop the parameters in the notation when they are not pertinent. With

$$R_i = (R_{i1}, \ldots, R_{iq})$$

we define the missing pattern, where

$$R_{ij} = \begin{cases} 0 & \text{if data entry } Y_{ij} \text{ is missing (that is, } Y_{ij} = \text{NA)} \\ 1 & \text{otherwise.} \end{cases}$$

Hence, looking at Fig. 11.1, we express the missing pattern with the indicator variable R shown on the right. It is important to note that our missingness indicator R by definition has no missing data. In other words, the data of the right-hand plot in Fig. 11.1 is complete. A key property is now the distribution of the missingness pattern given the data, that is we aim to explore

$$P(R_i | Y_i).$$

The simplest distribution occurs when the absence of the data is completely at random and does not depend on Y_i at all.

Definition 11.1 If the distribution of R_i is independent of Y_i, this is called **Missing Completely at Random (MCAR)**. Hence we have

$$P(R_i | Y_i) = P(R_i).$$

MCAR is the simplest setting and, in this case, complete case analyses would still give unbiased results. This is easily demonstrated as follows. Looking at complete cases means that we condition on cases where $R_i = 1$, or, more precisely, $R_{ij} = 1$ for $j = 1, \ldots, q$. Because $P(R_i | Y_i) = P(R_i)$ we have

$$P(Y_i | R_i = 1) = \frac{P(R_i = 1 | Y_i) P(Y_i)}{P(R_i = 1)} = \frac{P(R_i = 1)}{P(R_i = 1)} P(Y_i) = P(Y_i).$$

In other words, if we include only complete observations, we may not incorporate all of the available information, but at least we will not produce biased results. To summarise, missing completely at random means that the missingness has no relation to the data. That is, the reason for this missingness is independent of both what is observed and not observed.

A more problematic, but still manageable, circumstance is if the missing pattern depends only on the observed variables. To demonstrate, let Y_i be divided into $Y_i =$

(Y_{iO_i}, Y_{iM_i}) where O_i is the set of observed and M_i the set of indices of missing variables for an observation. That is,

$$O_i = \{j : R_{ij} = 1\} \text{ and } M_i = \{j : R_{ij} = 0\}.$$

For simplicity of notation, we drop the observation index i from O_i and M_i and simply write Y_{iO} and Y_{iM}, bearing in mind that the missing pattern may be different for different observations.

Definition 11.2 If the distribution of R_i depends only on the observed variables, this is called **Missing at Random (MAR)**. Hence, we have

$$P(R_i|Y_i) = P(R_i|Y_{iO}).$$

Missing at random means that the missingness indicator R is <u>not</u> independent of the observations Y, but conditional on the data that has been observed it is independent on the data not observed. That is, missingness of the unobserved data does not depend on their (unobserved) values. In case of MAR, one needs to take special measures to obtain unbiased results. Complete case analysis will certainly lead to biased results, because

$$P(Y_i|R_i = 1) = \frac{P(R_i = 1|Y_i)P(Y_i)}{P(R_i = 1)} = \frac{P(R_i = 1|Y_{iO})P(Y_i)}{P(R_i = 1)} \neq P(Y_i)$$

and the ratio

$$\frac{P(R_i = 1|Y_{iO})}{P(R_i = 1)} \tag{11.1.1}$$

is almost certainly not equal to 1. This implies that complete case analysis will lead to biased results.

Let us demonstrate how to deal with MAR data with a concrete example. Assume we have data (Y_i, X_i, Z_i) and are interested in the probability model $P(Y|X, Z)$. This mirrors the classical regression setting, with Y as the response variable and X and Z as the explanatory variables. We then fit a regression model for Y given X and Z, as in Chap. 7. We assume that Z_i is always observed, but Y_i and X_i can be missing. Let R_{Y_i} and R_{X_i} be the corresponding missingness indicators. We will explore a number of possible scenarios, but let us first assume that the covariates X_i are always observed. That is, $R_{X_i} = 1$ for all $i = 1, \ldots, n$ and only our response variable Y_i can be missing. We constrain our analysis to the complete data, i.e. we condition on observations with $R_{Y_i} = 1$. The complete case analysis in this case yields

$$P(Y_i|X_i, Z_i, R_{Y_i} = 1) = \frac{P(Y_i, X_i, Z_i, R_{Y_i} = 1)}{P(X_i, Z_i, R_{Y_i} = 1)} = \frac{P(R_{Y_i} = 1|Y_i, X_i, Z_i)P(Y_i, X_i, Z_i)}{P(R_{Y_i} = 1|X_i, Z_i)P(X_i, Z_i)}$$

$$= \frac{P(R_{Y_i} = 1|X_i, Z_i)}{P(R_{Y_i} = 1|X_i, Z_i)}P(Y_i|X_i, Z_i) = P(Y_i|X_i, Z_i),$$

where the simplification follows from the assumed MAR condition. This implies that complete case analysis does not violate the estimation and we can unbiasedly estimate the conditional distribution of Y_i given (X_i, Z_i). Hence, the MAR assumption allows us to make use of a complete case analysis, if we are interested in the conditional distribution of one variable given the others and the missing pattern occurs only in the response variable Y_i.

Let us again examine the MAR scenario, this time assuming that only the explanatory variable can be missing and all response variables are observed. That is, we have $R_{Y_i} = 1$ for $i = 1, \ldots, n$, but R_{X_i} can take values 1 and 0. If we constrain our analysis to complete cases, this gives

$$P(Y_i | X_i, Z_i, R_{X_i} = 1) = \frac{P(Y_i, X_i, Z_i, R_{X_i} = 1)}{P(X_i, Z_i, R_{X_i} = 1)} = \frac{P(R_{X_i} = 1 | Y_i, X_i, Z_i) P(Y_i, X_i, Z_i)}{P(R_{X_i} = 1 | X_i, Z_i) P(X_i, Z_i)}$$

$$= \frac{P(R_{X_i} = 1 | Y_i, Z_i)}{P(R_{X_i} = 1 | Z_i)} P(Y_i | X_i, Z_i).$$

Hence,

$$P(Y_i | X_i, Z_i) = \frac{P(R_{X_i} = 1 | Z_i)}{P(R_{X_i} = 1 | Y_i, Z_i)} P(Y_i | X_i, Z_i, R_{X_i} = 1). \qquad (11.1.2)$$

The ratio in (11.1.2) is crucial and, if it is not equal to 1, we induce a bias with complete case analysis. To demonstrate this more clearly, let $P(Y_i | X_i, Z_i)$ depend on some parameter θ and define with $s_i(\theta)$ the corresponding score function (see Chap. 4). Letting $f(y_i | x_i, z_i; \theta)$ denote the density function (or probability function for discrete Y_i), we set

$$s_i(\theta) = \frac{\partial \log f(Y_i | x_i, z_i; \theta)}{\partial \theta} \quad \text{and} \quad s_{\text{compl}}(\theta) = \sum_{i=1}^{n} R_{X_i} s_i(\theta). \qquad (11.1.3)$$

If we restrict ourselves to complete data, calculating the conditional expectation of the score given x_i and z_i gives

$$E\left\{ s_{\text{compl}}(\theta) | x, z \right\} = \sum_{i=1}^{n} E\left\{ R_{X_i} s_i(\theta) \right\} = \sum_{i=1}^{n} \int P(R_{X_i} = 1 | y_i, x_i, z_i) s_i(\theta) f(y_i | x_i, z_i; \theta) dy_i$$

$$= \sum_{i=1}^{n} \int P(R_{X_i} = 1 | y_i, z_i) s_i(\theta) f(y_i | x_i, z_i; \theta) dy_i$$

which does not simplify to zero. With the results of Chap. 4, we can see that setting the complete case score to zero, i.e. $\sum_{i=1}^{n} R_i s_i(\hat{\theta}) = 0$ will yield a biased estimate $\hat{\theta}$. A simplification occurs if R_X depends only on Z_i and is conditionally independent of Y_i given Z_i. In this case $P(R_{X_i} = 1. | Y_i, Z_i) = P(R_{X_i} = 1 | Z_i)$ and the ratio in (11.1.2) takes value 1.

A correction of the above MAR bias can be obtained with **inverse probability weighting.** This was proposed by Robins et al. (1994), and even dates back to Horvitz and Thompson (1952). Assume that we know the probability of observing X_i, that is, we know

$$P(R_{X_i} = 1 | y_i, z_i) =: \pi_x(y_i, z_i).$$

We can then replace the complete case score by weighting its summands with the inverse of $\pi_x(y_i, z_i)$. This gives the weighted score function

$$s_{w,\text{compl}}(\theta) = \sum_{i=1}^{n} \frac{R_{X_i}}{\pi_x(y_i, z_i)} s_i(\theta). \tag{11.1.4}$$

As $E(R_{X_i} | y_i, x_i, z_i) = \pi(y_i, z_i)$, it is not difficult to show that

$$E\left\{ s_{w,\text{compl}}(\theta) | x, z \right\} = 0.$$

This in turn means that the modified complete case estimator $\hat{\theta}_w$, defined by

$$s_{w,\text{compl}}(\hat{\theta}_w) = 0$$

is asymptotically unbiased. Clearly, $\pi_x(y_i, z_i)$ is usually unknown and needs to be estimated. Fortunately enough, this is actually often possible. Note that R_{X_i} is observed for *all* data and we assumed that both Y_i and Z_i are also observed. As R_{X_i} is a binary variable, we can, for example, use logistic regression models as discussed in Chap. 7. That is, we fit a regression model with R_{X_i} as response variable and Y_i and Z_i as covariates. We could also use any other method for estimating $\pi_x(y_i, x_i)$, as discussed at the end of Chap. 7. With any suitable estimate available for $\pi_x(y_i, z_i)$, we insert this in (11.1.4) and obtain the approximate weighted score function

$$\hat{s}_{w,\text{compl}}(\theta) = \sum_{i=1}^{n} \frac{R_{X_i}}{\hat{\pi}(y_i, z_i)} s_i(\theta).$$

It can be shown that the resulting estimate $\hat{\theta}$, which solves $\hat{s}_{w,\text{compl}}(\hat{\theta}) = 0$, is asymptotically unbiased.

In reality, one often has both missing cases occurring in the same data. That is, we have observations where Y_i is missing or X_i is missing or both. This is visualised in the following table

		R_{Y_i}	
		1	0
R_{X_i}	1	(a)	(b)
	0	(c)	(d)

Field (a) has no missing data and we discussed fields (b) and (c) above. Clearly, field (d) is a combination of the (b) and (c). It is worth noting, as we showed for (b), that whether we observe Y_i or not makes no difference when fitting the model using $f(y_i|x_i, z_i; \theta)$. But it does matter when we derive an estimate for the weight $\pi_x(y_i, x_i)$, because here y_i is considered as a covariate and not as the response variable. In this case, we need to adapt the methods derived for case (c), but now for regression of R_{X_i} on Y_i and Z_i. This can be done, but is a little complicated and for simplicity we refer instead to McLeish and Struthers (2006) for a review or to the books Little and Rubin (2002) and Schafer (1997).

So far we have looked at the MCAR and MAR settings. But what happens if the missingness of the data depends upon the value that the missing data should have taken. In other words, whether we observe a value for some variable or not depends itself on the value of this variable, which we define as follows

Definition 11.3 If the missing pattern is not independent of the missing variables, even after conditioning on all observed variables, this is called **Missing Not At Random** (MNAR). In this case $P(R_i|Y_i)$ does not simplify to $P(R_i|Y_{iO})$.

Because Y_i is not observed, it can simply not be modelled. It appears that we are trapped. Data analysis in the presence of MNAR cannot be corrected to be unbiased. The situation is sometimes called inaccessible, see Graham and Donaldson (1993). A very common manifestation of this case is as follows. A survey is run and the respondents are asked to report their monthly salary. People with a high income might not be willing to answer this question, the same holds for those with a very low income. Hence, whether the question is answered or not, and consequently whether the data are observed, depends on the value of the variable. This is MNAR. We must conclude that, if there is sign of MNAR, the analysis of the data is questionable and validity of the results can neither be guaranteed nor validated. This does not sound promising, but at least we can clearly define the limits where statistical data analytics becomes risky and might be fundamentally invalid. From now on, we will therefore assume that the missing data either occur in an MCAR or MAR setting.

11.1.2 EM Algorithm

So far, we have discussed missing data patterns and their implications for complete case analysis. Let us now move on to more advanced approaches. We begin by examining how we can estimate a parameter by maximising the likelihood function under missing data. Let $Y_i = (Y_{i1}, \ldots, Y_{iq})$ be the observations for $i = 1, \ldots, n$ and, as before, assume that

$$Y_i \sim F(y; \theta) \quad i.i.d..$$

If we observe all of the data, the log-likelihood is given by

$$l(\theta) = \sum_{i=1}^{n} \log f(y_i; \theta) = \sum_{i=1}^{n} l_i(\theta).$$

But how can we calculate the likelihood if some data are missing? To address this, let us start by setting $Y_i = (Y_{iO_i}, Y_{iM_i})$ where $O_i = \{j : R_{ij} = 1\}$ is the index of the observed variables and $M_i = \{j : R_{ij} = 0\}$ that of the missing variables. We again drop the subscript i for O and M and just write Y_{iO} and Y_{iM}. Because we do not observe the missing variables, the log-likelihood for the observed values is given by

$$l_O(\theta) = \sum_{i=1}^{n} \log f(y_{iO}; \theta) = \sum_{i=1}^{n} l_{iO}(\theta), \qquad (11.1.5)$$

where

$$l_{iO}(\theta) = \log f(y_{iO}; \theta) = \log \int f(y_i; \theta) dy_{iM}. \qquad (11.1.6)$$

This means that we are calculating the marginal density of y_{iO} by integrating out all possible values of the unobserved Y_{iM}. In principle, the only thing we have to do now is to maximise this log-likelihood using the results from Chap. 4. Unfortunately, this can be rather complicated, or even infeasible, as $f(y_{iO}; \theta)$ can have a very complex form and may depend in a complicated way on the parameter θ. This is because the marginal distribution needs to be calculated by integration, as shown in (11.1.6), which can be cumbersome and lead to a complex distribution. Consequently, the log-likelihood (11.1.5) with missing data can be too complex to work with it. Moreover, if the missingness pattern varies from observation to observation, as shown in Fig. 11.1, this means even more numerical effort, because each observation requires its own set of calculations. We will therefore propose different strategies that avoid the computationally heavy calculation of the marginal log-likelihood (11.1.5).

A convenient solution is possible if the calculation of the conditional mean of the log-likelihood is simple, or at least numerically manageable. In this case, the Expectation Maximisation (EM) algorithm, one of the key algorithms in statistics, can prove very useful. Importantly, it makes use of the entire data and not just the complete cases. The algorithm was introduced in this seminal paper by Dempster et al. (1973), but was preceded by a number of implementations for specific applications, see Little and Rubin (1987), and has been shown to converge under quite broad regularity conditions, see Vaida (2005). The algorithm consists of two steps, the Expectation step and the Maximisation step, which are applied iteratively. The two steps can be comprehended in the following way. In the E-step we replace all missing data with their expected values, given the current fit of the model. This produces a full dataset, i.e. no missing data. The data can then be used to fit the model, by maximising the likelihood. This gives the M-step. More formally, let $\theta_{(t)}$ be the parameter value in the t-th iteration of the algorithm or any starting value for $t = 1$. The EM algorithm proceeds as follows:

1. E(Expectation)-step
   Calculate

$$Q(\theta; \theta_{(t)}) = \sum_{i=1}^{n} \int l_i(\theta) f(y_{iM} | y_{iO}; \theta_{(t)}) dy_{iM}, \qquad (11.1.7)$$

   where $l_i(\theta) = \log f(y_i; \theta)$, with $y_i = (y_{iO}, y_{iM})$
2. M(Maximisation)-step
   Maximise $Q(\theta; \theta_{(t)})$ with respect to θ, that is solve

$$s(\theta_{(t+1)}; \theta_{(t)}) = 0,$$

   where

$$s(\theta; \theta_{(t)}) = \frac{\partial Q(\theta; \theta_{(t)})}{\partial \theta}.$$

3. Return to Step 1 until convergence.

The EM algorithm is most easily applied if $l_i(\theta)$ depends linearly on y_i or a statistic of y_i. This is because the E-step for a linear function in y_i is just the expectation of the random variable, which is usually easy to calculate. This is the case for exponential family distributions and we give an example for a bivariate normal distribution below. Before doing so, let us demonstrate the interesting fact that, while the EM algorithm may not find the Maximum Likelihood, it is guaranteed to increase it with every iteration step.

Proof Here we prove that the EM algorithm increases the likelihood with each step. Note that

$$f(y_i; \theta) = f(y_{iO}, y_{iM}; \theta) = f(y_{iO}; \theta) f(y_{iM} | y_{iO}; \theta)$$

such that the likelihood contribution of the i-th observation decomposes to

$$l_i(\theta) = l_{iO}(\theta) + \log f(y_{iM} | y_{iO}; \theta),$$

where, as above,

$$l_i(\theta) = \log f(y_i; \theta) \quad \text{and} \quad l_{iO}(\theta) = \log f(y_{iO}; \theta).$$

Because only y_{iO} is observed, $l_{iO}(\theta)$ is the contribution of the likelihood given in (11.1.5) and with the decomposition above we can write

$$l_O(\theta) = l(\theta) - \sum_{i=1}^{n} \log f(y_{iM}|y_{iO}; \theta), \tag{11.1.8}$$

where

$$l_O(\theta) = \sum_{i=1}^{n} l_{iO}(\theta) \text{ and } l(\theta) = \sum_{i=1}^{n} l_i(\theta).$$

The two quantities on the right-hand side of (11.1.8) depend on y_{iM}, which are the missing observations and hence unknown. The left-hand side of (11.1.8), however, does not depend on y_{iM}. Consequently, (11.1.8) holds for all possible values of y_{iM} and we may therefore take the expectation of both sides of (11.1.8) with respect to $f(y_{iM}|y_{iO}; \theta_{(t)})$. This gives

$$l_O(\theta) = Q(\theta|\theta_{(t)}) - H(\theta|\theta_{(t)}),$$

where $Q(\theta|\theta_{(t)})$ is defined in (11.1.7) and

$$H(\theta|\theta_{(t)}) = \sum_{i=1}^{n} \int \log f(y_{iM}|y_{iO}; \theta) f(y_{iM}|y_{iO}, \theta_{(t)}) dy_{iM}.$$

We can express component $H(\theta|\theta_{(t)})$ in terms of the Kullback–Leibler divergence discussed in Sect. 3.2.5. In fact, we have

$$KL(\theta; \theta_{(t)}) = H(\theta_{(t)}|\theta_{(t)}) - H(\theta|\theta_{(t)}).$$

As the Kullback–Leibler divergence is always positive, we can conclude that

$$H(\theta_{(t)}|\theta_{(t)}) \geq H(\theta|\theta_{(t)}).$$

If we now maximise $Q(\theta|\theta_{(t)})$ (or at least increase it) and set $\theta_{(t+1)}$ as the new value, we have

$$Q(\theta_{(t+1)}|\theta_{(t)}) \geq Q(\theta_{(t)}|\theta_{(t)}) \text{ and } -H(\theta_{(t+1)}|\theta_{(t)}) \geq -H(\theta_{(t)}|\theta_{(t)})$$

such that

$$l_O(\theta_{(t+1)}) \geq l_O(\theta_{(t)}).$$

This proves that the likelihood increases with each iteration step of the EM algorithm. □

Example 48 Let us consider the simple but demonstrative example of the bivariate normal distribution. Assume that $Y_i = (Y_{i1}, Y_{i2})$ comes from a bivariate normal distribution, i.e.

$$Y_i = \begin{pmatrix} Y_{i1} \\ Y_{i2} \end{pmatrix} \sim N\left(\begin{pmatrix} \mu_1 \\ \mu_2 \end{pmatrix}, \begin{pmatrix} \sigma_{11} & \sigma_{12} \\ \sigma_{21} & \sigma_{22} \end{pmatrix}\right) = N(\mu, \Sigma).$$

For simplicity, let us assume that Σ is known. We have *i.i.d.* data with $i = 1, \ldots, n$, but observe three missing patterns. For n_0 cases we have complete observations, for n_1 cases we only observe variable Y_{i1} and for n_2 of the cases we only observe Y_{i2}. Remember that the bivariate normal model implies the conditional model (see Definition 2.14)

$$Y_{i1}|y_{i2} \sim N\left(\mu_1 + \beta_{1.2}(y_{i2} - \mu_2); \sigma_{1.2}^2\right),$$

where

$$\sigma_{1.2}^2 = \sigma_{11}^2 - \frac{\sigma_{12}^2}{\sigma_{22}} \quad \text{and} \quad \beta_{1.2} = \frac{\sigma_{12}}{\sigma_{22}}$$

which analogously also holds for $Y_{i2}|y_{i1}$. The log-likelihood for all data being observed is given by

$$l(\mu) = -\frac{1}{2} \sum_{i=1}^{n} (y_i - \mu)^T \Sigma^{-1}(y - \mu) + \text{const}$$

$$= -\frac{1}{2} \sum_{i=1}^{n} \{(y_{i1} - \mu_1)^2 \sigma^{11} + (y_{i2} - \mu_2)^2 \sigma^{22} + 2(y_{i1} - \mu_1)\sigma^{12}(y_{i2} - \mu_2)\} + \text{const}$$

$$= \sum_{i=1}^{n} \{\sigma^{11}(y_{i1}\mu_1 - \frac{1}{2}\mu_1^2) + \sigma^{22}(y_{i2}\mu_2 - \frac{1}{2}\mu_2^2) + \sigma^{12}(y_{i1}\mu_2 + y_{i2}\mu_1 - \mu_1\mu_2)\} + \text{const}.$$

where the superscripts indicate the elements of Σ^{-1}. We now choose $\mu_1 = \mu_{1(t)}$ and $\mu_2 = \mu_{2(t)}$ as initial values and begin the first step of the EM algorithm. Applying the E step of the EM algorithm means that, for the n_1 cases where y_{i1} is missing, we need to calculate $E(y_{i1}|y_{i2})$, which is $\mu_1 + \beta_{1.2}(y_{i2} - \mu_2)$. These values replace the missing values y_{i1}. Similarly, the n_2 cases where y_{i2} is missing will be replaced with $\mu_2 + \beta_{2.1}(y_{i1} - \mu_1)$. This gives a complete dataset for which we can maximise the resulting likelihood, which is the M step. Let $\mu_{(t)}$ be the current estimate, then

$$\hat{\mu}_{1(t+1)} = \frac{1}{n} \sum_{i=1}^{n} \tilde{y}_{1i},$$

where

$$\tilde{y}_{1i} = \begin{cases} y_{i1} \text{ if variable is observed} \\ \mu_{1(t)} + \beta_{1.2}(y_{i2} - \mu_{2(t)}) \text{ if } y_{i1} \text{ is missing.} \end{cases}$$

We can calculate $\hat{\mu}_{2(t+1)}$ in a similar way. Let us move from this rather theoretical example into a more concrete simulated scenario, which is shown in Fig. 11.2. We simulated bivariate normal data with correlation 0.8, standard deviation 1 and mean 0. We only observe the full data points if $y_{1i} > 0$ and $y_{2i} > 0$, which are shown in black. In the other quadrants, the points are only partly observed. In the top left quadrant, we observe y_{i2} but y_{i1} is missing. The opposite holds in the bottom right quadrant. Both, y_{i1} and y_{i2} are missing in the bottom left quadrant. We show three

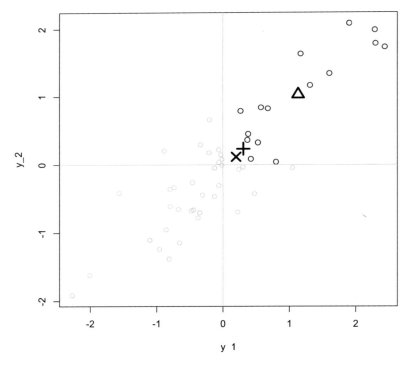

Fig. 11.2 Simulated normed data where only values in the top right quadrant are fully observed. The triangle represents the mean estimate for complete data, the vertical cross the estimate for available data, and the diagonal cross the EM estimate

estimates for the mean in the plot. The triangle is the estimated mean if we only make use of complete cases. The vertical cross is the estimate if we make use of available data, that is, μ_1 is estimated with the arithmetic mean of all observed y_{i1} values and the same with μ_2. This certainly is a plausible approach, but one can see that the EM algorithm estimate, shown with a diagonal cross, is better. In general, when observations are correlated, it is useful to not only use available data, but to also apply the EM algorithm.

▷

Example 49 Another very different application of the EM algorithm is for estimation in mixture models. A mixture model is appropriate when a random variable Y is assumed to come from a mixture of distributions, instead of a single distribution. When only two distributions are combined, we can model the variable as follows. Assume that Z is a binary random variable with

$$P(Z = 1) = \pi, \quad P(Z = 0) = 1 - \pi.$$

Given Z, we draw random variable Y from

$$Y|Z = z \sim \begin{cases} f_0(y) \text{ if } z = 0 \\ f_1(y) \text{ if } z = 1, \end{cases}$$

where $f_0(.)$ and $f_1(.)$ may be two distributions from the same distributional family or even from totally different families. Let us explore the former and assume

$$f_0(y) = f(y; \theta_0) \text{ and } f_1(y) = f(y; \theta_1).$$

If we could observe z_i, then, given the data (z_i, y_i) for $i = 1, \ldots, n$, the log-likelihood becomes

$$l(\theta_0, \theta_1, \pi) = \sum_{i=1}^{n} \left[(1 - z_i) \{ \log(1 - \pi) + \log f(y_i; \theta_0) \} + z_i \{ \log(\pi) + \log f(y_i; \theta_1) \} \right].$$

This is maximised by differentiation, e.g.

$$s_{\theta_0}(\theta_0, \theta_1, \pi) = \sum_{i=1}^{n} (1 - z_i) s_0(\theta_0) \tag{11.1.9}$$

$$s_{\theta_1}(\theta_0, \theta_1, \pi) = \sum_{i=1}^{n} z_i s_i(\theta_1) \tag{11.1.10}$$

$$s_{\pi}(\theta_0, \theta_1, \pi) = \sum_{i=1}^{n} \left\{ \frac{z_i - 1}{1 - \pi} + \frac{z_i}{\pi} \right\}. \tag{11.1.11}$$

We assume now that the indicator variable Z_i is not observed and we do not know whether Y_i was drawn with parameter θ_0 or θ_1. We can still write the log-likelihood, which is given by the sum of the possible values of Z_i, weighted by their corresponding probability. That is

$$l(\theta_0, \theta_1, \pi) = \sum_{i=1}^{n} \log\{(1 - \pi)f(y_i; \theta_0) + \pi f(y_i; \theta_1)\}.$$

Clearly, this likelihood looks much more complicated and should not be maximised directly. Instead, it can be maximised more easily by treating variable Z_i as a missing value. Hence, we artificially construct a dataset with an additional column of missing values for Z, which allows us to now use the EM algorithm. To do so, we need to calculate

$$E(Z_i|y_i) = P(Z_i = 1|y_i) = \frac{\pi f(y_i; \theta_1)}{(1 - \pi)f(y_i; \theta_0) + \pi f(y_i; \theta_1)}$$

which is easy with the current parameter estimates. We have now completed our E-step. The next step is to replace z_i in the score functions (11.1.9), (11.1.10) and (11.1.11) with $E(Z_i|y_i)$. This allows us to easily solve the score functions and get parameter updates for θ_0, θ_1 and π. This is the M-step. In Fig. 11.3, we show simulated data from a mixture of two normal distributions. The data are fitted with

$$Y_i|Z_i = 0 \sim N(\mu_0, \Sigma_0) \text{ and } Y_i|Z_i = 1 \sim N(\mu_1, \Sigma_1),$$

where both, mean μ and covariance matrix Σ, may differ between $Z = 0$ or $Z = 1$. Estimation in the M-step is easily carried out by taking weighted arithmetic means and weighted sample covariance matrices, where the weights are determined by the E-step with $P(Z_i = 1|y_i)$ and $P(Z_i = 0|y_i)$. After a few steps, we can decompose the two groups. This approach to mixture model estimation is quite flexible and is used frequently in classification. We refer to Böhning (1999) or to McLachlan (2000) for a more thorough exploration of the topic.

▷

When applying the EM algorithm, it is worth bearing in mind that it provides an estimate in the presence of missing data by maximising (or increasing) the likelihood. However, it does not automatically allow for correct variance estimates. That is to say, any Fisher information which is (or may be) obtained in the M-step cannot be used to estimate the variance. This is because the M-step relies on missing observations being replaced by their conditional expectation, given all observed data. Overall, we can conclude that the EM algorithm provides a simple way to obtain parameter estimates by increasing the likelihood of the observed data, but variance calculation of the variance can be cumbersome. We will close this section

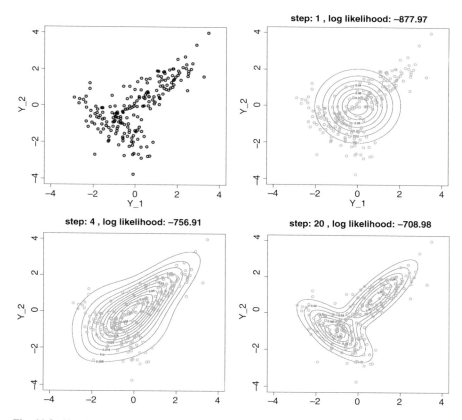

Fig. 11.3 Simulated data from a mixture of normal distributions (top row, left plot) and the estimate of the EM algorithm (starting value in top right plot, Iteration 4 in the bottom left plot and after convergence in the bottom right plot)

by motivating and mathematically deriving the Louis' formula, which is the most commonly used approach for estimating the variance of the estimate.

Proof Let us look more closely into an accurate estimate of the Fisher information that takes the presence of missing data into account and allows for variance estimates. The likelihood for the observed data is written as

$$l_O(\theta) = \sum_{i=1}^{n} \log \int f(y_i; \theta) dy_{iM}.$$

This gives the score function

$$
s_O(\theta) = \frac{\partial l_O(\theta)}{\partial \theta} = \sum_{i=1}^{n} \frac{\int \frac{\partial}{\partial \theta} f(y_i; \theta) dy_{iM}}{\int f(y_i; \theta) dy_{iM}}
$$

$$
= \sum_{i=1}^{n} \frac{\int \frac{\frac{\partial}{\partial \theta} f(y_i; \theta)}{f(y_i; \theta)} f(y_i; \theta) dy_{iM}}{f(y_{iO}; \theta)}
$$

$$
= \sum_{i=1}^{n} \int \frac{\partial}{\partial \theta} \log f(y_i; \theta) \frac{f(y_i; \theta)}{f(y_{iO}; \theta)} dy_{iM} \qquad (11.1.12)
$$

$$
= \sum_{i=1}^{n} \int \frac{\partial}{\partial \theta} \log f(y_i; \theta) f(y_{iM} | y_{iO}; \theta) dy_{iM}
$$

$$
= \sum_{i=1}^{n} E(s_i(\theta) | y_{iO}) := \sum_{i=1}^{n} s_{iO}(\theta), \qquad (11.1.13)
$$

where $s_i(\theta)$ is the full data score for the i-the observation, i.e.

$$
s_i(\theta) = \frac{\partial}{\partial \theta} \log f(y_i; \theta)
$$

and $s_{iO}(\theta)$ is the score of the observed data

$$
s_{iO}(\theta) = \int s_i(\theta) f(y_{iM} | y_{iO}) dy_{iM}.
$$

In order to derive the Fisher information, we need to calculate the second order derivative. This gives the observed Fisher information

$$
J_O(\theta) = -\frac{\partial s_O(\theta)}{\partial \theta}
$$

$$
= -\sum_{i=1}^{n} \frac{\partial}{\partial \theta} \frac{\int \frac{\partial}{\partial \theta} f(y_i; \theta) dy_{iM}}{\int f(y_i; \theta) dy_{iM}}
$$

$$
= \sum_{i=1}^{n} \frac{-\int \frac{\partial^2}{(\partial \theta)^2} f(y_i; \theta) dy_{iM}}{\int f(y_I; \theta dy_{iM})} + \sum_{i=1}^{n} s_{iO}(\theta) s_{iO}(\theta). \qquad (11.1.14)
$$

Note that

$$
\frac{\frac{\partial^2}{(\partial \theta)^2} f(y_i; \theta)}{f(y_i; \theta)} = \frac{\partial^2 \log f(y; \theta)}{\partial \theta \partial \theta} + \frac{\partial \log f(y_i; \theta)}{\partial \theta} \frac{\partial \log f(y_i; \theta)}{\partial \theta}
$$

such that the first component in (11.1.14) can be rewritten as

$$J_O(\theta) = \sum_{i=1}^{n} \{E(J_i(\theta)|y_{iO}) - E(s_i(\theta)s_i(\theta)|y_{iO}) + s_{iO}(\theta)s_{iO}(\theta)\},$$

where

$$J_i(\theta) = -\frac{\partial^2}{\partial\theta\,\partial\theta} \log f(y_i;\theta)$$

is the contribution for the observed full data. The above derivative is also known as Louis' formula and was introduced in Louis (1982). These quantities need to be evaluated in order to obtain a reliable variance estimate. Calculation might not be possible in analytic form, but simulation-based methods are often applicable. A different strategy for calculating the Fisher Information was proposed by Oakes (1999), which directly relies on the use of the function $Q(\theta;\theta')$ given in (11.1.7).

$\square$

11.1.3 Multiple Imputation

The basic principle of the EM algorithm can be extended to also incorporate estimation variability. The idea of the EM algorithm was to replace the missing values with their conditional expectation, given the observed data. Hence, broadly speaking, we set the missing values y_{iM} to $E(Y_{iM}|y_{iO})$, which is then considered as the observed value in the M-step. Note that the E-step gives a single fixed value for the missing data, at least if the parameters are kept fixed. We know, however, that Y_{iM} is a random value and by setting Y_{iM} to its mean value $E(Y_{iM}|y_{iO})$, this randomness is ignored. This can clearly be improved upon and an alternative strategy would be helpful here. Note that we will never be able to recover the missing value Y_{iM}, but we are often able to learn something about the posterior distribution $f(y_{iM}|y_{iO})$. If this is the case, we can estimate the conditional distribution and then simulate from it.

To formalise this idea, let us start by being a bit more precise with our notation. We already defined $M_i = \{j : R_{ij} = 0\}$ and $O_i = \{j : R_{ij} = 1\}$. Assume, as before, that $R_{ij} = 0$ implies that Y_{ij} is unobserved. We can now take observations with $R_{ij} = 1$ to learn more about the distribution of Y_{ij}. In the next step, we draw a random variable Y_{ij}^* from this posterior distribution and treat this as a (random) proxy for the missing observation. This idea is carried out in multiple imputation, where we impute (simulate) the missing values to obtain multiple complete datasets. The multiple imputation procedure is as follows.

1. Create K complete datasets by replacing all missing data with simulated values Y_{iM}^* drawn from $f(y_{iM}|y_{iO})$.

2. With the K complete datasets, fit the model $Y_i \sim f(y; \theta)$ using standard Maximum Likelihood, or any other suitable estimation routine leading to K estimates $\hat{\theta}^*_{(1)}, \ldots, \hat{\theta}^*_{(K)}$.
3. Obtain the final parameter estimate with the mean of the K estimates and an estimate of its variance with Rubin's rule.

Let us first discuss the three steps in more depth. We will discuss the first step at the end, because we have already covered the core idea and it is the least relevant to understanding the true idea behind Multiple Imputation.

Once we have generated a complete dataset, we no longer have missing data problems and Step 2 simply involves using standard results and methods for estimation. If we, for instance, use Maximum Likelihood estimation, we obtain $\hat{\theta}_{(k)}$ as the Maximum Likelihood estimate for the k-th imputed dataset and correspondingly $I(\hat{\theta}_{(k)})$ as a plug-in estimate for the Fisher information. We know that if the k-th completed dataset was the true data set, that is, if the imputed values were identical to the missing values, then $I^{-1}(\hat{\theta}_{(k)})$ would be the estimate for the variance of $\hat{\theta}_{(k)}$. However, the k-th completed dataset consists of simulated values $Y^*_{iM(k)}$ drawn from $f(y_{iM}|y_{iO})$ for $i = 1, \ldots, n$, which means that $I^{-1}(\hat{\theta}_{(k)})$ is in fact not a valid estimate for the true variance. We denote with

$$\hat{V}_{(k)} = I^{-1}(\hat{\theta}_{(k)})$$

the simulation-based estimate of the variance for a single dataset. Hence, from Step 2, we not only get parameter estimates $\hat{\theta}^*_{(1)}, \ldots, \hat{\theta}^*_{(K)}$, but also variance estimates $\hat{V}_{(1)}, \ldots, \hat{V}_{(K)}$.

These variance estimates are now combined in the third step of the imputation algorithm from above with Rubin's rule. Firstly, we get our parameter estimate by averaging the estimates of each individual case

$$\hat{\theta}_{MI} = \frac{1}{K} \sum_{k=1}^{K} \hat{\theta}^*_{(k)}. \tag{11.1.15}$$

Rubin's rule now gives an estimate of the variance of $\hat{\theta}_{MI}$. Its components are given by

$$\bar{\hat{V}} = \frac{1}{K} \sum_{k=1}^{K} \hat{V}_{(k)}$$

$$\bar{B} = \frac{1}{K-1} \sum_{k=1}^{K} (\hat{\theta}_{(k)} - \hat{\theta}_{MI})(\hat{\theta}_{(k)} - \hat{\theta}_{MI})^T.$$

Rubin's variance estimate itself is given by

$$\widehat{Var}(\hat{\theta}_{MI}) = \bar{V} + (1 + K^{-1})\bar{B}. \qquad (11.1.16)$$

In other words, we use (11.1.16) as a variance estimate for the multiple imputation estimate (11.1.15).

Proof We will justify Rubin's rule from a Bayesian perspective. We start by assuming that the posterior of θ given the complete data y takes the form

$$\theta - \hat{\theta}|y \sim N\left(0, I^{-1}(\hat{\theta})\right). \qquad (11.1.17)$$

Here, $\hat{\theta}$ is the posterior mean $E(\theta|y)$, which is equivalent to the Maximum Likelihood estimate given the complete data with a flat prior. To simplify notation slightly, let y represent the complete data, which decomposes to $y = (y_M, y_O)$. Note that we do not observe y, but only y_O and therefore we need to derive the conditional quantities $E(\theta|y_O)$ and $Var(\theta|y_O)$. The first can be obtained with

$$E(\theta|y_O) = E_{Y_M}(E(\theta|y)) = E_{Y_M}(\hat{\theta})$$

using Property 2.2. A reasonable estimate of the above is the Multiple Imputation estimate $\hat{\theta}_{MI}$, which simply serves as the arithmetic mean. Let us now look at the variance, which is given by

$$Var(\theta|y_O) = E_{Y_M}(Var(\theta|y)) + Var_{Y_M}(E(\theta|y))$$
$$= E_{Y_M}(I^{-1}(\hat{\theta})) + Var_{Y_M}(\hat{\theta}). \qquad (11.1.18)$$

The first component can be estimated by taking the average $\bar{V} = \frac{1}{K}\sum_{k=1}^{K} \hat{V}_{(k)} = \frac{1}{K}\sum_{k=1}^{K} I^{-1}(\hat{\theta}_{(k)})$. The second component of (11.1.18) expresses the variability of the imputed estimates, which can be estimated with $\bar{B}$. Hence, if y is not observed, but only y_O is given, the variance of $\theta - \hat{\theta}$ in (11.1.17) changes to (11.1.18) which can be estimated with $\bar{V} + \bar{B}$. Note that $\hat{\theta}$ is defined as the Maximum Likelihood estimate for the full data y. However, we do not know y because y_M is missing. In other words, the above statement is not yet of practical use. To make it applicable, we need to replace the unknown Maximum Likelihood estimate $\hat{\theta}$ with the Multiple Imputation estimate $\hat{\theta}_{MI} = \sum_{k=1}^{K} \hat{\theta}_{(k)}/K$. With the simple decomposition

$$\theta - \hat{\theta}_{MI} = (\theta - \hat{\theta}) + (\hat{\theta} - \hat{\theta}_{MI})$$

we get

$$Var(\theta - \hat{\theta}_{MI}|y_O) = Var(\theta - \hat{\theta}|y_O) + Var(\hat{\theta}_{MI} - \hat{\theta}|y_O)$$

because the distribution of θ is independent under multiple imputation. The first component has been derived already and the second component is estimated by the sample variance of the imputations, that is

$$\frac{1}{K}\frac{1}{K-1}\sum_{k=1}^{K}(\hat{\theta}_{(k)} - \hat{\theta}_{MI})(\hat{\theta}_{(k)} - \hat{\theta}_{MI}) = \bar{B}/K.$$

Combining the above components justifies Rubin's rule and $Var(\theta|y_{obs})$ can be estimated with (11.1.16). A non-Bayesian formal proof is more complicated but comes to the same conclusion and the above variance formula can also be used in Frequentist settings.   □

We have now discussed Steps 2 and 3 of the multiple imputation algorithm from above. Let us now go back to Step 1, that is, how to simulate y_{iM} from the conditional distribution given y_{iO}. In fact, there are various approaches and we will only give a brief example here. In principle, this is a prediction problem, in that we want to predict y_{iM} given y_{iO}. In other words, all suitable prediction and regression models can be used here and we can rely on the results derived in Chap. 7, with

$$Y_i = \begin{pmatrix} Y_{iO} \\ Y_{iM} \end{pmatrix} \sim N\left(\begin{pmatrix} \mu_O \\ \mu_M \end{pmatrix}, \Sigma = \begin{pmatrix} \Sigma_{OO} & \\ \Sigma_{MO} & \Sigma_{MM} \end{pmatrix}\right).$$

In this case

$$Y_{iM}|y_{iO} \sim N(\mu_{M.O}, \Sigma_{\mu M.O}),$$

where

$$\mu_{M.O} = \mu_M + \Sigma_{MO}\Sigma_{OO}^{-1}(y_{iO} - \mu_O)$$

$$\Sigma_{MM.O} = \Sigma_{MM} - \Sigma_{MO}\Sigma_{OO}^{-1}\Sigma_{OM}.$$

Given the available data, one can first estimate the parameters in the model and then apply the Multiple Imputation step. Alternative methods are classification trees or any other numerically more advanced prediction routines (including Bayesian approaches).

Example 50 Let us demonstrate the usefulness of multiple imputation with a small example. We consider the rental guide data from Chap. 7, focusing on 1000 apartments. The covariates are: x_1 = floor space, x_2 = bath.good (= 1 if bathroom is above standard, 0 otherwise), x_3 = bath.simple (= 1 if bathroom is below standard, 0 otherwise) and x_4 = kitchen.good (= 1 if kitchen has special appliances, 0 otherwise). We fit the model

$$Y = \beta_0 + x^T\beta + \varepsilon,$$

Fig. 11.4 Pattern of
missingness in rental data

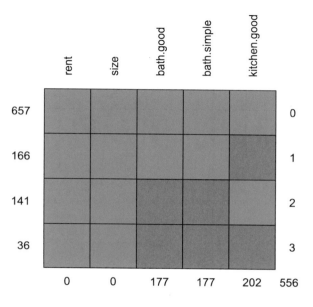

where Y is the rent and $x = (x_1, x_2, x_3, x_4)$. We then artificially remove some
values from x_2, x_3 and x_4, as visualised in Fig. 11.4. Out of the 1000 observations,
657 are complete cases (number on left-hand side) and hence have 0 missing items
(number on right-hand side). Moreover, for 166 observations the state of the kitchen,
i.e. 1 item, is missing. For 144 observations we have 2 missing items and for 38
observations we have 3 missing items and only rent and size are observed. We model
three cases:

1. "Full model"—use all of the data. This is available to us because we artificially
   induced missing data.
2. "Complete model"—use only complete cases. All cases with missing values are
   ignored.
3. "Impute model"—use multiple imputation for parameter estimation. This is
   calculated using the `mice` package in R.

The corresponding parameter estimates and their standard deviations are shown in
Fig. 11.5. The complete case analysis clearly has the largest variability, which is
indicated by the confidence intervals. Secondly, for all coefficients, the multiple
imputation estimate is closer to the full data estimate. This demonstrates that
multiple imputation helps us to obtain more precise results as compared to a
complete case analysis.

▷

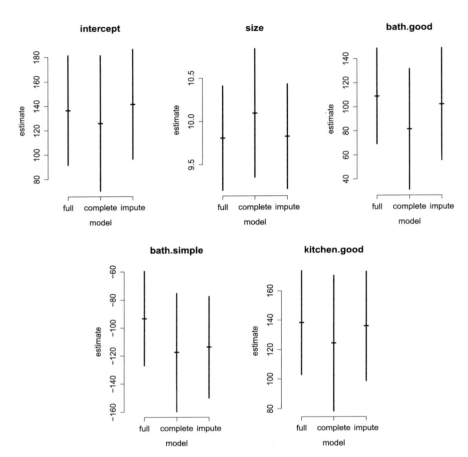

Fig. 11.5 Estimated coefficients in rent data given the full data with no missing values (full), the complete case data (complete) and the imputed data (impute). The vertical bars show the estimated standard errors

11.1.4   Censored Observations

In some situations we are not able to observe what we are interested in, because some observations are censored. Examples could include technical measurements, where we are not able to measure the quantity of interest if it falls below (or above) the precision level of the measuring device. The problem of censoring also regularly occurs in duration analysis, where we do not observe the duration, but just that it has exceeded some threshold. If we take the original measurements and ignore that their values are censored, this induces a bias and the results of the data analysis can be very wrong. The likelihood approach, however, allows us to rather simply accommodate censored observations. To exemplify this process, let Y_i be the measurement of particulate matter in the air at measuring stations in a city. The measurement devices can record particles above a threshold c, which determines the accuracy of the system. If the particulate matter measurement falls below this threshold, the observation is set to level c and a censoring indicator in the data is given, stating that this measurement is below the accuracy level. Hence, a variable R_i is included, which now states

$$R_i = \begin{cases} 1 & \text{if } Y_i > c \\ 0 & \text{if } Y_i \leq c, \text{i.e. censored.} \end{cases}$$

Assuming that

$$Y_i \sim f(.; \theta) \quad i.i.d.$$

we can directly write down the log-likelihood in case of censored observations:

$$l(\theta) = \sum_{i=1}^{n} R_i \log f(y_i; \theta) + \sum_{i=1}^{n} (1 - R_i) \log F(c; \theta),$$

where $F(.; \theta)$ is the cumulative distribution function of $f(.; \theta)$. In other words, for censored variables we only know that $Y_i \leq c$, which provides the contribution to the log-likelihood. We see that Maximum Likelihood estimation quite easily allows us to accommodate censored observations.

11.1.5 Omitting Variables (Simpson's Paradox)

If variables depend upon each other in a complex way, the dependence structure can change if some, but not all, variables are observed. Such situations can underlie Simpson's Paradox, which is best described with an example. To demonstrate, we make use of the data published by Radelet and Pierce (1991), which gives the number of death penalty verdicts in Florida for a 12-year period. The verdicts, ordered by the race of the defendant, are given in Table 11.1. A naive analysis of the data reveals that there are a lower proportion of Afro-Americans given the death sentence than Caucasian Americans. We have, however, omitted information about the *victim's* race. The data with this included is shown in Table 11.2. We can now see that if the victim is Caucasian, Afro-Americans are given the death penalty in 22.9% of cases, which is about double the rate of Caucasian defendants. The same holds for Afro-American victims, where no death penalty was recorded for Caucasian defendants, while in 2.8% of the trials the Afro-American defendant received the death penalty. Overall, Afro-Americans have a higher death penalty rate, which gets reversed if we ignore the victim's race. This phenomenon was described in Yule (1903), but was coined with reference to the discussion of the topic in Simpson (1951). For further detail we refer to Pavlides and Perlman (2009).

Table 11.1 The number of death penalty sentences in Florida, separated by race

Defendant's race	Penalty		
	Death	No death	Percent death
Caucasian	53	430	11.0
Afro-American	15	176	7.9

Table 11.2 The number of death penalty sentences in Florida, given both the defendant's and victim's race

Defendant's race	Victim's race	Penalty		
		Death	No death	Percent death
Caucasian	Caucasian	53	414	11.3
Afro-American	Caucasian	11	37	22.9
Caucasian	Afro-American	0	16	0.0
Afro-American	Afro-American	4	139	2.8

11.2 Biased Data

We saw in the previous section that ignoring missing data can lead to biased results. This section will be devoted to exactly this problem. A fundamental assumption for any reasonable data analysis is that the data are "representative", that is, that analysis of the data allows us to draw valid conclusions. But what happens if this is not the case and the data themselves are biased? Let us make this clear with a very simple example. Assume that we are interested in the average number of children in a family, which would allow us to estimate the reproduction rate of a population. We collect data by asking randomly selected children how many siblings they have. When added to the child being asked, this gives an estimate for the number of children living in family households. This process inevitably leads to a biased estimate, as the number of children per household is overestimated. The explanation is quite simple: households with no children are not accounted for. Such biases can often be corrected, which will be later demonstrated. Before doing so, we emphasise that it is often not so obvious that data are biased. Usually, the bias is hidden and one needs to think carefully about how it can be quantified and then corrected.

A second and rather common example of biased data occurs with the friendship paradox, observed by Scott (1991). In a social network, it can be observed that your friends, on average, have more friends than you. This sounds like a paradox. However, it can be explained in terms of biased data. Assume a social network where individuals are nodes connected by edges, which express mutual friendship. If we randomly select individuals in the network and count how many friends they have, this yields unbiased data and allows us to estimate the average degree, i.e. number of edges or friends, of actors in a network. If we do not record these numbers, but instead randomly select a friend of one of the originally sampled individuals, we obtain a biased sample. This is because individuals with no edges (friends) are excluded from the data and, similarly, individuals with many friends have a higher chance of being selected. This shows us that before we analyse data we need to question whether the data are biased. In the previous section we assumed missingness for individual entries relating to a single unit or individual in the database, which we labelled as item non-response. In this section we will focus on biased data and therefore on the situation where *all* observations from some units are missing, i.e. unit non-response.

The question of how biased samples can be used in order to obtain unbiased results is related to the field of unequal probability sampling. For a summary, see Berger and Tillé (2009). The methods trace back to Hansen and Hurwitz (1943) and Horvitz and Thompson (1952) and are frequently used in survey statistics. In this case, a biased sample can be drawn deliberately in order to increase accuracy and reduce statistical variability. In other words, it can be recommendable to draw a sample which is biased, but one needs to take the unequal inclusion probabilities into account to obtain unbiased results. Let us make this clear with a simple example. Assume we aim to run a survey to measure the amount of timber in a district. We could divide the area of the district in grid squares of, for example, 100 by 100 m. We could then randomly select some of the squares and go there to measure the amount of timber. While this may give an unbiased sample for the timber per hectare (i.e. 100 times 100 m), this sampling strategy is a little awkward. This is clear, because if we select hectares in urban areas, we know in advance that the amount of timber is low, while grid hectares in rural areas or in forests are likely to provide more information. Hence, we might be tempted to oversample rural areas and forests, but by doing so we produce a biased sample, which then needs to be corrected before analysing the data.

Let us formalise this concept. For each individual i, we collect data Y_i, where Y_i may be multivariate. As before, we define with the indicator R_i whether we have observed data on individual i. We assume a random sample, that is, we assume that the data are collected based on some random selection of individuals. We define with

$$\pi_i = P(R_i = 1),$$

the inclusion probability, where π_i might depend on other quantities, which are usually correlated with Y_i. In other words, we assume that R_i and Y_i are correlated, which is omitted in the notation. We further assume that π_i is known, or can at least be approximated reasonably well. The correction for biased data has been proposed already and it is strikingly simple. If we know π_i, we can replace the likelihood of the model which we aim to fit with its weighted counterpart. To demonstrate, assume that we have data $y_i \sim f(y; \theta)$, for all $i = 1, 2, \ldots$, and unit i is observed with probability π_i. The likelihood of the observed data is given by

$$l(\theta) = \sum_i R_i \log f(y_i; \theta).$$

As in the previous section, we calculate the score function as

$$E\left(\sum_i R_i \frac{\partial \log f(Y_i; \theta)}{\partial \theta}\right) = E_Y\left(\sum_i E(R_i|Y_i)\frac{\partial \log f(Y_i; \theta)}{\partial \theta}\right)$$

which does not equal zero as long as R_i depends on Y_i. However, if we replace the likelihood with its weighted counterpart

$$l_w(\theta) = \sum_i \frac{R_i}{\pi_i} \log f(y_i; \theta)$$

we can easily determine that the score is unbiased, i.e. taking the expectation from the weighted score gives zero for the true parameter. Hence, correcting for biased data is very straightforward, as long as we know the source of the biasedness. This, however, is generally not easy and an example should help to see how this works in practice.

Example 51 Web-scraping is a common tool in data science, but the resulting data can carry a bias, as the following example demonstrates. We consider data from an online platform for real estate advertisements, that focus on the rent per square metre of rental properties. The intention is to produce a rental atlas, showing the average rents for each district. If we assume that apartments are offered at random timepoints and that the advertisement period for each apartment is the same, the durations would follow the left-hand plot of Fig. 11.6. The horizontal line shows the time period that apartments are offered. If we now sample data, as indicated by the vertical line, we get an unbiased sample and can determine an unbiased market assessment. However, in practice, the data are different. Once an apartment is rented out, the advertisement is removed from the platform. In other words, the data that we achieve with web-scraping actually are closer to the right-hand plot of Fig. 11.6. Clearly, apartments that are longer on the market have a higher inclusion probability. That per se is not a problem, but now we need to question whether

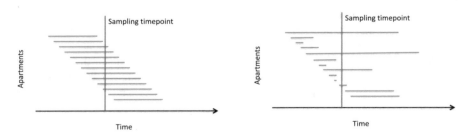

Fig. 11.6 Different assumptions about the duration of real estate advertisements. The diagram on the left shows sampling with identical durations, while the diagram on the right shows a sample with varying advertisement durations

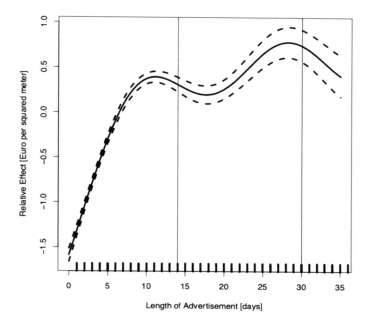

Fig. 11.7 Change in rent per square metre with length of advertisement on a real estate internet platform

the length of the advertisement is correlated with the rent of the apartment. The answer is most certainly yes, as inexpensive apartments find a new tenant faster than expensive apartments. Hence, more expensive apartments have a higher chance of being included in this data collected through web-scraping. In fact, a more thorough analysis of apartments in Munich using data from immoscout.de reveals that if we regress the rent per square metre on the advertisement length, we get the price change as shown in Fig. 11.7. We can now easily correct the bias by additionally recording the length of advertisement while web-scraping, which can be used to construct π_i and determine a weighted likelihood.

▷

Thus, biased data can be corrected, if we know or can approximate the inclusion probability. We want to emphasise, however, that biased data are in fact not a rarity, but occur rather frequently in data science. The reason is that data are often recorded as the by-product of a process. For example, the process of buying a product from a retailer or the process of making a booking. Such data are not originally collected for data analytic purposes, hence, one needs to be careful that any bias in the data is accounted for before a data analysis is pursued. The good news is, however, that often the inclusion probability π_i can be reasonably approximated through a simple weighted analysis and we can address the issue of bias directly.

11.3   Quality Versus Quantity

This section is of a more theoretical nature and has no clear practical application in particular. We have included it, however, to demonstrate that the quality of data is the utmost importance and arguments in the line of "if I have enough data, quality doesn't matter that much" are void and will motivate why data quality cannot be compensated by quantity. We already discussed above that Missing Not At Random data (MNAR) takes us to the limits of proper data analysis. To explore this point, let us focus on the balance between data quality and data quantity. In the big data era, it is often argued that the quality of data does not matter as long as there is lots of it available. To refute this claim, let us follow the line of argument proposed by Meng (2018) (see also the corresponding video youtu.be/8YLdIDOMEZs).

For simplicity of notation and calculation, we assume a finite population of size N and intend to determine some quantity of interest of this population. To be specific, we assume that the units in the population have a property, denoted by Y_i, and we are interested in the population mean

$$\mu_g = \sum_{i=1}^{N} g(Y_i)/N.$$

The function $g(.)$ can be any function, but we assume it to be known. We now take a sample of individuals, not necessarily at random, from the population and adapt our missingness indicator R_i to denote whether each individual in the population is either present or absent in our sample, that is

$$R_i = \begin{cases} 1 \text{ if data for } i\text{-th individual are recorded} \\ 0 \text{ otherwise.} \end{cases}$$

Taking the available data, we obtain our estimate for μ_g with

$$\hat{\mu}_g = \sum_{i=1}^{N} R_i g(Y_i) / \sum_{i=1}^{N} R_i,$$

that is, the sum of sampled individuals over the number of sampled individuals, i.e. the sample mean. Because we are dealing with a finite population, we can interpret the above sums as expectations and write

$$\hat{\mu}_g = \frac{\sum_{i=1}^{N} R_i g(Y_i)/N}{\sum_{i=1}^{N} R_i/N} = \frac{E(R_j g(Y_j))}{E(R_j)},$$

where j corresponds to a single observation uniformly drawn from $\{1, \ldots, N\}$. Looking now at $\hat{\mu}_g - \mu_g$, we obtain

$$\hat{\mu}_g - \mu_g = \frac{E(R_j g(Y_j))}{E(R_j)} - E(g(Y_j))$$

$$= \frac{E(R_j g(Y_j)) - E(R_j)E(g(Y_j))}{E(R_j)}$$

$$= \frac{Cov(R_j, g(Y_j))}{E(R_j)}$$

$$= \varrho_{Rg} \times \sigma_g \times \sqrt{\frac{Var(R_j)}{E^2(R_j)}}, \tag{11.3.1}$$

where ρ_{Rg} is the (empirical) correlation between R_j and $g(Y_j)$, σ_g is the standard deviation of $g(Y_j)$ and $Var(R_j)$ is the variance of R_j. As R_j is a $\{0, 1\}$ variable, we get that $Var(R_j) = E(R_j)(1 - E(R_j))$, which simplifies the final component of (11.3.1). Denoting the resulting sample size with $n = N \cdot E(R_j)$, we obtain from (11.3.1)

$$\hat{\mu}_g - \mu_g = \varrho_{Rg} \times \sigma_g \times \sqrt{\frac{N - n}{n}}. \tag{11.3.2}$$

This can be seen as one of the fundamental equations in data analytics. It states that if we make use of the data and calculate $\hat{\mu}_g$ to estimate the population quantity μ_g, the accuracy of the estimate depends on the product of three quantities:

$$\varrho_{Rg} = \text{quality of the data}$$

$$\sigma_g = \text{variability of the quantity of interest}$$

$$\sqrt{\frac{N - n}{n}} = \text{quantity of the data.}$$

In the big data era, one is tempted to set $n = N$, that is, to attempt to take observations of *all* individuals. However, this is nearly never the case and we will never be able to sample the entire population, i.e. $n < N$. This happens, in particular, if we have a missing not at random situation (MNAR), which is the case if there is correlation between R_i and $g(Y_i)$.

Before we look at this setting, we first look at the missing completely at random case (MCAR). In this case, R_i and $g(Y_i)$ are uncorrelated, which corresponds to taking a random sample. We now take any distribution of R_i, such that $\sum_{i=1}^{N} R_i = n$, i.e. we condition on the sample size. We denote the resulting expectation and variance with $E_R(.)$ and $Var_R(.)$. If R_i and $g(Y_i)$ are uncorrelated, we have $E_R(\varrho_{Rg}) = 0$, such that $\hat{\mu}_g$ is an unbiased estimate for μ_g, and the variance is

$$Var_R(\hat{\mu}_g - \mu_g) = E_R(\varrho_{Rg}^2) \times \sigma_g^2 \times \frac{N - n}{n},$$

where

$$E_R(\varrho_{Rg}^2) = Var_R(\varrho_{Rg}^2) + E^2(\varrho_{Rg}) = \frac{1}{N-1} + 0$$

is given by finite population arguments (see e.g. Thompson (2002)). Putting this together gives

$$MSE_R(\hat{\mu}_g) = \frac{N-n}{N-1}\frac{\sigma_g^2}{n} \le \frac{\sigma_g^2}{n}. \qquad (11.3.3)$$

The first factor is usually called the finite population correction factor and the second component is the sample variance. We see that as n increases, the variance gets smaller and, in fact, if $n = N$ we have complete information and the variance vanishes. However, this only holds for MCAR values. What happens if the data are biased and R_i and $g(Y_i)$ are correlated, that is, we have a MNAR situation? With (11.3.2), the mean squared error is

$$MSE_R(\hat{\mu}_g) = \{Var_R(\varrho_{Rg}) + E_R^2(\varrho_{Rg})\}\frac{N-n}{n}\sigma_g^2$$

$$= E_R(\varrho_{Rg}^2)\frac{N-n}{n}\sigma_g^2. \qquad (11.3.4)$$

This depends on the correlation of R and $g(Y)$. In other words, it depends on the varying probability of recording $g(Y_i)$, given the value of $g(Y_i)$. When working with big data, one usually accepts that the recorded data might be biased. But the argument seems to be that as long as we have enough data, this bias can be ignored. This can definitely be misleading, which we can even quantify with (11.3.3) and (11.3.4).

   Formula (11.3.3) gives the mean squared error of a simple random sample of size n. This applies for the classical statistically valid setting, where a proper (unbiased) sample is drawn from N individuals and the quantity μ_g is estimated. Formula (11.3.4), in contrast, gives the Mean Squared Error if a large, possibly biased, database is available. We can now relate these two formulae by calculating an "effective sample size" n_{eff}. That is, we calculate the size of a proper simple random sample that would provide the equivalent level of accuracy as can be obtained from the large, but biased, database. More formally, we set n in (11.3.3) to n_{eff} and ignore the finite population correction factor for simplicity. We then set (11.3.3) to equal (11.3.4). To demonstrate, let $p = n/N$ be the proportion of observed individuals of the population in the database. We now solve

$$\frac{\sigma_g^2}{n_{eff}} = E_R(\varrho_{Rg}^2)\frac{1-p}{p}\sigma_g^2 \Leftrightarrow n_{eff} = \frac{p}{1-p}\frac{1}{E_R(\varrho_{Rg}^2)}.$$

Table 11.3 Effective sample size n_{eff} for different values of the proportion of available data p and the correlation ϱ_{Rg}

p	$\sqrt{E(\varrho_{Rg}^2)}$	
	0.05	0.01
80%	1600	40,000
90%	3600	90,000
95%	7600	190,000

Table 11.3 gives the effective sample size for different values of p and $E(\varrho_{Rg}^2)$. For example, imagine a scenario where we have access to 90% of a database containing 1,000,000 customers, which is a very large sample. However, if we further assume that there is a small correlation of order $\sqrt{E(\varrho_{Rg}^2)} = 0.05$ between the information we want $g(Y_i)$ and the availability of the data R_i, then the information we obtain from our 900.000 individuals is as good as a random sample of only 3600 individuals. This is quite a discrepancy and it reflects the fact that the quality of the data needs to be questioned, even when working with big data. In particular, it teaches the lesson that quantity does not replace quality and that we therefore need to carefully statistically assess and model the quality of our data.

11.4 Measurement and Measurement Error

11.4.1 Theory of Measurement

So far, we have discussed problems with missing data, contrasted quality and quantity and proposed methods to deal with missing data entries. Now, let us focus on errors in the data, that is, what to do when the data entry itself is erroneous. Usually, data entries come from measurements and taking a measurement is an essential step in their generation. Before analysing data, it is therefore advisable to examine the quality of the measurement itself. An excellent and detailed discussion of this issue is provided by Hand (2004) in his book "Measurement. Theory and Practice. The world through quantification". In general, a measurement is formally defined as the assignment of a number to an object, which describes a specific property. One essential aspect is that we can compare objects using measurements. Physical measurements may be the first that come to mind, for example, the height of a person, length of a tool, speed of a car, number of clicks on a web page, etc. These measurements are called **representational measurements**, as they relate to existing attributes of an object. However, a measurement can also be seen more generally. In the humanities, measurements like intelligence, pain and xenophobia are of interest and likewise in economics, measurements of inflation, welfare, subjective well being and quality of a product. One common strategy for obtaining measurements of such nebulous properties is to build a score from many variables, e.g. from a questionnaire. The use of these kinds of measurements requires much more deliberation and, in many cases, different strategies for measurement are required. These type of measurements are also called **pragmatic measurements**, where the attribute is, in effect, defined by the measurement process itself (Hand 2004). As a consequence, variability in the observed measurement is often driven by variability in the measurement process itself and consequently is an inherent property of the data.

We define measurement error as follows:

Definition 11.4 A Measurement error is defined as the difference between a measured value of a quantity and its true value.

To make this more concrete, let us use the following notation: Let X be the true variable and X^* be its measured value. The most simple stochastic model for a measurement error is the additive model

$$X^* = X + U \text{ with } E(U) = \mu_U \text{ and } Var(U) = \sigma_U^2,$$

where U is an additive, stochastic measurement error. The expected value μ_U is the systematic measurement error, which is zero if the measurement process is well calibrated. This is also called a valid measure. If μ_U is 0 and U is independent of X, on average we capture the true value with our measurement process. The measurement error variance σ_U^2 relates to its precision. For a physical measurement device, its precision is sometimes given by the value $2\sigma_U$, i.e. twice the standard deviation. Assuming that U is normally distributed and $\mu_U = 0$, this implies that X^* is in the range of $X \pm 2\sigma_U$ with a probability of 95% and within the range of $X \pm 3\sigma_U$ with probability 99.7%.

11.4.2 Effect of Measurement Error in Regression

The consequences and mitigation of measurement error in statistical inference have been extensively discussed in the literature. For an excellent overview see Carroll et al. (2006) or see Gustafson (2003) for a Bayesian perspective. We will now concentrate on the effect of measurement error in regression models. We first assume that the response variable Y is measured with error, i.e. that we observe Y^* instead of the Y. Assuming an additive measurement error we get

$$Y = \beta_0 + \beta_1 X + \varepsilon \text{ and } Y^* = Y + U$$

such that

$$Y^* = \beta_0 + \beta_1 X + \varepsilon + U. \tag{11.4.1}$$

Equation (11.4.1) indicates the relationship between the observed variable Y^* and X. We still have a regression, but the error term now consists of both the

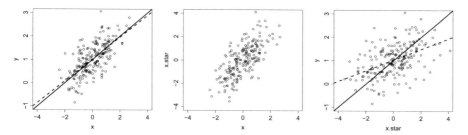

Fig. 11.8 Original data (left-hand plot), measurement error (middle plot) and Y plotted against X^* (right-hand plot)

residual error ε and the measurement error U. Assuming that the measurement error is independent of ε and X, we can simply reduce (11.4.1) to a regression model with a new error term $\varepsilon^* = \varepsilon + U$. As the variance of ε^* is $\sigma_{\varepsilon^*}^2 = \sigma_\varepsilon^2 + \sigma_U^2$, the relationship between X and Y is less pronounced, that is, we have an increased residual variance. However, because the assumptions of linear regression are still met, one can apply valid statistical inference as usual with the data (Y^*, X). In particular, this means that all procedures (estimates, tests, etc.) can be used exactly as presented in previous chapters. Interestingly, the estimate of the parameter β_x is still unbiased, even if Y is measured with a systematic error, i.e. $Y^* = Y + U$ with $E(u) = \mu_U \neq 0$. The observed regression can then be written as

$$E(Y^*|X) = E(\beta_0 + \beta_1 X + U) = \beta_0 + \mu_U + \beta_1 X. \tag{11.4.2}$$

From (11.4.2) we can see that using Y^* and X for estimation gives a biased estimator of the intercept, but unbiased and hence valid estimation of β_1.

Let us now consider the case of a covariate X, which we only observe through the erroneous measurement X^*. We again assume an additive error model, i.e.

$$Y = \beta_0 + \beta_1 X + \epsilon \text{ and } X^* = X + U, \tag{11.4.3}$$

where $U \sim N(0, \sigma_U^2)$ and $\varepsilon \sim N(0, \sigma_\varepsilon^2)$ are independent from each other and X.

This is sometimes called the "error in variables" model. We will see that this kind of error most certainly influences the estimation of β_1 in (11.4.3) and we will obtain biased estimates of β_x if the measurement error is ignored. To demonstrate this visually, we give an example in Fig. 11.8. We simulated 200 values of X from a standard normal distribution and set $Y = 0.5X + \varepsilon$, where the residual is simulated $N(0, \sigma_\varepsilon = 0.5)$. The left-hand plot shows simulated Y plotted against X. The true regression line is represented by the solid line (slope 0.5) and the fitted parameters by the dotted line. The middle plot visualises the simulated measurement error. Instead of X (horizontal axis) we observe X^* (vertical axis), where $X^* = X + U$ and U is standard normally distributed. If we now plot the observed Y against these observed values X^*, we obtain the right-hand plot. If we then fit a regression model of Y on X^*, we obtain the fitted dashed line, which is clearly biased compared to the solid true line with slope 0.5. In fact, we will show below, that the bias leads to a general underestimation of the slope parameter. This effect is called **attenuation** and frequently occurs in the presence of measurement error of a regressor variable. One reason for the attenuation is that, if our measurement of X carries measurement error, the information in X is lower and we find a weaker relationship between X^* and Y. Let us investigate this effect by further investigating this relationship. The observed model in linear regression is $E(Y|X^*) = \beta_0 + \beta_1 E(X|X^*)$, which is different to model (11.4.3). To investigate the bias in more depth, we assume a random covariate X with $X \sim N(\mu_x, \sigma_x^2)$. Then, following the measurement error model from above, we get $X^*|X \sim N(X, \sigma_U^2)$ and with the results of Sect. 2.1.6 we can see that (X, X^*) has a bivariate normal distribution

$$
\begin{pmatrix} X \\ X^* \end{pmatrix} \sim N \left(\begin{pmatrix} \mu_X \\ \mu_X \end{pmatrix}, \begin{pmatrix} \sigma_X^2 & \sigma_X^2 \\ \sigma_X^2 & \sigma_U^2 + \sigma_X^2 \end{pmatrix} \right).
$$

Exchanging the role of X and X^*, we can derive the conditional distribution of X given X^* with (see Definition 2.14)

$$
X|X^* \sim N \left(\mu_X + \frac{\sigma_X^2}{\sigma_X^2 + \sigma_U^2} (X^* - \mu_X), \sigma_X^2 - \frac{\sigma_X^4}{\sigma_U^2 + \sigma_X^2} \right).
$$

This leads us to the observed model:

$$
E(Y|X^*) = \beta_0^* + \beta_1^* X^*,
$$

where the regression coefficients are given by

$$
\beta_1^* = \frac{\sigma_X^2}{\sigma_X^2 + \sigma_U^2} \beta_1 \text{ and } \beta_0^* = \beta_0 + \left(1 - \frac{\sigma_X^2}{\sigma_X^2 + \sigma_U^2} \right) \beta_1 \mu_X. \tag{11.4.4}
$$

Assuming normality for the residuals gives the modified regression model

$$
Y = \beta_0^* + X^* \beta_1^* + \varepsilon^* \text{ with } \varepsilon^* \sim N \left(0, \sigma_\varepsilon^2 + \frac{\beta_1^2 \sigma_u^2 \sigma_x^2}{\sigma_x^2 + \sigma_u^2} \right). \tag{11.4.5}
$$

From the above equation, we see that, for a normally distributed covariate X, the relationship between Y and X^* can again be described as a simple linear regression model, but now with parameters $\beta^* = (\beta_0^*, \beta_1^*)$ and error term ε^* with $\sigma_{\varepsilon^*}^2 = \sigma_\varepsilon^2 + \frac{\beta_1^2 \sigma_U^2 \sigma_X^2}{\sigma_X^2 + \sigma_U^2}$. The least squares estimator that uses the observed data (Y_i, X_i^*), i.e. that simply ignores measurement error, is called the **naive estimator** $\hat{\beta}^*$. It is clear from the results in Chap. 7 that $\hat{\beta}^*$ is an unbiased estimator of β^*, but is clearly a biased estimator of β. In fact, we get

$$E(\hat{\beta}_1^*) = \beta_1^* = \beta_1 \frac{\sigma_X^2}{\sigma_X^2 + \sigma_U^2}. \tag{11.4.6}$$

From Eq. (11.4.6), we can see that the slope β_1 is attenuated by the factor $\frac{\sigma_X^2}{\sigma_X^2 + \sigma_U^2} = r$. This factor r is often called the **reliability ratio**, as it relates the variance of the true value X to the variance of X^*, namely $\sigma_{X^*}^2 = \sigma_X^2 + \sigma_U^2$. If there is no measurement error, i.e. $\sigma_U^2 = 0$, then $r = 1$ and for $\sigma_U^2 \to \infty$, the reliability decreases to zero, i.e. $r \to 0$. Hence, the factor r can be comprehended as the proportion of the information in X^* about X which is reliable.

11.4.3 Correction for Measurement Error in Linear Regression

Now that we know that measurement error on X induces a bias, we want to correct for it, i.e. we want to find an unbiased estimate. Using Eq. (11.4.6), we could simply estimate β_1 with

$$\hat{\beta}_1 = \hat{\beta}_{1LS}^* \cdot \frac{\sigma_X^2 + \sigma_U^2}{\sigma_X} = \hat{\beta}_{1LS}^* \cdot \frac{\sigma_{X^*}^2}{\sigma_{X^*}^2 - \sigma_U^2}.$$

This correction for attenuation can only be applied if we have information about $\sigma_{X^*}^2$ and σ_U^2. Obviously $\sigma_{X^*}^2$ can be estimated from the data using the empirical variance of the observed values of X^*. However, as the true variable X is not known, we can estimate neither σ_X^2 nor σ_U^2 from data (y, x^*). Consequently, we are faced with a problem of identification. We cannot identify β_1 and σ_U^2 from available data, even if our sample size increases. To overcome this problem, we need extra information about the measurement error variance, i.e. about the measurement process. There are three main ways of getting this information:

a) **Validation Data.** This is the situation where we can observe both X and X^* correctly to gather extra data. This is often called a validation sample and clearly requires that such a sample can even be constructed or recorded. More precisely, we obtain additional data (X_i, X_i^*) for $i = 1, \ldots, n^*$ from which we can fit the measurement error model

$$X_i^* = X_i + U$$

or any such variants, to obtain an estimate for σ_U^2.

b) **Replication Data.** The above approach is not always possible, because exact measurements of X are sometimes not available or even possible. We might, however, be able to take two or more measurements X^* (replications) for the same unit. That is, we take repeated measurements X^* to get more information about X. Assuming independence between any two measurements, we can deduce the measurement error variance by looking at the differences between pairs of measurements. More formally, let X_{i1}^* and X_{i2}^* be two measurements taken from the same unit. Following the measurement model, we assume

$$X_{i1}^* = X_i + U_{i1} \text{ and } X_{i2}^* = X_i + U_{i2}.$$

If we further assume independence of U_{i1} and U_{i2}, the difference follows

$$X_{i1}^* - X_{i2}^* = U_{i1} - U_{i2}.$$

Following a normal model for the measurement error U gives

$$Var(X_{i1}^* - X_{i2}^*) = 2\sigma_U^2.$$

With data pairs (X_{i1}^*, X_{i2}^*) for $i = 1, \ldots, n^*$, we can estimate σ_U.

c) **Assumptions.** If there is no information available about the measurement error, Eq. (11.4.6) can be used for a sensitivity analysis, i.e. we can look at the result while making a plausible assumption for σ_U^2. Simply ignoring measurement error assumes, perhaps unreasonably, that $\sigma_U^2 = 0$.

The results presented above were deduced from an additive measurement error with joint normality of X and X^*. They can be extended in various ways. Firstly, the assumption of normality of X can be dropped. It can be shown that the least squares estimator converges to $\beta_1^* = \frac{\sigma_X^2}{\sigma_X^2 + \sigma_U^2}$ for all X distributions and also for fixed values of X, see Carroll et al. (2006). The analysis can also be extended to the multivariate case where variances correspond to variance matrices and correction formulae are given by matrix multiplication. We do not want to go into depth here and instead refer to Carroll et al. (2006) for details.

11.4.4 General Strategies for Measurement Error Correction

The procedures described above to account for measurements errors can be seen in a more general setting. In the regression context, we have the parameter of interest, e.g. the relation between covariates X, Z and outcome variable Y. We assume that Z can be observed without measurement error. Instead of X, we observe X^*, which contains measurement error. We are interested in the parameters β of the "main model"

$$E(Y|X, Z) = m_Y(X, Z; .), \tag{11.4.7}$$

where $m_Y(\cdot)$ refers to some regression model which is parameterised by β. A simple example is the linear model

$$Y = \beta_0 + \beta_1 X + \beta_2 Z + \varepsilon.$$

To take measurement error into account, we also need a statistical model for the measurement process that describes the relationship between X^* and X and other observed variables. One common strategy is to set up a "measurement model"

$$E(X|X^*, Z) = m_X(X^*, Z; \gamma), \tag{11.4.8}$$

where $m_X(\cdot)$ is a second regression model but now with X as response variable. Given these two models, statistical inference is now possible with a range of different approaches, including Bayes and Likelihood. We restrict ourselves here to a simple and popular method called regression calibration. It includes three steps.

1. Estimate a model for $E(X|X^*, Z, \gamma)$ using validation data (X_j, X_j^*, Z_j).
2. Calculate the estimate $\tilde{X}_i := E(X_i|X^*, Z, \hat{\gamma}_i)$ and fit the main model given the data $(Y_i, \tilde{X}_i, Z_i)$.
3. Calculate variance-estimators and confidence intervals taking Step 1 and Step 2 into account.

11.5 Exercises

Exercise 1 (Use R Statistical Software)
Give a short summary of the different possible missing data patterns. Add the relevant formulas, an intuitive explanation and give examples for all three cases.

Now, let us take a look at the different approaches to dealing with missing data. To be able to compare the various methods we will take the GermanCredit_ms dataset, which was created from the GermanCredit_com dataset by deleting values at random. Keep in mind that dealing with missing values is just the first step towards further analysis of the data. In our case, we ultimately want to fit a logistic model to predict users with bad credit scores.

1. Use the VIM package to visualise the missingness pattern with the function aggr. Take a look at the function md.pattern in mice as well.
2. The simplest way of overcoming the problem of missing values is by doing a complete case analysis. Fit a regression model on the complete cases. What are the drawbacks of such an analysis? Another simple approach is (unconditional) mean imputation, meaning that we set all values of missing data to the mean of the respective observed values. Fit once again a logistic regression model after performing mean imputation using the mice package. Compare the imputed values with the real ones and discuss the problems of this approach as well.

3. Briefly describe the framework of multiple imputation, and take a look at the different approaches to imputation that are available in the `mice` package. Pick some methods according to the variable type (continuous, etc.) and finally perform multiple imputation for our problem.
4. Use the function `mice` to get a sample of completed datasets. Compare the imputed values with the real ones, then fit a model on the completed data sets, by using the functions `with.mids` and `pool`. How are the results for the different completed datasets combined in this scenario? What is the theoretical backbone of the variance estimate? Compare the model fit for different numbers of completed data sets.
5. Perform (in R) a final comparison of all the different approaches covered in this exercise.

Exercise 2 (Use R Statistical Software)
What is the advantage of the EM algorithm for fitting a model in the presence of missing data over simply maximising the marginal likelihood? Pin your explanation down to formulas in a concrete example.

The EM algorithm can be useful to fit a model on an incomplete data set, or if the desired model involves latent variables, both of which we have already introduced in this chapter. In this exercise, we will take a look at another frequent use for the EM algorithm: imputation. You are given the following table, which has one missing entry:

		B		
		j=1	j=2	j=3
A	i=1	8	12	19
	i=2	17	26	?

Assume for the cells x_{ij} an additive model

$$x_{ij} = \mu + \alpha_i + \beta_j + e_{ij},$$

with $\sum_{i=1}^{2} \alpha_i = \sum_{j=1}^{3} \beta_j = 0$. Here μ is the overall constant, α_i the effect of the rows, β_j the effect of the columns and e_{ij} an error term.

1. Based on the results in this chapter, would it be possible to adapt the EM algorithm for imputation? If so, how? Give an intuitive explanation.
2. Write down both steps of the EM algorithm for our specific example. Compute Q and s.
3. Finally, implement the algorithm in R. Plot the estimate of the missing value against the iteration of the EM algorithm.

Chapter 12
Experiments and Causality

A central question in statistical data analysis is when and how one can draw causal conclusions. In the typical setting, we want to ascertain how a covariate X influences an outcome Y. However, we also want to be certain that our statistical model represents a true causal process and not just some observed and possibly spurious correlation. A common example is the strong correlation between the number of storks sighted and the number of births in a country, which obviously is related to the size of the country in question, see Hunter et al. (1978). Other examples also spring to mind, such as the strong correlation between shoe size and reading ability in children (the older children have larger feet have better reading ability). Unfortunately, not all spurious associations are this obvious. Thus, it is imperative that we keep causal effects in mind when analysing data and develop methods that allow us to identify "real" correlations and not spurious ones.

In philosophy of science and many other scientific branches, the discussion of causality is a central, and often controversial, issue. For example, there is much debate over what exactly suggests that a causal relationship is at all present. Chambliss and Schutt (2003) list conditions like the association between variables, proper time order or existence of a causal mechanism. In a recent book on causal inference, Hernan and Robins (2020) give an excellent review on the issue for applications in medicine, economics and social sciences. They start with the approach that true causality can be deduced if an outcome Y depends on an action X. For example, the health status of a person i ($Y_i = 0$ healthy, $Y_i = 1$ ill) depends on the treatment X ($X = 1$ medication, $X = 0$ no medication). There is an individual causal effect if $Y_i(X = 0) \neq Y_i(X = 1)$. Instead of looking at the individual level we may look at the population level where we have $P(Y = 1|X = 0) = p_0$ and $P(Y = 1|X = 1) = p_1$. Then, $p_1 - p_0$ is called the average causal effect of X on Y. We could easily estimate the effect if we would know the outcome Y_i on the individual level for both $X = 1$ and $X = 0$. The practical problem is that we usually cannot observe $Y_i(X = 1)$ and $Y_i(X = 0)$ for each person, since we have to decide for one treatment for every person. This is called a counterfactual

© The Author(s), under exclusive license to Springer Nature Switzerland AG 2021
G. Kauermann et al., *Statistical Foundations, Reasoning and Inference*,
Springer Series in Statistics, https://doi.org/10.1007/978-3-030-69827-0_12

Table 12.1 Table of hypothetical average weekly sales for a given year, broken down by price level

		Our price level	
		"cheap"	"expensive"
Competitor's price level	"cheap"	60	40
	"expensive"	70	50

Table 12.2 The co-occurrence of different price levels for the competitor and our company

		Our price level		
		"cheap"	"expensive"	
Competitor's price level	"cheap"	0.4	0.1	0.5
	"expensive"	0.1	0.4	0.5
		0.5	0.5	1

setting (counter the fact). In practice, we only observe Y_i for one choice of X, the individual causal treatment effect can therefore not be observed. If we just look at $P(Y_i = 1 | X_i = 1)$ for the individuals who received the treatment and compare it to $P(Y_i = 1 | X_i = 0)$ for the individuals who did not receive medication, there are many sources of bias. For example, if only severely ill individuals get the medication, the observed difference does not equal the true causal effect. We do not intend to explore this topic too deeply—this is a large and rapidly developing field—but instead focus more on practical issues which might appear in data science applications. To motivate a typical problem we begin our exploration of the topic with a simple example, which will be extended later in the chapter.

Assume that we are a retailer selling a product and are interested in its price elasticity and the price sensitivity of our customers. In the market we have a single competitor and, to keep things simple, we assume two fixed prices for both us and our competitor, "cheap" and "expensive". Table 12.1 gives last year's average weekly sales, broken down by price. Sales increase by 20 if we decrease our price, which holds regardless of the competitor's price level. Moreover, if our competitor increases his price, we obtain more customers and our sales increase by 10. It is reasonable to assume that our price level and the competitor's price level are dependent. If we decrease our price, the competitor is tempted to do so as well and vice versa. Table 12.2 shows the dependence between our price level and the competitor's price level.

The table states that in 40% of the weeks both of the products were cheap, while only 10% of the time our product was "expensive" while the competitor's was "cheap" and also 10% when ours was "cheap" and his "expensive". The setting is simple but still demonstrates a real-life problem that can occur when analysing sales data. Assume that we do not have the competitor's price in our database, i.e. we cannot distinguish our sales with reference to the competitor's price. In this case, we observe average sales for each price by taking the margins of the first table and weighting with the second table. Our sales if the product is cheap are

$$\frac{1}{0.5}(0.4 \cdot 60 + 0.1 \cdot 70) = 62$$

and when it is expensive, they are

$$\frac{1}{0.5}(0.1 \cdot 40 + 0.4 \cdot 50) = 48.$$

Consequently, when looking only at our sales data, we conclude that decreasing our price only increases the number of sales by 14, which clearly gives a biased result when compared with the first table, where the increase is clearly 20.

Faced with this problem, we want to present statistical tools to obtain unbiased results for the influence of our price (i.e. the quantity of interest) on the number of sales (i.e. the response variable). We will discuss three possibilities in this chapter for obtaining valid information in the above setting:

1. *Include the competitor's price in the data analysis*
   Clearly, as we learned in the last chapter, omitting relevant input variables can give biased results, as can be seen in the previous example. We do not emphasise this again in this chapter, but instead simply refer to the material on missing data in the previous chapter.
2. *Make our price level independent of the competitor's price level*
   If Table 12.2 had equal probability for all entries, then our price level and the competitor's price level would be independent and the calculated average sales for the cheap and expensive price levels would be 65 and 45, respectively. Hence, if we can consider our price to be independent of the competitors' price, we no longer calculate biased results. This can be achieved with (statistical) experiments.
3. *Include an instrumental variable in our analysis.*
   If we are able to find a third quantity that influences our own price but is independent of the competitor's price level, we can also obtain unbiased results.

In the first half of the chapter, we discuss experiments as a classical way of assessing causality. The second half of the chapter is dedicated to sketching more fundamental approaches to statistical causality, including propensity scores and directed acyclic graphs.

12.1   Design of Experiments

12.1.1   Experiment Versus Observational Data

A time-honoured and historically important field of statistics is the design of experiments, which can be explored in more detail in the textbooks Montgomery (2013) and Box et al. (2005) or the classical Fisher (1990). Assume we are interested in investigating the influence of some quantity X on a response variable Y. This could be sketched graphically as

$$\text{\textcircled{X}} \rightarrow \text{\textcircled{Y}}.$$

Unfortunately, we are not often able to directly investigate the influence of X on Y, as numerous other quantities may exist that also influence Y. Let us denote these quantities with W. Furthermore, W can be decomposed into quantities that are observable and those that are fundamentally uncontrollable and unobservable. We denote the former with Z and the latter with U, such that $W = (Z, U)$. We now assume that X, Z and U potentially influence Y and are possibly mutually dependent. We sketch this as follows:

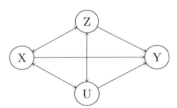

This can also be written in probabilistic terms as $f_{Y|X,Z,U} \cdot f_{X,Z,U}$. Here, $f_{X,Z,U}$ denotes the distribution of X, Z and U while $f_{Y|X,Z,U}$ expresses the conditional distribution of Y given X, Z and U. It is worth noting that Y depends on X, Z and U and $f_{Y|X,Z,U}$ expresses how samples of Y are generated. Coming back to the original question, we want to know how Y depends on X, without taking Z and U into account. That is we focus on Y and X only and aim to quantify how Y depends on X. This is expressed in the conditional distribution of Y given X denoted as $f_{Y|X}$. Omitting the quantities Z and U corresponds to them being integrated out. Hence, if we consider only data on X and Y, we obtain the following probability model

$$\int f_{Y|X,Z,U} f_{Z,U} dz du = \int \frac{f_{Y,X,Z,U}}{f_{X,Z,U}} f_{Z,U} dz du$$

$$= \int \frac{f_{Y,X,Z,U}}{f_{X|Z,U} f_{Z,U}} f_{Z,U} dz du = \int \frac{f_{Y,X,Z,U}}{f_{X|Z,U}} dz du \neq f_{Y|X}.$$

This is clearly different to the model $f_{Y|X}$ that we are interested in. Hence, we will potentially draw incorrect conclusions if we ignore the possible influence of Z and U on Y. This was demonstrated with the example above, where we considered sales data (Y), which depend on our price (X) but may also depend on the competitor's price (Z) and possibly other quantities (U).

An exception to the calculations above occurs if X is independent of both, Z and U. If this is the case, $f_{X|Z,U} = f_X$ so

$$\int f_{Y|X,Z,U} f_{Z,U} dzdu = \int \frac{f_{Y,X,Z,U}}{f_{X|Z,U}} dzdu = \int \frac{f_{Y,X,Z,U}}{f_X} dzdu = f_{Y|X}.$$

This shows an interesting and important property. Namely, it demonstrates that our goal should be to have (or make) X independent of Z and U to assess how X influences Y without Z and U. One simple, though not always possible, approach to achieve such independence is to apply **randomisation**. This leads to experimental designs, meaning that we actively set the value of X and record the value of Y. In the most simple scenario, we have X as a binary variable which describes **A/B testing**, see e.g. Kohavi and Longbotham (2017). For instance, in a study to evaluate website layouts, X could describe which of two websites is shown to the visitor and Y measures the conversion rate, i.e. the percentage of page views that lead to a purchase. Randomisation now states that we randomly assign a value X and then observe Y. In practice, each incoming visitor is randomly assigned one of the two layouts. By doing so, we force X to be independent of any other influencing quantities (like time of the day, weekday, browser type, etc.), such that the resulting data allow us to measure the influence of X on Y, irrespective of Z and U.

Full randomisation is not always possible for technical reasons or as a result of other constraints. For instance, if one wants to evaluate price elasticity, one can hardly vary the price of a product randomly for each new request. If randomisation is not possible, we try to achieve independence (or approximate independence) by applying what is called a **balanced design**. In principle, a balanced design means that we make X empirically independent of Z, but not necessarily independent of the unobserved variables U. To explain, let us assume that both Z and X are discrete valued. A balanced design is then achieved if X and Z are empirically uncorrelated. This means each combination of X and Z values is observed (or recorded) the same number of times. Assume, for instance, that Z is the day of the week and X represents two prices for a product. A balanced design for a four-week period could then be as shown in Table 12.3.

Table 12.3 Values of X in a balanced design with $Z =$ weekday

Week	z						
	Mon	Tue	Wed	Thu	Fri	Sat	Sun
1	1	2	1	2	1	2	1
2	2	1	2	1	2	1	2
3	1	2	1	2	1	2	1
4	2	1	2	1	2	1	2

In such experiments, the experimenter usually has full knowledge of the experimental setup, which remains unknown to the participants. This is sometimes called **single-blind**, where one party (the participant) is "blind" to the experimental setup. On the other hand, in a **double-blind** experiment, neither the experimenter nor the participant knows the value of X. Experiments of this type are often found in medical research, where X represents the medication (e.g. the new treatment versus placebo) and Y is the health status (e.g. disease cured or not). In this case, neither the experimenters (the doctors) nor the patients know which medication is given. It is the data analyst that **unblinds** the data before analysis. Double-blind experiments are necessary if subjective measures are taken by the experimenter, who might influence results towards the outcome that they expect or desire.

12.1.2   ANOVA

For comparing two groups in a randomised design, the two-sample t-test as discussed in Sect. 6.2.1 can be used to perform a comparison of the two group means. For more than two groups and complex designs the central statistical method for analysing experimental data is the **analysis of variance** approach, abbreviated with **ANOVA**. We will demonstrate the key concepts here, but direct the more interested reader to Montgomery (2013). Let us start with the simple scenario where we want to investigate how our response variable Y depends on a single discrete-valued input variable X, which takes values $\{1, \ldots, K\}$. In the field of experimental design, X is referred to as a factor in this setting. We would usually index our observations as pairs (Y_i, X_i) for observations $i = 1, \ldots, n$. In the field of experimental design a slightly modified notation is more common, which we use here for the sake of clarity and ease of comparison. To be specific, we introduce the following multi-index notation:

1. y_{kj} is the j-th observation of Y, that has X set to its k-th value
2. $k = 1, \ldots, K$ represents the possible values of X
3. $j = 1, \ldots, n_k$ represents the observation index, with n_k being the number of observations with $X = k$
4. $n = \sum_{k=1}^{K} n_k$ is the total number of observations.

We assume that y_{kj} is a realisation of a normally distributed random variable, that is

$$Y_{kj} = \mu + \beta_k + \varepsilon_{kj}, \tag{12.1.1}$$

where $\varepsilon_{kj} \sim N(0, \sigma^2)$ are i.i.d., μ represents a common mean among all observations, usually called the **grand mean**, and β_k represents how much the expected value of Y_{kj} deviates from μ. Model (12.1.1) is not identifiable, because subtracting a constant from all β_k and adding this to μ does not change the model.

We therefore need an additional linear constraint on the β coefficients, which is conventionally chosen as

$$\sum_{k=1}^{K} n_k \beta_k = 0. \tag{12.1.2}$$

Different constraints are possible as well, e.g. setting $\beta_K = 0$ or $\sum_{k=1}^{K} \beta_k = 0$, however, there is no meaningful difference in their interpretation and we will stick with (12.1.2) because it leads to simpler Maximum Likelihood estimates.

Interestingly, Model (12.1.1) can also be seen as a linear regression model. To demonstrate, we rewrite Model (12.1.1) in matrix form, which gives

$$\begin{pmatrix} y_{11} \\ \vdots \\ y_{1n_1} \\ y_{21} \\ \vdots \\ y_{2n_2} \\ \vdots \\ y_{K1} \\ \vdots \\ y_{Kn_k} \end{pmatrix} = \begin{pmatrix} 1 & 1 & \cdots & 0 \\ \vdots & \vdots & & \vdots \\ 1 & 1 & \cdots & 0 \\ 1 & 0 & \cdots & 0 \\ \vdots & \vdots & & \vdots \\ 1 & 0 & \cdots & 0 \\ \vdots & \vdots & & \vdots \\ 1 & -\frac{n_1}{n_K} & \cdots & -\frac{n_{K-1}}{n_K} \\ \vdots & \vdots & & \vdots \\ 1 & -\frac{n_1}{n_K} & \cdots & -\frac{n_{K-1}}{n_K} \end{pmatrix} \begin{pmatrix} \mu \\ \beta_1 \\ \vdots \\ \beta_{K-1} \end{pmatrix} + \begin{pmatrix} \varepsilon_{11} \\ \vdots \\ \varepsilon_{1n_1} \\ \varepsilon_{21} \\ \vdots \\ \varepsilon_{2n_2} \\ \vdots \\ \varepsilon_{K1} \\ \vdots \\ \varepsilon_{Kn_K} \end{pmatrix}$$

$$\Leftrightarrow Y = X\beta + \varepsilon.$$

Therefore, in principle we are not discussing a new topic here and could just refer to the results and approaches presented in Chap. 7. However, the analysis of experimental data is different from the material discussed in Chap. 7 in that we focus primarily on *whether* X influences Y, while *how* X influences Y remains of secondary interest. To do so, we focus on sums of squared (fitted) residuals.

Let us start by explicitly writing down the estimates in the model

$$\hat{\mu} = \sum_{k=1}^{K} \sum_{j=1}^{n_k} y_{kj}/n =: \bar{y}_{..}$$

$$\hat{\beta}_k = \sum_{j=1}^{n_k} (y_{kj} - \bar{y}_{..})/n_k = \bar{y}_{k.} - \bar{y}_{..},$$

where $\bar{y}_{k.} = \sum_{j=1}^{n_k} y_{kj}/n_k$. Note that these are Maximum Likelihood estimates and are equivalent to the matrix valued estimate $\hat{\beta} = (X^T X)^{-1} X^T Y$ derived in

Chap. 7. The next step is to test whether X has an influence on Y. Formulating this as a hypotheses on the parameters gives the following hypothesis test:

$$H_0 : \beta_k = 0 \text{ for } k = 1, \ldots, K$$
$$H_1 : \beta_k \neq 0 \text{ for at least two values of } k \in \{1, \ldots, K\}.$$

Note that under H_1, at least two components of β need to be non-zero, as the sum of the β-coefficients sums up to zero. We focus on the residual sum of squares and define the following terms:

1. Sum of Squares of the residuals in the H_1 model with X:

$$RSS_R := \sum_{k=1}^{K} \sum_{j=1}^{n_k} (y_{kj} - \bar{y}_{k.})^2$$

2. Total sum of Squares:

$$RSS_T = \sum_{k=1}^{K} \sum_{j=1}^{n_k} (y_{kj} - \bar{y}_{..})^2$$

The two quantities can be related with

$$RSS_T = \sum_{k=1}^{K} \sum_{j=1}^{n_k} (y_{kj} - \bar{y}_{k.} + \bar{y}_{k.} - \bar{y}_{..})^2$$

$$= RSS_R + \sum_{k=1}^{K} n_k (\bar{y}_{k.} - \bar{y}_{..})^2$$

$$= RSS_R + \sum_{k=1}^{K} n_k \hat{\beta}_k^2$$

$$= RSS_R + RSS_X.$$

The quantity $\sum_{k=1}^{K} n_k \hat{\beta}_k^2$ expresses the treatment effects and gives the variation of $\bar{y}_{k.}$ around $\bar{y}_{..}$, that is, the squared differences of group means around the grand mean. The larger this value, the more evidence exists that H_0 does not hold. We need, however, to normalise this properly to incorporate the effect of group size and the residual variance. This is addressed in the F-statistic

$$F = \frac{(RSS_T - RSS_R)/(df_T - df_R)}{RSS_R/df_R} = \frac{RSS_X/df_X}{RSS_R/df_R},$$

where df_T and df_R represent the degrees of freedom in the indexed model. This is defined as

$$df = \text{(number of observations)} - \text{(number of estimated parameters)}.$$

Hence, $df_T = n - 1$, because under H_0, only the single parameter μ is required, while $df_R = n - K$. Consequently, $df_X = df_T - df_R$ gives the number of treatment effects, which is $K - 1$. The F-statistic was proposed by Fisher (1925), who derived the distribution of F under the hypothesis H_0. If H_0 holds, then

$$F \sim \mathcal{F}_{df_X, df_R},$$

where $\mathcal{F}_{df_1, df_2}$ stands for the F-distribution with df_1 and df_2 degrees of freedom. Note that the statistic can be rewritten as

$$F = \left(\frac{RSS_T}{RSS_R} - 1 \right) \cdot \frac{df_R}{df_T - df_R}.$$

Because $RSS_R \geq 0$, we have $F \geq 0$ and the estimate for the residual variance is given by

$$\hat{\sigma}_\varepsilon^2 = \frac{RSS_R}{df_R}.$$

The actual shape of the F-distribution is of little practical use or interest, however, its quantiles are eminently useful in testing. These quantiles are implemented in any standard statistics package, such that the test decision is given by

$$F > F_{1-\alpha, df_X, df_R} \Leftrightarrow \text{"}H_1\text{"},$$

where $F_{1-\alpha, df_X, df_R}$ is the $(1 - \alpha)$-quantile of the F-distribution with df_X and df_R degrees of freedom. In the same way, we can derive p-values from the F-distribution. The results of this test are commonly provided in an "ANOVA table", as shown in Table 12.4. ANOVA tables are standard when analysing experimental data and their structure is always very similar, regardless of the software used to calculate the numbers. The most important number is the p-value, which indicates whether X

Table 12.4 An overview of the various values that are displayed in an ANOVA table

Source of error	Sum of squares	Degrees of freedom	Mean squared error	F-statistics	p-value
X	RSS_X	df_X	$MS_X = \frac{RSS_X}{df_X}$	MS_X / MS_R	$P(F > MS_X / MS_R)$
Residual	RSS_R	df_R	$MS_R = \frac{RSS_R}{df_R}$		
Total	RSS_T	df_T			

has a significant effect on Y. This effect can be interpreted as a causal effect for a randomised design, where the levels of the factor X are assigned randomly to the experimental units. The means for each group can also be inspected to gain further information.

12.1.3   Block Designs

The above setting is very simple and can be extended in various ways. In Table 12.3, we showed a balanced design for the case that an observable quantity Z also influences Y. Assume, for instance, that there is a weekday effect that is not of any particular interest to us. All we want to know (and test), is whether covariate X has an influence, while accounting for a possible weekday effect. In experimental design, this is called a **block effect**. It means that the experiment is pursued in each block (e.g. weekday) and the model needs to take the block effect into account. To demonstrate, let Z be categorical with B possible outcomes, that is $Z \in \{1, \ldots, B\}$. We need to guarantee that we have at least one observation for all possible combinations of X and Z. The observations are now indexed by

1. $y_{kbj} = j$-th observation of Y with X set to its k-th value and the block variable Z set to its b-th value
2. $k = 1, \ldots, K$ represents the value of X
3. $b = 1, \ldots, B$ represents the value of block variable Z
4. $j = 1, \ldots, n_{kb}$ represents the observation index, with n_{kb} being the number of observations with $X = k$ and $Z = b$
5. $n = \sum_{k=1}^{K} \sum_{b=1}^{B} n_{kb}$ is the total number of observations.

It is important to stress that n_{kb} can be equal to 1, that is, we can analyse data where no replicates are available for any or all combinations of X and Z. In fact, in many experimental settings one aims to limit the number of replicates to save time and money. The question in these cases is: how small (in terms of the sample size) can an experiment be so that one still can estimate and test the effect of the variable of interest? The model for the analysis of variance is now given by

$$Y_{kbj} = \mu + \beta_k + \alpha_b + \varepsilon_{kbj}, \tag{12.1.3}$$

where $\varepsilon_{kbj} \sim N(0, \sigma^2)$ i.i.d. and the α- and β-coefficients fulfil the constraints

$$\sum_{k=1}^{K} n_{k.} \beta_k = 0 \text{ and } \sum_{b=1}^{B} n_{.b} \alpha_b = 0,$$

where $n_{k.} = \sum_{b=1}^{B} n_{kb}$ and $n_{.b} = \sum_{k=1}^{K} n_{kb}$. We will test the null hypothesis that X has no influence on Y, which is expressed through constraints on the β-coefficients

$$H_0 : \beta_k = 0 \text{ for all } k = 1, \ldots, K$$
$$H_1 : \beta_k \neq 0 \text{ for at least two } k.$$

More formally, this can be seen as testing the simpler model

$$Y_{kbj} = \mu + \alpha_b + \varepsilon_{kbj} \tag{12.1.4}$$

against (12.1.3). In (12.1.4), only the block variable B, but not X, has an influence on Y. With a little bit of calculation, we can write down the estimates in analytic form as

$$\hat{\beta}_k = \bar{y}_{k..} - \bar{y}_{...}$$
$$\hat{\alpha}_b = \bar{y}_{.b.} - \bar{y}_{...}$$
$$\hat{\mu} = \bar{y}_{...}$$

with the common notation of using dots as indices, e.g.

$$\bar{y}_{k..} = \sum_{b=1}^{B} \sum_{j=1}^{n_{kb}} y_{kbj} / \sum_{b} n_{kb}.$$

To compare and test the model with block effects against the model that also includes X, we will again make use of residual sums of squares and extend the ANOVA table from above. Note that

$$\sum_{k=1}^{K} \sum_{b=1}^{B} \sum_{j=1}^{n_{kb}} (y_{kbj} - \bar{y}_{...})^2 = \sum_{k=1}^{K} \sum_{b=1}^{B} \sum_{j=1}^{n_{kb}} (y_{kbj} - \bar{y}_{k..} + \bar{y}_{k..} - \bar{y}_{.b.} + \bar{y}_{.b.} - \bar{y}_{...})^2$$

$$= \sum_{k=1}^{K} \sum_{b=1}^{B} \sum_{j=1}^{n_{kb}} \{(\bar{y}_{k..} - \bar{y}_{...}) + (\bar{y}_{.b.} - \bar{y}_{...})$$

$$+ (y_{kbj} - (\bar{y}_{k..} - \bar{y}_{...}) - (\bar{y}_{.b.} - \bar{y}_{...}) - \bar{y}_{...})\}^2$$

$$= \sum_{k=1}^{K} n_{k.} \underbrace{(\bar{y}_{k..} - \bar{y}_{...})^2}_{\hat{\beta}_k} + \sum_{b=1}^{B} n_{.b} \underbrace{(\bar{y}_{.b.} - \bar{y}_{...})^2}_{\hat{\alpha}_b} + \sum_{k=1}^{K} \sum_{b=1}^{B} \sum_{j=1}^{n_{kb}} (y_{kbj} - \hat{\mu}_{kb})^2.$$

$$\tag{12.1.5}$$

We define the following sums of squares:

$$RSS_T = \sum_{k=1}^{K} \sum_{b=1}^{B} \sum_{j=1}^{n_{kb}} (y_{kbj} - \bar{y}_{...})^2$$

$$RSS_X = \sum_{k=1}^{K} n_k \cdot \hat{\beta}_k^2$$

$$RSS_Z = \sum_{b=1}^{B} n_{\cdot b} \hat{\alpha}_b^2$$

$$RSS_R = \sum_{k=1}^{K} \sum_{b=1}^{B} \sum_{j=1}^{n_{kb}} (y_{kbj} - \bar{y}_{kb\cdot})^2.$$

This allows to write

$$RSS_T = RSS_X + RSS_Z + RSS_R.$$

When putting these numbers into an ANOVA table, we should bear in mind that our focus is exclusively on testing the influence of X. In other words, whether the block effect is evident in the data is of little interest. The hypothesis we aim to test is the same as before, i.e.

$$H_0 : \beta_k = 0 \text{ for } k = 1, \ldots, K$$
$$H_1 : \beta_k \neq 0 \text{ for at least two values of } k \in \{1, \ldots, K\}.$$

This leads to Table 12.5. The degrees of freedom are defined as above:

$$df_T = n - 1$$
$$df_X = K - 1$$
$$df_Z = B - 1$$
$$df_R = n - K - B + 1.$$

Table 12.5 An overview of the various values displayed in an ANOVA table for block design

Source	Sum of squares	df	Mean square error	F	p
X	RSS_X	df_X	$MS_X = \frac{RSS_X}{df_X}$	MS_X/MS_R	$P(F > MS_X/MS_R)$
Z	RSS_Z	df_Z	$MS_Z = \frac{RSS_Z}{df_Z}$	MS_Z/MS_R	
Residual	RSS_R	df_R	$MS_R = \frac{RSS_R}{df_R}$		
Total	RSS_T	df_T			

Table 12.6 An ANOVA
table displaying the result of
the example analysis

	Sum Sq	df	Mean Sq	F value	Pr(>F)
Chemical	12.95	4	4.32	2.375	0.121
Bolt	157.00	4	39.25	21.606	0.0001
Residuals	21.80	12	1.82		

Note that in the ANOVA table in Table 12.5, we only include the p-value for X (first row), but not for a possible block effect (second row). This is reasonable, as our focus is exclusively on testing whether X has an influence on Y, accounting for the block effect Z, but not assessing its validity.

Example 52 Let us explore a small example to demonstrate interpretation of the above ANOVA table. We measure the thickness of the fabric produced at a factory, where the thread used to produce the fabric is treated with four different chemicals our value of interest. The thread itself comes from five different bolts, which are treated as blocks. For each setting we have one observation, giving a sample size of $n = 10$. The corresponding ANOVA table is shown in Table 12.6 and the data are plotted in Fig. 12.1.

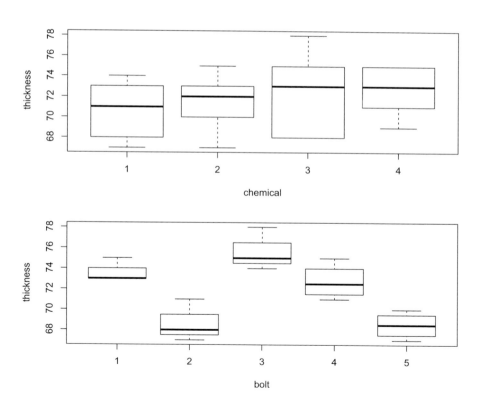

Fig. 12.1 Thickness of fabric plotted against treatment and bold

The table shows a p-value of 0.12 for the effect of the chemical and is not significant. The second p-value of 0.0001 is not of primary interest, as it merely confirms our assumption that the thickness of the fabric depends on the bolt used to make it.

12.1.4 More Complex Designs

It lies beyond the scope of this book to provide a complete introduction to experimental design, as there exist many extensions and highly sophisticated methods. Instead, we refer to available and well-established literature in this field, such as Montgomery (2013). Nevertheless, let us at least discuss one design which is helpful for the analysis of multiple variables $X_1, X_2, X_3, \ldots$. Assume, for instance, that we are interested in designing a product with 3 different properties X_1: colour, X_2: shape and X_3: weight. It can be very expensive and time consuming to run an experiment with all possible combinations of X_1, X_2 and X_3. One possible approach in this case is the **Latin square**. To demonstrate, we assume additive effects in the model

$$Y_{k_1 k_2 k_3 j} = \mu + \beta_{k_1}^1 + \beta_{k_2}^2 + \beta_{k_3}^3 + \varepsilon_{k_1 k_2 k_3 j}$$

with $k_1 = 1, \ldots, K_1; k_2 = 1, \ldots, K_2; k_3 = 1, \ldots, K_3$ and the index of observations with the same combination of covariates given by $j = 1, \ldots, n_{k_1 k_2 k_3}$. Let us assume, as usual, $\varepsilon_{k_1 k_2 k_3 j}$ is normal and, for simplicity, that $K_1 = K_2 = K_3 = 4$. This allows us to construct the following design. Instead of running the experiment for all $4 \cdot 4 \cdot 4 = 64$ possible combinations of X_1, X_2 and X_3, one can reduce the design to the 16 combinations shown in Fig. 12.2. The values in the table are those of X_3. This layout is called a Latin square. It has the property that in each row and column, we have all numbers from 1 to 4 for X_3. A Latin square can therefore be considered as the technical concept behind SUDOKU puzzles. The approach can be extended to higher dimensions and is often used in experimental analyses where numerous variables are of interest and need to be tested. The

Fig. 12.2 A 4x4x4 Latin square describing the possible combinations of the three variables to test

		X_2			
		1	2	3	4
X_1	1	$X_3{=}1$	2	3	4
	2	2	3	4	1
	3	3	4	1	2
	4	4	1	2	3

resulting data can be analysed with a linear multiple regression model, leading to ANOVA tables as previously discussed.

12.2   Instrumental Variables

It is often taken for granted that the best way of getting evidence is to conduct an experiment. However, there can be drawbacks. For example, the artificial conditions required can complicate the analysis or conducting an experiment can simply be too expensive, unethical or even impossible. Furthermore, it is often of interest to use observational data to draw causal conclusions. While this is not always possible, as motivated above, different strategies have been developed in different scientific disciplines to address exactly this question. One approach is to make use of what are called instrumental variables. To demonstrate the idea, let us start by assuming that available data have not been recorded under experimental conditions. Can we still obtain information about how X influences Y with this data? As before, we are interested in quantifying the effect of X on Y with regression and only have data on X and Y. This would allow us to fit a regression model

$$Y = \beta_0 + x\beta_x + \tilde{\varepsilon}. \tag{12.2.1}$$

A fundamental assumption in regression models is accordingly that the residual $\tilde{\varepsilon}$, to which we deliberately added a tilde, is uncorrelated with X. This sounds like a plausible assumption, which can nonetheless be violated, as the following scenario demonstrates. Assume that there is an unobserved quantity U which also influences Y and assume that X and U are correlated. This is similar to Sect. 12.1.1, but we have ignored the observable quantities Z for the sake of simplicity. To be specific, we assume the following dependence structure, visualised by the directed graph:

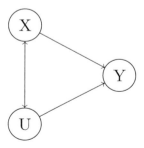

This diagram visualises that X and U are mutually dependent and both have an influence on Y. The question is whether only observing X and Y gives valid information on the model $f_{Y|X}$, that is, on the model only for the variables Y and X. We will see that this is not the case for dependent X and U.

To demonstrate this, we use the regression model (12.2.1) but let the true model be

$$Y = \beta_0 + X\beta_X + U\beta_U + \varepsilon, \tag{12.2.2}$$

where ε is uncorrelated to X and U. As U is not observable, we define $\tilde{\varepsilon}$ in (12.2.1) as $U\beta_U + \varepsilon$ from (12.2.2). We now can relate the two models with

$$Y = \beta_0 + X\beta_X + \underbrace{U\beta_U + \varepsilon} \tag{12.2.3}$$

$$= \beta_0 + X\beta_X + \quad \tilde{\varepsilon}. \tag{12.2.4}$$

If X and U are not independent, then, by extension, X and $\tilde{\varepsilon}$ are also not independent and hence not necessarily uncorrelated. This can have severe consequences. For instance, if we wanted to quantify the influence of X on Y with our model, we would calculate

$$\frac{\partial Y}{\partial X} = \beta_X + \frac{\partial \tilde{\varepsilon}}{\partial X} =: \tilde{\beta}_X.$$

Hence, instead of coefficient β_X we are faced with a bias $\partial\tilde{\varepsilon}/\partial X$ and fitting model (12.2.1) to data on Y and X will yield an estimate of $\tilde{\beta}_X$, but not an estimate of β_X. Hence, we induce a bias which itself can be quite problematic. In statistical terms this is referred to as **endogeneity** problem. In regression, covariates need to be exogeneous, that is, uncorrelated with ε and are referred to as endogenous if this is not the case.

This brings us back to the problem from the beginning of the chapter. To formalise this, let X be the price of the product and Y the resulting sales. We are trying to fit Model (12.2.1). However, it is reasonable to assume the actual process is more akin to (12.2.2), with U indicating our competitor's price. In this scenario, we would expect β_X to be negative, i.e. increasing our price would lead to a decrease in our sales, and β_U to be positive, i.e. increasing the competitor's price would lead to an increase in our sales. It is also plausible to assume that X and U are dependent, which is even considered a law of economics in such a market with multiple suppliers. It is not possible to estimate the price elasticity coefficient β_X if we ignore the competitor's price and just use our own price X and sales data Y, because the data only allow us to estimate $\tilde{\beta}_X$, which is clearly biased. Hence, a corrected analysis needs to take the competitor's price U into account, that is, we need to extend our dataset with observations of U. Now that web scraping is prevalent, this might be possible, depending upon the product and market. If we are able to get information about the competitor's price, we definitely need to include this in our model and hence in our data analysis.

However, not every market is this transparent and data on U might not be accessible to us. If we continue to work with Model (12.2.3), we have residuals $\tilde{\varepsilon}$ that are correlated with covariate X. In practice, this means that our dataset, even if it is very large, will not allow us to unbiasedly estimate the price elasticity. One possible way to circumvent this dilemma is to make use of an **instrumental variable** Z, which has the following two properties:

(i) Z is uncorrelated with the residual $\tilde{\varepsilon}$.
(ii) Z is correlated with X.

Let us again visualise the dependence structure with a few graphs. We assume that both X and some quantity U have an influence on Y.

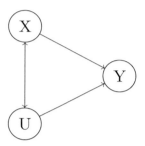

We are only interested in the effect of X on Y for a given value of U. But ignoring U will give residuals correlated with X. The instrumental variable Z is now constructed such that it fulfils the following dependence structure:

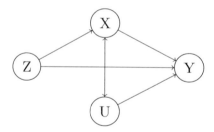

Hence, Z has an influence on X, but is independent of U (and hence independent of ε). Let us now motivate how we can use Z to obtain the effect of X on Y. Note that we are interested in

$$\frac{\partial Y}{\partial X}|(U = u)$$

meaning that we want to know how Y changes with X, while keeping $U = u$ fixed. Note that, with U fixed in Model (12.2.2), we get $\partial Y/\partial X|(U = u) = \beta_x$. Extending this with Z yields

$$\frac{\partial Y}{\partial X}|(U = u) = \frac{\partial Y/\partial Z}{\partial X/\partial Z}|(U = u) = \frac{\partial Y/\partial Z}{\partial X/\partial Z}.$$

Because Z is independent of U, we can omit the condition $U = u$. Consequently, the right-hand side of the equation allows us to estimate the effect of X, even if data on U is not available. That is, in the example, we can estimate price elasticity even if we do not know the competitor's price level. We need, however, to find or construct an instrumental variable that fulfils the two required properties, which is often no mean feat. Assuming that we have found an appropriate instrumental variable, this suggests the following estimation strategy. We fit a regression of Y on Z and X on Z, giving the two estimates $\hat{\beta}_{YZ}$ and $\hat{\beta}_{XZ}$. The estimate for $\hat{\beta}_X$ is then given by the ratio $\hat{\beta}_{YZ}/\hat{\beta}_{XZ}$. In the example, we could, for instance, take Z as the price of our product in a different market, which is sometimes called a Hausman-type instrument (see Hausman (1996)). In this case, the prices X and Z are correlated, as they depend on the company's pricing policy, but assuming that the markets are different, we may assume the necessary independence (or missing correlation) between Z and U. An alternative is to use the market power of our company, e.g. taking Z as the number of products produced. This was proposed by Stern (1996). In this case, the market size Z certainly influences our price X, but we may assume that it is independent of the competitors' price. Clearly, a combination of different instrumental variables is also possible. Estimation can also be pursued with a two stage procedure, by first regressing X on Z, taking the predicted value of X and further regressing this on Y. To be specific, a linear regression of X on Z gives

$$\hat{X} = Z(Z^T Z)^{-1} Z^T X$$

on which we now regress Y to obtain the estimate for β_X, i.e.

$$\begin{aligned} \hat{\beta}_X &= (\hat{X}^T \hat{X})^{-1} \hat{X}^T Y \\ &= (X^T Z(Z^T Z)^{-1} Z^T X)^{-1} X^T Z(Z^T Z)^{-1} Z^T Y. \end{aligned}$$

We refer to Cameron and Trivedi (2005) for more details, see also Angrist et al. (1986) or Stock and Trebbi (2003). Overall, with internet price comparison portals and webscaping methods becoming readily available in many aspects of the economy, it is becoming much easier to find or construct instrumental variables.

Example 53 Let us demonstrate the idea of instrumental variables with a simple example. Let Y be the sales of a product offered at price X. We assume that our price and the competitor's price U are correlated. Finally, let Z be a Hausman-type instrumental variable, such as the price of a different product from our company in a different market. We simulate X, U and Z from a normal distribution with the following covariance matrix

$$Var \begin{pmatrix} X \\ U \\ Z \end{pmatrix} = \begin{pmatrix} 1.25 & . & . \\ 1 & 1 & . \\ 0.25 & 0 & 0.25 \end{pmatrix}.$$

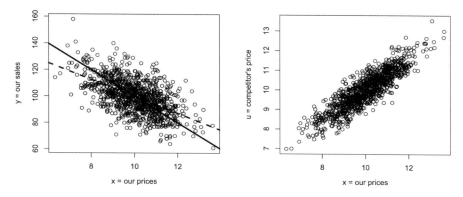

Fig. 12.3 Simulated sales plotted against price (left plot) and price plotted against competitor's price (right plot)

Finally, we simulate Y from

$$Y = \lceil 100 - 10x - 5u + \varepsilon \rceil = \lceil 100 - 10x + \tilde{\varepsilon} \rceil,$$

where $\lceil \rceil$ is the ceiling operator which rounds the Y up to the nearest whole number and $\varepsilon \sim N(0, 10)$. We plot the simulated data in Fig. 12.3, where we show Y plotted against X. We also plot U against X. The true effect is shown as a solid line ($\beta_X = -10$) and the biased fitted effect is given as a dashed line ($\hat{\tilde{\beta}} = -6.3$). We clearly underestimate price elasticity if we omit the endogeneity of X. If we take the instrumental variable Z into account and regress both Y and X on Z we obtain the estimate

$$\frac{\hat{\beta}_{YZ}}{\hat{\beta}_{XZ}} = -10.4$$

which is close to the true value -10. The same outcomes also result from the two stage procedure.

$\triangleright$

12.3   Propensity Score Matching

Let us continue the discussion about drawing inference from observational data. **Propensity scores** allow us to separate the likelihood of a subject choosing a treatment from their response to said treatment, which can be especially useful when attempting to draw causal conclusions from data. To motivate the use of a propensity score, imagine that a company offers customers a special product.

Perhaps an internet based retailer offers a fixed-price annual subscription for free shipping and related benefits. The customers can decide whether they want to take the offer or not. The general question for the retailer is whether this offer increases sales for a given customer. To answer this question, we could compare the sales of the customers that took the offer to the sales of those that did not. However, this carries the risk that we overestimate (or underestimate) the effect of the offer and induce a bias. This bias occurs because customers that take advantage of such offers might have different purchasing behaviour to those that do not. In other words, more regular customers are likely to find a flat rate for free shipping more attractive than occasional shoppers. Let us try to formalise this concept. Let $D_i = 1$ indicate that person i is attracted to the special offer, while $D_i = 0$ indicates that the person is not attracted. Variable D_i indicates whether the customer can be treated or not. Let us further denote with C_i the indicator whether the i-th person receives the offer ($C_i = 1$) or not ($C_i = 0$). That is, C_i indicates whether the i-th was treated or not. We can conceptualise this through the following matrix:

	Participate	
Willingness to participate	Yes: $C = 1$	No: $C = 0$
Yes: $D = 1$	$Y(1)$	$Y(0)$
No: $D = 0$	0	$Y(0)$

Customers that took the offer are located in cell ($D = 1, C = 1$), while customers that did not take the offer could in principle be attracted but did not sign up ($D = 1, C = 0$), but could also be generally not attracted by such offers ($D = 0, C = 0$). If the offer is sent to all customers, then usually cells ($D = 1, C = 0$) and ($D = 0, C = 1$) are empty, because we expect interested users to take the offer and disinterested users not to. We define with $Y_i(1)$ the sales of a person i if he or she took to the annual subscription and with $Y_i(0)$ the sales if they did not. Note that we only observe $Y_i(0)$ and $Y_i(1)$ and the naïve approach would be to compare the average sales of $Y_i(1)$ with the average sales of $Y_i(0)$. But if we think about it, it is clear that this is in fact not what we are really looking for. We are interested in the **treatment effect** (i.e. the increase in sales) among the customers that can be treated, i.e. the customers that feel attracted by such offers and sign up. Formally, we are only interested in the first row of the above matrix, which is defined as the "Average Treatment effect on the Treated (ATT)". This can be written as

$$\tau = E(Y(1)|D = 1) - E(Y(0)|D = 1).$$

Clearly, since variable D is unknown, the quantity $E(Y(0)|D = 1)$ is neither observed nor observable unless we run an experiment, which we aim to avoid. Note that $E(Y(0)|D = 1)$ represents the average sales among the customers that are attracted by sale deals but did not sign up. Let us see what happens if we replace $E(Y(0)|D = 1)$ with $E(Y(0)|D = 0)$, meaning that we classify all customers that did not take the offer as those that are generally not attracted to such offers. Note that $E(Y(0)|D = 0)$ represents the average sales of customers that did not feel attracted by an annual subscription and hence did not take it, which is readily available in the data. It is, however, clear that $E(Y(0)|D = 1)$ is likely to differ from $E(Y(0)|D = 0)$. This is demonstrated in the following equation

$$E(Y(1)|D = 1) - E(Y(0)|D = 0) = \tau + \underbrace{E(Y(0)|D = 1) - E(Y(0)|D = 0)}_{\text{bias}},$$

where the bias is referred to as self-selection bias. Only in the absence of self-selection bias can we draw valid conclusions. This occurs, for instance, in experiments where the treatment is assigned randomly. In observed (non-experimental) data, we need to determine the bias or, to be more specific, we need to determine $E(Y(0)|D = 1)$.

To do so, we define a score based on a set of observable covariates X. In the above example, X could be metadata for each customer, such as age, gender, place of residence or previous sales, etc. We assume now that $Y(0)$ and $Y(1)$ are conditionally independent of D if we know X. We can express this as

$$f_{Y(0),Y(1),D|X} = f_{Y(0),Y(1)|X} f_{D|X} \qquad (12.3.1)$$

or as a sketch

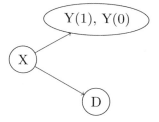

This implies that the self-selection expressed in D depends on observable quantities and hence, given X, D has no influence on $Y(0)$ and $Y(1)$. This property is called **unconfoundedness**. We now define the propensity score as the probability $\pi(x_i)$, that customer i feels attracted by the offer, i.e. he or she will take the treatment, i.e. $D_i = 1$. The propensity score is formally defined as

$$\pi(x) = P(D = 1|x)$$

and expresses the probability of receiving the treatment given the covariates x. Given that the offer is made to all individuals and assuming that individuals that feel attracted ($D_i = 1$) will take it ($C_i = 1$), while those who do not feel tempted by such offers ($D_i = 0$) will not take them ($C_i = 0$), we can conclude that the off-diagonals in the table above are empty. This in turn allows us to estimate $\pi(.)$, as we have observed C_i and hence D_i for the data. That is, we have observed whether the customers took the offer on an annual subscription (i.e. the treatment) or not. This is a binary indicator variable, which we can incorporate into a logistic regression model as discussed in Sect. 7.5. For each treated individual i, we then obtain (an estimate of) $\pi(x_i)$. If unconfoundedness holds, we can replace (12.3.1) with

$$f_{Y(0),Y(1),D|\pi(x)} = f_{Y(0),Y(1)|\pi(x)} f_{D|\pi(x)}.$$

The idea is now to match individuals based on their propensity scores $\pi(x_i)$. To clarify with an example, let i be an individual that has taken the offer. We then match this individual to a customer j, that has not taken the offer, such that $\pi(x_i)$ and $\pi(x_j)$ is as small as possible. Since the choice of customer j depends on the customer index i, we notate this as $j(i)$. This matching can be done with nearest neighbour or other clustering methods, which can also be performed on the original X values instead of $\pi(x)$. Once the matching has been performed a simple approach to estimating τ would be to take

$$\hat{\tau} = \sum_{i=1}^{n} (Y_i(1) - Y_{j(i)}(0)).$$

Instead of just taking the closest individual, one can also use weights expressing their similarity. We refer to Pearl (2009) for more details.

The basic idea of this matching is to find a control group which is comparable to the group of treated individuals. The idea of using covariates X and doing the matching with $\pi(x) = P(D = 1|x)$ was proposed in the pioneering paper of Rosenbaum and Rubin (1983). Propensity score matching is often applied in economics, but also plays an important role in medical statistics, where the effects of different therapies are assessed with the help of observational data. An essential part of using the method is the conditional independence assumption (12.3.1). We need a set of covariates X that describes the decision of $D = 1$ or $D = 0$ and is related to Y. This aspect has been discussed intensively in the economic and medical literature, see e.g. Kuss et al. (2016), Stuart (2010) or Caliendo and Kopeinig (2008).

12.4 Directed Acyclic Graphs (DAGs)

Without explicitly introducing the subject we have already made use of directed graphical models in this chapter and introduced undirected graphical models in Sect. 10.1.2. We want to extend this here and present Directed Acyclic Graphs (DAGs) as a convenient approach to modelling, analysis and visualisation of causal relations. The foundation of this method is the directed graph. As with graphical models, we take variables or measurements as nodes in the graph, which are connected by edges. The edges are directed but cycles are forbidden. A simple example is shown in Fig. 12.4. In fact, all visualisations shown in the previous sections make use of DAGs, which express the dependence structure of the causal

Fig. 12.4 Simple DAG

relation. In Fig. 12.4, it means that Y_3 depends on Y_1 and Y_2, but Y_1 and Y_2 are independent. As in Sect. 10.1, we notate independence with the symbol $\perp\!\!\!\perp$, such that the DAG shown in Fig. 12.4 represents

$$Y_1 \perp\!\!\!\perp Y_2 \text{ or in short } 1 \perp\!\!\!\perp 2.$$

The DAG induces an ordering in the variables and the probability of (Y_1, Y_2, Y_3) factorises to

$$f_{Y_1 Y_2 Y_3} = f_{Y_3 | Y_1, Y_2} \cdot f_{Y_1} \cdot f_{Y_2}$$

or in short

$$f_{123} = f_1 f_2 f_{3|1,2}.$$

We define with $pa(i)$ the parent set of node i. That is, for each $j \in pa(i)$, there is a directed edge from j to i. In the simple DAG in Fig. 12.4 we have

$$pa(1) = \emptyset$$
$$pa(2) = \emptyset$$
$$pa(3) = \{1, 2\}.$$

With this notation, we can express the dependence structure induced by a DAG. To this end, let $V = \{1, \ldots, q\}$ be the index set of a set of variables. Based on the DAG, we define the joint probability as

$$f_V = \prod_{j \in V} f_{j | pa(j)}.$$

Hence, the DAG is a simple visualisation of the dependence structure.

The DAG allows us to correctly process information taking the causal structure of the data into account. Looking at Fig. 12.4, we see that Y_1 and Y_2 are independent. But what happens if Y_3 is observed? Or, to put it differently, we question whether Y_1 and Y_2 are conditionally independent given Y_3. In fact, if we condition on Y_3, we find that Y_1 and Y_2 are dependent. To demonstrate this effect, assume a technical system where Y_1 and Y_2 are two independent components. Only if both Y_1 and

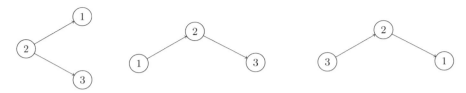

Fig. 12.5 Three DAGs expressing the same conditional independence properties

Y_2 are functioning, the system works, which is indicated with Y_3. If the system is not running, it could be a failure of either Y_1 or Y_2 and by checking whether Y_2 is running, we obtain information about Y_1. For instance, if we find Y_2 is working when Y_3 indicates a failure, then we know that Y_1 caused the disruption. In other words, conditioning on Y_3 we find that Y_1 and Y_2 are not any longer independent.

Conditional independences are not uniquely confined to a DAG structure as can be easily demonstrated with the DAGs shown in Fig. 12.5, where $1 \perp\!\!\!\perp 3|2$ in all graphs.

Despite the fact that conditional (and marginal) independences do not lead to unique DAGs, they can nevertheless help to uncover causal relations in the data. Assume, for instance, that we observe the following conditional and marginal independences between the variables $V = \{1, 2, 3, 4\}$

$$2 \perp\!\!\!\perp 3|1 \text{ and } 4 \perp\!\!\!\perp 1|\{2, 3\},$$

where all other pairwise conditional and marginal dependencies are present. We can then conclude that the dependence structure can be visualised with the following DAG.

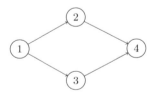

Note that the DAG expresses both conditional and marginal independences. If we look at all four variables, we see that the DAG expresses $1 \perp\!\!\!\perp 4|2, 3$, as

$$f_{4|\{1,2,3\}} = f_{4|pa(4)} = f_{4|\{2,3\}}.$$

If we integrate out Y_4, we get the marginal conditional independence statement $2 \perp\!\!\!\perp 3|1$, as

$$f_{3|\{1,2\}} = f_{3|pa(3)} = f_{3|1}.$$

The idea is now to investigate the conditional and marginal independence structure in the data to derive a DAG. Taking into account the non-uniqueness of DAGs, as shown in Fig. 12.5, we can easily see that this is not always possible. Still, DAGs are an important tool to model complex and high dimensional dependence structure in a simple directed graph.

A general introduction to causality and its representation with DAGs is given by Spirtes et al. (1993) or Pearl (2009). A novel field of application is Neuroinformatics for brain imaging (see e.g. Weichwald et al. (2015)). More recent work is given in e.g. Janzing et al. (2013).

12.5 Exercises

Exercise 1 (Use R Statistical Software)

Begin by stating and explaining the general formula that makes 'Design of Experiments' necessary in the first place.

A company has produced three tv ads ('A', 'B' , 'C'). They perform an experiment to find out which one is the most promising among them. The ads get shown to 15 people, who rate the ads with a score from 1 to 11. The experiment uses a balanced design with respect to the following age groups: 1 : [0, 18], 2 : (18, 30], 3 : (30, 45], 4 : (45, 60], 5 : (60, ∞]. The experiment yields the following ratings:

	A	B	C
1	9	10	6
2	8	11	6
3	11	3	6
4	10	2	6
5	10	2	5
sum	48	28	29

1. State a useful linear model for the data, adapting the one that is given in the lecture notes.
2. Compute the estimates for μ, α_1, α_2 and α_3 as well as the block effects $\beta_1, \ldots, \beta_5$.

   Hint: You can solve the rest of this exercise with R. To insert the data, you can use the following lines of code:

   ```
   score <- c(10,8,11,10,9,2,11,3,2,10,5,6,6,6,6)
   spot <- factor(c(rep("A",5),rep("B",5),rep("C",5)))
   age <- factor(rep(c(5,2,3,4,1), times=3))
   advertising <- data.frame(score=score, spot=spot, age=age)
   advertising <- advertising[order(age),]
   ```

3. Compute the various RSS and create an ANOVA table.

4. Perform an F-test ($\alpha = 0.05$) to decide whether the feedback for the different ads is significantly different.

Exercise 2 (Use R Statistical Software)

Consider the following data where $Y = 1$ if a patient recovers from a disease without any late effects of the disease while $Y = 0$ if a patient suffers from late effects of the disease. $X = 1$ denotes patients treated with a new promising drug while $X = 0$ are patients which get a treatment with a standard drug. As additional confounder, the categorised age of the patient is available, where $L = 1$ if the patient is < 60 years old while $L = 0$ stands for elder patients (≥ 60 years).

$L = 0$			$L = 1$		
X	$Y = 0$	$Y = 1$	X	$Y = 0$	$Y = 1$
0	153	102	0	29	65
1	36	209	1	16	390

Estimate the treatment effect $E(Y|X = 1) - E(Y|X = 0)$ ignoring the confounder by collapsing the data into a 2×2-table for X and Y and using the observed frequencies in that table. Why is this estimate probably incorrect?

References

Akaike, H. 1973. Information theory as an extension of the maximum likelihood principle. In *Proceedings of the 2nd International Symposium on Information Theory*, 267–281.

Akaike, H. 1983. Information measures and model selection. *International Statistical Institute* 44: 277–291.

Aldrich, J. 1997. R. A. Fisher and the making of maximum likelihood 1912–1922. *Statistical Science* 12: 162–176.

Anderson, T.W. 2003. *An Introduction to Multivariate Statistics*. 3rd ed. New York: Wiley.

Ando, T. 2010. *Bayesian Model Selection and Statistical Modeling*. Boca Raton: Chapman and Hall/CRC.

Angrist, J.D., D.B. Rubin, and G.W. Imbens. 1986. Identification of causal effects using instrumental variables. *Journal of the American Statistical Association* 91: 444–455.

Arlot, S., and A. Celisse. 2010. A survey of cross-validation procedures for model selection. *Statistical Survey* 4: 40–79.

Baird, D. 1983. The Fisher/Pearson chi-squared controversy: A turning point for inductive inference. *The British Journal for the Philosophy of Science* 34(2): 105–118.

Bauer, A., A. Bender, A. Klima, and H. Küchenhoff. 2018. Koala: A new paradigm for election coverage in multi-party electoral systems, July 2018

Beirlant, J., Y. Goegebeur, J. Teugels, and J. Segers. 2004. *Statistics of Extremes: Theory and Applications*. New York: Wiley.

Benjamini, Y., and Y. Hochberg. 1995. Controlling the false discovery rate: A practical and powerful approach to multiple testing. *Journal of the Royal Statistical Society* 57: 289–300.

Beran, R., and G.R. Ducharme. 1991. *Asympotic Theory for Bootstrap Methods in Statistics*. Centre De Recherches Mathématiques, hal-01826943, version1. https://hal.archives-ouvertes.fr/hal-01826943.

Berger, J., J. Bernardo, and D. Sun. 2009. The formal definition of reference priors. *The Annals of Statistics* 37(2): 905–938.

Berger, J., and L. Pericchi. 2001. Objective bayesian methods for model selection: Introduction and comparison (with discussion). In *Model Selection*, ed. P. Lahiri, 135–207.

Berger, Y.G., and Y. Tillé. 2009. Sampling with unequal probabilities. *Handbook of Statistics* 29: 39–54.

Berk, R., L. Brown, A. Buja, K. Zhang, and L. Zhao. 2013. Valid post-selection inference. *Annals of Statistics* 41: 802–837.

Bernardo, J., and A. Smith. 1994. *Bayesian Theory*. Wiley Series in Probability and Statistics. New York: Wiley

© The Author(s), under exclusive license to Springer Nature Switzerland AG 2021
G. Kauermann et al., *Statistical Foundations, Reasoning and Inference*,
Springer Series in Statistics, https://doi.org/10.1007/978-3-030-69827-0

Bickel, P.J., and D.A. Freedman. 1981. Some asymptotic theory for the bootstrap. *The Annals of Statistics* 9(6): 1196–1217.

Böhning, D. 1999. *Computeer-Assisted Analysis of Mixture and Applications*. Boca Raton: Chapman & Hall.

Bonferroni, C.E. 1936. Teoria statistica delle classi e calcolo delle probabilità. *Pubblicazioni del R Istituto Superiore di Scienze Economiche e Commerciali di Firenze* 8: 3–62.

Box, G.E.P., J.S. Hunter, and W.G. Hunter. 2005. *Statistics for Experimenters*. New York: Wiley.

Box, G.E.P., and G.C. Tiao. 1973. *Bayesian Inference in Statistical Analysis*. Wiley Classics Library Edition. New York: Wiley.

Breiman, L. 1992. The little bootstrap and other methods for dimensionality selection in regression: X-fixed prediction error. *Journal of the American Statistical Association* 87: 738–754.

Breiman, L. 2001. Random forests. *Machine Learning* 45: 5–32.

Burnham, K.P., and D.R. Anderson. 2002. *Model Selection and Multimodel Inference*. Berlin: Springer.

Caliendo, M., and S. Kopeinig. 2008. Some practical guidance for the implementation of propensity score matching. *Journal of Economic Surveys* 22: 31–72.

Cameron, A.C., and P.K. Trivedi. 2005. *Microecconometrics*. Cambridge: Cambridge University Press.

Cantelli, F.P. 1933. Sulla determinazione empirica delle leggi di probabilità. *Giorn. Ist. Ital. Attuari* 4: 421–424.

Carroll, R.J., D. Ruppert, L.A. Stefanski, and C.M. Crainiceanu. 2006. *Measurement Error in Nonlinear Models*. Boca Raton: Chapman and Hall/CRC.

Chambliss, D.F., and R.K. Schutt. 2003. *Making Sense of the Social World*.

Claekens, G., and N.L. Hjort. 2008. *Model Selection and Model Averaging*. Cambridge: Cambridge University Press.

Clarke, B., and A. Barron. 1994. Jeffreys prior is asymptotically least favorable under entropy risk. *Journal of Statistical Planning and Inference* 41(1): 37–60.

Clausnitzer, C., H. Küchenhoff, A. Goldhammer, and O. Adam. 2004. Organoleptische wirkung von sauerstoff in kohlensäurehaltigem mineralwasser. *Ern'ahrung & Medizin* 19: 184–187.

Cleveland, W.S. 2001. Data science: An action plan for expanding the technical areas of the field of statistics. *International Statistical Review* 69: 21–26.

Clopper, C.J., and E.S. Pearson. 1934. The use of confidence or fiducial limits illustrated in the case of the binomial. *Biometrika* 26(4): 404–413.

Coles, S. 2001. *An Introduction to Statistical Modeling of Extreme Values*. Berlin: Springer.

Congdon, P. 2003. *Applied Bayesian Modelling*. New York: Wiley.

Cui, W., and E. George. 2008. Empirical Bayes vs. fully Bayes variable selection. *Journal of Statistical Planning and Inference* 138: 888–900.

Czado, C. 2010. *Pair-Copula Constructions of Multivariate Copulas*. Berlin: Springer.

Danzig, G.B., and M.N. Thapa. 1997. *Linear Programming 1: Introduction*. Berlin: Springer.

de Boor, C. 1972. On calculation with B-splines. *Journal of Approximation Theory* 6: 50–62.

De Finetti, B. 1974. *Theory of Probability: A Critical Introductory Treatment*. Number Bd. 1 in Probability and Statistics Series. New York: Wiley.

de Haan, L. 1970. *On Regular Variation and Its Application to the Weak Convergence of Sample Extremes*. Mathematical Centre Tracts, 32. Amsterdam: Mathematisch Centrum.

Dempster, A.P., N.M. Laird, and D.B. Rubin. 1973. Maximum likelihood from incomplete data via the EM algorithm. *Journal of the Royal Statistical Society* 39: 1–38.

Dudewicz, E.J., and S.N. Mishra. 1988. *Modern Mathematical Statistics*. New York: Wiley.

DuMouchel, W., and G. Duncan. 1983. Using sample survey weights in multiple regression analyses of stratified samples. *Journal of the American Statistical Association* 78: 535–548.

Edwards, A.W.F. 1974. The history of likelihood. *International Statistical Review/Revue Internationale de Statistique* 42: 9–15.

Efron, B. 1979. Bootstrap methods: Another look at the jackknife. *The Annals of Statistics* 7(1): 1–26.

Efron, B. 2010. *Large-Scale Inference: Empirical Bayes Methods for Estimation, Testing, and Prediction*. Cambridge: Cambridge University Press.

Eilers, P., and B.D. Marx. 1996. Flexible smoothing with b-splines and penalties. *Statistical Science* 11: 89–121.

Fahrmeir, L., T. Kneib, S. Lang, and B. Marx. 2015. *Regression—Models, Methods and Applications*. Berlin: Springer.

Fan, J., and I. Gijbels. 1996. *Local Polynomial Regression*. Boca Raton: Chapman and Hall/CRC.

Fisher, R.A. 1912. An absolute criterion for fitting frequency curves. *Messenger of Mathematics* 41: 155–160.

Fisher, R.A. 1915. Frequency distribution of the values of the correlation coefficients in samples from an indefinitely large population. *Biometrika* 10(4): 507–521.

Fisher, R.A. 1922. On the mathematical foundations of theoretical statistics. *Philosophical Transactions of the Royal Society of London. Series A, Containing Papers of a Mathematical or Physical Character* 222: 309–368.

Fisher, R.A. 1925. *Statistical Methods for Research Workers*. Edinburgh: Olive and Boyd.

Fisher, R.A. 1990. *Statistical Methods Experimental Design and Scientific Inference*. Oxford: Oxford Science Publications.

Fisher, R.A., and L. Tippett. 1928. Limiting forms of the frequency distribution of the largest or smallest member of a sample. *Proceedings of the Cambridge Philosophical Society* 24: 180–190.

Fox, C.W., and S.J. Roberts. 2012. A tutorial on variational Bayesian inference. *Artificial Intelligence Review* 38(2): 85–95.

Friedman, J., T. Hastie, and R. Tibshirani. 2009. *Elements of Statistical Learning*. Berlin: Springer.

Gelfand, A., and A. Smith. 1990. Sampling based approaches to calculating marginal densities. *Journal of the American Statistical Association* 85: 398–409.

Gelman, A. 2007. Struggles with survey weighting and regression modeling. *Statistical Science* 22: 153–164.

Geyer, C.J. 1992. Practical Markov chain Monte Carlo. *Statistical Science* 7(4): 473–483.

Gilks, W.R., and P. Wild. 1992. Adaptive rejection sampling for Gibbs sampling. *Journal of the Royal Statistical Society, Series C. Applied Statistics* 41(2): 337–348.

Gillies. 2000. *Philosophical Theories of Probability*. London: Routledge.

Glivenko, V. 1933. Sulla determinazione empirica delle leggi di probabilità. *Giorn. Ist. Ital. Attuari* 4: 92–99.

Good, P. 2005. *Permutation, Parameter and Bootstrap Tests of Hypotheses*. 3rd ed. Berlin: Springer.

Goodman, S. 2008. A dirty dozen: Twelve p-value misconceptions. *Seminars in Hematology* 45: 135–140.

Graham, J.W., and S.I. Donaldson. 1993. Evaluating interventions with differential attrition: The importance of nonresponse mechanisms and use of follow-up data. *Journal of Applied Psychology* 78: 119–128.

Grimmett, G., and D. Stirzaker. 2001. *Probability and Random Processes*, 3rd ed. Oxford: Oxford University Press.

Gumbel, E.J. 1958. *Statistics of Extremes*. New York: Columbia University Press.

Gustafson, P. 2003. *Measurement Error and Misclassification in Statistics and Epidemiology Impacts and Bayesian Adjustments*. Boca Raton: Chapman and Hall/CRC.

Hand, D. 2004. *Measurement. Theory and Practice. The World Through Quantification*. New York: Wiley.

Hansen, M.H., and W.N. Hurwitz. 1943. On the theory of sampling from finite populations. *Annals of Mathematical Statistics* 14(4): 333–362.

Härdle, W.K., and O. Okhrin. 2010. De copulis non est disputandum. *Advances in Statistical Analysis* 94: 1–31.

Härdle, W.K., and L. Simar. 2012. *Applied Multivariate Statistical Analysis*. Berlin: Springer.

Harford, T. 2014. Big data: Are we making a big mistake. *Significance* 11(5): 14–19.

Hastie, T., R. Tibshirani, and J. Friedman. 2009. *The Elements of Statistical Learning*. Berlin: Springer.

Hastie, T., R. Tibshirani, and M. Wainwright. 2015. *Statistical Learning with Sparsity: The Lasso and Generalizations*. Boca Raton: CRC Press.

Hastie, T., R. Tibshirani, D. Witten, and G. James. 2015. *An Introduction to Statistical Learning*. Berlin: Springer.

Hastie, T.J., and R.J. Tibshirani. 1990. *Generalized Additive Models*. Boca Raton: Chapman and Hall/CRC.

Hastings, W. 1970. Monte Carlo sampling methods using Markov Chains and their application. *Biometrika* 57(19): 97–109.

Hausman, J.A. 1996. *Valuation of New Goods Under Perfect and Imperfect Competition*, 209–248. Chap. 5. Chicago: University of Chicago Press.

Held, L., and M. Ott. 2015. How the maximal evidence of p-values against point null hypotheses depends on sample size. *The American Statistician* 70(4): 335–341.

Held, L., and D. Sabanés Bové. 2014. *Applied Statistical Inference*. Berlin: Springer.

Hernan, M. A., and J.M. Robins. 2020. *Causal Inference: What If*. Boca Raton: Chapman and Hall/CRC.

Heumann, C., and M. Schomaker. 2016. *Introduction to Statistics and Data Analysis*. Berlin: Springer.

Hoeting, J.A., D. Madigan, A.E. Raftery, and C.T. Volinsky. 1999. Bayesian model averaging: A tutorial (with comments by M. Clyde, David Draper and E. I. George, and a rejoinder by the authors). *Statistical Science* 14(4): 382–417.

Højsgaard, S., D. Edwards, and S. Lauritzen. 2012. *Graphical Models with R*. Berlin: Springer.

Holm, S. 1979. A simple sequentially rejective multiple test procedure. *Scandinavian Journal of Statistics* 6: 65–70.

Horowitz, J. L. 2001. The bootstrap. In *Handbook of Econometrics*, ed. J. Heckman, and E. Leamer, 3159–3228. 1st ed. Vol. 5, Chap. 52. San Diego: Elsevier.

Horvitz, D.G., and D.J. Thompson. 1952. A generalization of sampling without replacement from a finite universe. *Journal of the American Statistical Association* 47: 663–658.

Hu, F., and J. Zidek. 2002. The weighted likelihood. *Canadian Journal of Statistics* 30: 347–371.

Huber, P.J. 1967. The behaviour of maximum likelihood estimates under nonstandard condistions. In *Proceedings of the Fifth Berkeley Symposium on Mathematical Statistics and Probability*, 221–233.

Hunter, W.G., J.S. Hunter, and G.E.P. Box. 1978. *Statistics for Experimenters*. Hoboken, NJ: Wiley.

Hurvich, C.M., and C.-L. Tsai. 1989. Regression and time series model selection in small samples. *Biometrika* 76(2): 297.

Janzing, D., D. Balduzzi, M. Grosse-Wentrup, and B. Schölkopf. 2013. Quantifying causal influences. *The Annals of Statistics* 41: 2324–2358.

Joe, H. 2014. *Dependence Modelling with Copula*. Boca Raton: CRC Press.

Kabaila, P., A.H. Welsh, and W. Abeysekera. 2016. Model-averaged confidence intervals. *Scandinavian Journal of Statistics* 43(1): 35–48.

Karr, A.F. 1993. *Probability*. Berlin: Springer.

Kass, R.E., and A.E. Raftery. 1995. Bayes factors. *Journal of the American Statistical Association* 90(430): 773–795.

Kauermann, G., and H. Kuechenhoff. 2011. Nach fukushima stellt sich die frage des risikos neu. *Frankfurter Allgemeine Zeitung*, N1.

Kendall, M., and A. Stuart. 1973. *The Advanced Theory of Statistics*. Griffin's Statistical Monographs and Courses. London: Griffin.

Kloek, T., and H. Dijk. 1978. Bayesian estimates of equation system parameters: An application of integration by Monte Carlo. *Econometrica* 46(1): 1–19.

Kneib, T. 2013. Beyond mean regression (with discussion and rejoinder). *Statistical Modelling* 13: 275–385.

Koenker, R. 1996. *Quantile Regression*. Cambridge: Cambridge University Press.

Kohavi, R., and R. Longbotham. 2017. Online controlled experiments and A/B testing. *Encyclopedia of Machine Learning and Data Mining*.

Kolmogorov, A. 1933. *Grundbegriffe der Wahrscheinlichkeitsrechnung*. Ergebnisse der Mathematik und ihrer Grenzgebiete. J. Berlin: Springer.

Konishi, S., and G. Kitagawa. 1996. Generalised information criteria in model selection. *Biometrika* 83(4): 875–890.

Krupskii, P., and H. Joe. 2015. Structured factor copula models: Theory, inference and computation. *Journal of Multivariate Analysis* 138: 53–73.

Kuss, O., M. Blettner, and J. Borgenmau. 2016. Propensity score matching: An alternative method of analyzing treatment effects. *Deutsches Ärzteblatt Int.* 113: 597–603.

Lauritzen, S.L. 1996. *Graphical Models*. Oxford University Press.

Leeb, H., and B. Pötscher. 2005. Model selection and inference: Facts and fiction. *Econometric Theory* 21: 21–59.

Leeb, H., B.M. Pötscher, and K. Ewald. 2015. On various confidence intervals post-model-selection. *Statistical Science* 30: 216–227.

Lehmann, E.L., and G. Casella. 1998. *Theory of Point Estimation*. Berlin: Springer.

Leuzinger-Bohleber, M., M. Hautzinger, G. Fiedler, W. Keller, U. Bahrke, L. Kallenbach, J. Kaufhold, M. Ernst, A. Negele, M. Schoett, H. Kuechenhoff, F. Guenther, B. Rueger, and M. Beutel. 2019. Outcome of psychoanalytic and cognitive-behavioural long-term therapy with chronically depressed patients: A controlled trial with preferential and randomized allocation. *Canadian Journal of Psychiatry-Revue Canadienne De Psychiatrie* 64(1): 47–58.

Little, R.J.A., and D.B. Rubin. 1987. *Statistical Analysis with Missing Data*. New York: Wiley.

Little, R.J.A., and D.B. Rubin. 2002. *Statistical Analysis with Missing Data*. New York: Wiley.

Lockhart, R., J. Taylor, and R. Tibshirani. 2014. A significance test for the Lasso (with discussion). *Annals of Statistics* 42: 413–468.

Louis, T.A. 1982. Finding the observed information matrix when using the EM-algorithm. *Journal of the Royal Statistical Society: Series B (Methodological)* 44: 226–233.

Mardia, K., J. Kent, and J. Bibby. 1979. *Multivariate Analysis*. London: Academic Press.

McGrayne, S.B. 2011. *The Theory that Would not die: How Bayes' Rule cracked the Enigma Code, Hunted Down Russian Submarines & Emerged Triumphant from Two Centuries of Co.*

McLachlan, G. 2000. *Finite Mixture Models*. New York: Wiley.

McLeish, D.L., and C.A. Struthers. 2006. Estimation of regression parameters in missing data problems. *Canadian Journal of Statistics* 34: 233–259.

Meng, X.-L. 2018. Statistical paradises and paradoxes in big data (i): Law of large populations, big data paradox, and the 2016 US presidential election. *The Annals of Applied Statistics* 12: 685–726.

Metropolis, N., M. Rosenbluth, A. Rosenbluth, A. Teller, and E. Teller. 1953. Equations of state calculations by fast computing machines. *The Journal of Chemical Physics* 21(6): 1087–1092.

Montgomery, D.C. 2013. *Design and Analysis of Experiments*. New York: Wiley.

Myers, R.H., D.C. Montgomery, G.G. Vining, and T.J. Robinson. 2010. *Generalized Linear Models: With Applications in Engineering and the Sciences*. New York: Wiley.

Nelder, J.A., and P. McCullagh. 1989. *Generalized Linear Models*. Boca Raton: Chapman and Hall.

Nelsen, R.B. 2006. *An Introduction to Copulas*. 2nd ed. New York: Springer.

Neyman, J., and E.S. Pearson. 1933. IX. On the problem of the most efficient tests of statistical hypotheses. *Philosophical Transactions of the Royal Society of London* 231: 289–337.

Oakes, D. 1999. Direct calculation of the information matrix via the EM. *Journal of the Royal Statistical Society* 61: 479–482.

Park, T., and G. Casella. 2008. The Bayesian Lasso. *Journal of the American Statistical Association* 103(482).

Pavlides, M., and M. Perlman. 2009. How likely is Simpsons paradoxon? *The American Statistician* 63: 226–233.

Pearl, J. 2009. *"Understanding Propensity Scores", Causality: Models, Reasoning and Inference.* 2nd ed. Cambridge: Cambridge University Press.

Pfister, N., P. Böhlmann, B. Schöllkopf, and J. Peters. 2018. Kernel-based tests for joint independence. *Journal of the Royal Statistical Society: Series B (Statistical Methodology)* 80(1): 5–31.

Politis, D.M., and J.P. Romano. 1994. Large sample confidence regions based on subsamples under minimal assumptions. *The Annals of Statistics* 22(4): 2031–2050.

Politis, D.M., J.P. Romano, and M. Wolf. 1999. *Subsampling*. New York: Springer.

Politis, D.N. 2015. *Model-Free Prediction and Regression*. Berlin: Springer.

Quenouille, M.H. 1956. Notes on bias in estimation. *Biometrica* 43: 353–360.

Radelet, M.L., and G.L. Pierce. 1991. Choosing those who will die: Race and the death penalty in florida. *Florida Law Review* 43: 1–34.

Robbins, H., and S. Monro. 1951. A stochastic approximation method. *The Annals of Mathematical Statistics* 22: 400–407.

Robert, C., and G. Casella. 2004. *Monte Carlo Statistical Methods*. Berlin: Springer.

Robert, C., and G. Casella. 2010. *Introducing Monte Carlo Methods with R*. Berlin: Springer.

Robins, J.M., A. Rotnitzky, and L.P. Zhao. 1994. Estimation of regression coefficients when some regressors are not always observed. *Journal of the American Statistical Association* 89(427): 846–866.

Rosenbaum, P.R. 2002. *Observational Studies (2nd Edition)*. New York: Springer Verlag.

Rosenbaum, P.R., and D.B. Rubin. 1983. The central role of the propensity score in observational studies for causal effects. *Biometrika* 70: 41–55.

Rossi, A., L. Pappalardo, P. Cintia, F.M. Iaia, J. Fernàndez, and D. Medina. 2018. Effective injury forecasting in soccer with GPS training data and machine learning. *PLoS ONE* 13, e0201264.

Rubin, D.B. 1981. The Bayesian bootstrap. *Annals of Statistics* 9(1): 130–134.

Rue, H., and L. Held. 2005. *Gaussian Markov Random Fields*. Boca Raton: CRC Press.

Rue, H., S. Martino, and N. Chopin. 2009. Approximate Bayesian inference for latent Gaussian models by using integrated nested laplace approximations. *Royal Statistical Society, Series B* 71: 319–392.

Ruppert, D., M.P. Wand, and R.J. Carroll. 2003. *Semiparametric Regression*. Cambridge: Cambridge University Press.

Schafer, J.L. 1997. *Analysis of Incomplete Multivariate Data*. Boca Raton: Chapman & Hall.

Scott, F.L. 1991. Why your friends have more friends than you do. *American Journal of Scociology* 96: 1464–1477.

Severini, T. 2000. *Likelihood Methods in Statistics*. Oxford: Oxford University Press.

Shibata, R. 1989. *Statistical Aspects of Model Selection*. Berlin: Springer.

Simpson, E.H. 1951. The interpretation of interaction in contingency tables. *JRSS (B)* 13: 238–241.

Sklar, A. 1959. Fonctions de répartition à n dimensions et leurs marges. *Publications de l'Institut de statistique de l'Université de Paris* 8: 229–231.

Spiegelhalter, D.J., N.G. Best, B.P. Carlin, and A. van der Linde. 2002. Bayesian measures of model complexity and fit. *Journal of Royal Statistical Society* 64: 583–639.

Spirtes, P., C. Glymour, and R. Scheines. 1993. *Causation, Prediction, and Search*. Vol. 81. New York: Springer.

Stern, H.S. 1996. Neural networks in applied statistics. *Technomatrics* 38: 205–214.

Stigler, S.M. 2007. The epic story of maximum likelihood. *Statistical Science* 22: 598–620.

Stock, J.H., and F. Trebbi. 2003. Retrospectives: Who invented instrumental variable regression? *Journal of Economic Perspectives* 17(3): 177–194.

Stoer, J., and R. Bulirsch. 2002. *Introduction to Numerical Analysis*, 3rd ed. Berlin: Springer.

Stuart, E.A. 2010. Matching methods for causal inference: A review and a look forward. *Statistical science: A review Journal of the Institute of Mathematical Sciences* 25: 1–21.

Symonds, M.R.E., and A. Moussalli. 2011. A brief guide to model selection, multimodel inference and model averaging in behavioural ecology using akaike's information criterion. *Behavioral Ecology and Sociobiology* 65(91): 13–21.

Takeuchi, K. 1979. Distribution of informational statistics and a criterion of model fitting. *Mathematical Sciences* 153: 12–18.

Thompson, S.K. 2002. *Sampling*. New York: Wiley.

Tibshirani, R. 1996. Regression shrinkage and selection via the lasso. *Journal of the Royal Statistical Society, Series B* 58, 267–288.

Tukey, J.W. 1958. Bias and confidence in not quite large samples. *The Annals of Mathematical Statistics* 29: 614–623.

Vaida, F. 2005. Parameter convergence for EM and MM algorithms. *Statistica Sinica* 15: 831–840.

Von Mises, R. 1928. *Wahrscheinlichkeit, Statistik und Wahrheit*. Number Bd. 3 in Schriften zur wissenschaftlichen Weltauffassung. J. Berlin: Springer.

Wald, A. 1943. Tests of statistical hypotheses concerning several parameters when the number of observations is large. *Transaction of American Mathematical Society* 54: 426–482.

Wasserstein, R.L., and N.A. Lazar. 2016. The asa statement on p-values: Context, process, and purpose. *The American Statistician* 70(2): 129–133.

Wedderburn, R.W.M., and J.A. Nelder. 1972. Generalized linear models. *Journal of Royal Statistical Society* 135: 370–384.

Weichwald, S., T. Meyer, O. Özdenizci, B. Schölkopf, T. Ball, and M. Grosse-Wentrup. 2015. Causal interpretation rules for encoding and decoding models in neuroimaging. *Neuroimage* 110: 48–59.

Wermuth, N., and D.R. Cox. 1996. *Multivariate Dependencies—Models, Analysis and Interpretation*. Boca Raton: Chapman and Hall.

Whittaker, J. 1989. *Graphical Models*. New York: Wiley.

Wood, S.N. 2017. *Generalized Additive Models: An Introduction with R*. Boca Raton: Chapman and Hall/CRC.

Wu, C.F.J. 1986. Jackknife, bootstrap and other resampling methods in regression analysis (with discussions). *Annals of Statistics* 14: 1261–1350.

Yule, G. 1903. Notes on the theory of association of attributes in statistics. *Biometrika* 2: 121–134.

Zelterman, D. 2015. *Applied Multivariate Statistics with R*. Berlin: Springer.

Zheng, T., and J.L. Gastwirth. 2010. On bootstrap tests of symmetry about an unknown median. *Journal of Data Science* 8: 413–427.

Zimmerman, D.W., B.D. Zumbo, and R.H. Williams. 2003. Bias in estimation and hypothesis testing of correlation. *Psicologica* 24: 133–158.

Index

© The Author(s), under exclusive license to Springer Nature Switzerland AG 2021
G. Kauermann et al., *Statistical Foundations, Reasoning and Inference*,
Springer Series in Statistics, https://doi.org/10.1007/978-3-030-69827-0

Printed in the United States
by Baker & Taylor Publisher Services